PROMOTING LOCAL GROWTH

Promoting Local Growth

Process, practice and policy

Edited by
DANIEL FELSENSTEIN
Hebrew University of Jerusalem, Israel

MICHAEL TAYLOR
The University of Birmingham, United Kingdom

Ashgate

Aldershot • Burlington USA • Singapore • Sydney

Published by
Ashgate Publishing Limited
Gower House
Croft Road
Aldershot
Hants GU11 3HR
England

Ashgate Publishing Company
131 Main Street
Burlington, VT 05401-5600 USA

Ashgate website: http://www.ashgate.com

British Library Cataloguing in Publication Data
Promoting local growth : process, practice and policy. -
(The organisation of industrial space)
1.Regional economics 2.Economic policy
I.Felsenstein, Daniel II.Taylor, Michael, 1946 Feb. 28-
337

Library of Congress Control Number: 2001091404

ISBN 0 7546 1686 X

Printed and bound in Great Britain by MPG Books Ltd, Bodmin, Cornwall

Contents

Part II Practice

Part III Policy

List of Figures

List of Tables

List of Contributors

Claes G. Alvstam is Professor of International Economic Geography, Department of Human and Economic Geography, School of Economics, Göteborg University, Sweden.

Raphael Bar-El is Head of the Department of Public Policy and Administration and teaches in the Department of Economics, Ben-Gurion University of the Negev, Be'er Sheva, Israel.

Itzhak Benenson is Senior Lecturer in the Department of Geography, Tel-Aviv University, Tel Aviv, Israel.

Oded Borenstein is in the Planning Office, Ministry of the Interior, Haifa District, Haifa, Israel.

Tony Eastham is Professor in the Departments of Civil and Electrical and Electronic Enginering and Associate Vice President for Research & Development, Hong Kong University of Science and Technology, Clearwater Bay, Kowloon, Hong Kong.

Ayda Eraydin is a Professor in the Department of City and Regional Planning, Faculty of Architecture, Middle East Technical University, Ankara, Turkey.

Daniel Felsenstein is Senior Lecturer in the Department of Geography, Hebrew University of Jerusalem, Mount Scopus, Jerusalem, Israel.

Hein van Gils is a Researcher in the Department of Economic Geography, Erasmus University, Rotterdam, The Netherlands.

Annelie Jönsson is in the Department of Human and Economic Geography, School of Economics, Göteborg University, Sweden.

Baruch A. Kipnis is Professor in the Department of Geography, University of Haifa, Haifa, Israel.

Lambert van der Laan, is Associate Professor of Economics in the Department of Economics, Erasmus University, Rotterdam, The Netherlands.

John Langdale is Associate Professsor in the School of Earth Science, Macquarie University, North Ryde, New South Wales, Australia.

Helen Lawton Smith is Reader in Local Economic Development, Coventry Business School, Coventry University and Senior Research Associate at the School of Geography, Oxford University, United Kingdom.

Richard Le Heron is Professor of Geography, Department of Geography, University of Auckland, Auckland, New Zealand.

Simon Leonard is Senior Lecturer in the Department of Geography, University of Portsmouth, Portsmouth, United Kingdom.

Edward J. Malecki is Director of the Centre for Urban and Regional Analysis and Professor of Geography at the Ohio State University, Columbus, Ohio, USA.

Peter Maskell is Professor of Business Economics in the Copenhagen Business School, Copenhagen, Denmark.

Philip McDermott is Director, McDermott Fairway Group Ltd., Takapuna, North Shore City, New Zealand.

Päivi Oinas is Research Fellow, CASBEC, Erasmus University, Rotterdam, The Netherlands.

Sam Ock Park is Professor and Head, Department of Geography, Seoul National University, Seoul, Korea.

Jerry Patchell is Assistant Professor in the Division of Social Science, Hong Kong University of Science and Technology, Clearwater Bay, Kowloon, Hong Kong.

Izhak Schnell is Associate Professor in the Department of Geography, Tel-Aviv University, Tel Aviv, Israel.

Michael Sofer is the Head of Geography at the Levinsky Teachers College and lectures in the Departments of Geography at Tel-Aviv and Bar-Ilan Universities, Tel Aviv, Israel.

Christine Tamásy is a Researcher and Lecturer in the Department of Economic and Social Geography, Faculty of Economics and Social Sciences, University of Cologne, Germany.

Michael Taylor is Professor of Human Geography in the School of Geography and Environmental Sciences at the University of Birmingham, United Kingdom.

Eirik Vatne is Professor in the Department of Geography, Norwegian School of Economics and Business Administration, University of Bergen, Bergen, Norway.

1 Introduction: Promoting Local Growth – A New Look at an Old Saga

Daniel Felsenstein and Michael Taylor

The paradox of the emergence of the region and the locality as the main arenas for growth in a globalising world is widely recognised, though it is implicit that the processes of re-scaling associated with globalisation have changed the circumstances of local growth. Nevertheless, the resurgence of interest in growth at the local level has led to a revisiting of past ideas on regional growth. Marshallian 'atmosphere' and 'agglomeration' have been rediscovered and recast by institutionalists as 'new industrial districts'. Externalities, increasing returns to scale and imperfect competition have resurfaced as the mainstays and stylised underpinnings of the economists' 'new economic geography'. And, staple exogenous ingredients of regional growth, such as technology and human capital, have been endogenised in the 'new growth theories'.

A common denominator of the renewed interest in the old story of regional growth relates to the openness of the locality and the region as primary initiator of localised change in a globalising system. Flows of capital, ideas, technology, tastes and culture transcend national and local boundaries homogenising place and space. In this environment, localities have to make strenuous efforts to promote growth, stressing their uniqueness and ensuring that economic development 'sticks' at their location and does not flow elsewhere (Markusen, 1996). However, this recognition of the 'open system' nature of places and regions is hardly new, either. Many of the classic treatises of regional growth, written over thirty years ago, open with this observation (see for example, Borts and Stein, 1964; Siebert 1969).

What then, is the justification for the resurgence of interest in the old saga of local and regional growth? One answer might lie in the fact that research from different national and regional contexts has shown that contrary to neo-classical growth theory, convergence rates between regions since the mid 1980s have slowed considerably (Barro and Sala-i-Martin, 1991; Armstrong and Vickerman, 1995). This suggests that rather than being self-correcting, local and regional growth dynamics may well be self-reinforcing, thereby perpetuating regional inequalities. Another answer might lie in the evidence that has emerged of an uneven regional landscape punctuated by clusters of both high growth and high deprivation – a global mosaic of successful and unsuccessful regional economies. This empirical evidence has led to a re-thinking of the neo-classical growth model and the development of 'new endogenous growth' theory with its stylised facts ('externalities', 'returns to scale'), on the one hand, and 'new institutional theory' with its equally stylised soft concepts ('embeddedness', 'social capital', 'learning regions'), on the other.

An irony of the current situation is that if interest in local and regional growth is due to the seemingly intractable spatial disparities that it promotes, we would expect more attention to be focussed on those individuals and households excluded from this growth, and on the mechanisms of exclusion. In practise, contemporary discussions of 'new industrial spaces', 'new territorial complexes', 'new' industries, regions, sectors or institutions, seem to have very little to say about the welfare of those individuals affected by growth (see Felsenstein, this volume). This applies to theorizing and empirical work on local growth from both the new endogenous growth and the new institutionalist perspectives. Even the ostensibly people-centred view of growth grounded in 'social capital' (and adopted by both schools) has been interpreted as describing an isolated world where trust and face-to-face relations are based on performance not community and commercial reputation rather than civic engagement (Cohen and Fields, 2000).

In part, this 'blind spot' can be ascribed to the way new theoretical perspectives on local growth have been developed, especially among the institutionalists. Here, the emphasis has been almost exclusively on growth and exploration of processes of growth but only in regions that are apparently 'successful'. This is a methodology that, as Staber (1996) has argued, selects on the dependent variable – falling foul of the non sequitor that because 'successful' regions are the product of positive growth processes, the processes that can currently be identified in a 'successful' region created that growth. The meaning of 'success' at a regional scale is rarely unpacked. Is it concerned with levels of employment, the welfare of individuals, rates of new firm formation, or rates of investment? Too often in analyses flavoured with neo-liberalism the unasked questions are, growth for what and growth for whom?

It would therefore seem that the analysis of local and regional growth needs to be reframed to convey a meaning that goes beyond simply describing 'something greater than existed before'. Only in this way will the current interest in the old saga of local growth, represent something essentially 'new'.

The thrust of the present volume reflects the foregoing observations. The selection of papers has been guided by a desire to first, clarify some of the key concepts underlying the interest in local growth, and second to refocus some of the attention on issues of labour, welfare and distribution.

The majority of the papers assembled for this book were originally presented at a residential conference of the IGU commission on The Organisation of Industrial Space that was held in Israel in June 1999. The title of that meeting was 'Promoting Growth: New Industries, Policies and Forms of Governance' and the spirit of that theme has been preserved here. Thirteen of the chapters were originally presented as papers at that meeting and have subsequently been revised as chapters for this book. A further seven chapters were specially commissioned for this volume.

While the conference focussed in detail on the new economic activities that act as the motors of local growth, the present book expands this view to gain a broader perspective on local growth. In doing so, we also address some of the debates at the forefront of current theory and practise in economic geography. In the 'process' section, key concepts underlying local growth are examined. The vocabulary of 'embeddedness', 'learning regions', 'social capital 'and 'industrial districts' used to invoke processes of growth is critically examined and anchored to the more basic concepts of the lexicon of growth viz, 'development', 'welfare' and 'power'. In this respect, some of the softer and fuzzier concepts popular in contemporary economic geography are subject to a further round of definition and tightening (Markusen, 1999).

The section dealing with local growth 'practise' assembles empirical evidence of local growth stories grounded in technological innovation, internet-based economic activity and tertiary-type activities. This collection of case-study knowledge also addresses a further contemporary concern in economic geography, that of 'thin empirics' (Martin and Sunley, 2001). While local growth is both time and place-specific, a critical mass of solid empirical work can suggest generic factors that appear time and again in different contexts. One recurrent issue is that of technology-based economic activity promoting growth and spatial disparities at one and the same time. These disparities are not just between the 'haves' and the 'have nots' but also overlap spatial divides, such as urban and rural populations, inner and outer-city communities and the like. Much empirical work suggests that the key in promoting growth is 'making it stick' in places where conditions may be adverse. Furthermore, the rise of internet-based growth and the concomitant tertiarisation of economic activity raises the spectre and challenge of 'placeless growth'.

The third section of the book addresses issues of 'policy'. Underlying the smorgasbord of approaches presented here is a pervading ethos that policy is needed in order to address market failures of different kinds and that in general, 'policy works'. The challenges of globalisation, competitiveness, long-term structural change and labour market imperfections cannot be left to any 'hidden-hand'. However, the necessary tuning, is place-specific. Policy relevance is again an area that has been targeted in recent critiques of economic geography (Lovering, 1999). Central to the re-emerging discourse (Peck, 1999; Pollard *et al.,* 2000; Banks and MacKain, 2000; Martin and Sunley, 2001) is the apparent relegation of policy relevant research in economic geography to a policy monitoring rather than a policy formulation role, with policy formulation being the preserve of economists. The contributions of this section of the book belie such a simplistic dichotomisation and demonstrate the full spectrum of policy relevant research from formulation to monitoring and critique based on theoretically-informed 'thick' empirics.

Without doubt, useful policy needs to be grounded in conceptual clarity and rich empirics. In light of the foregoing debate, we would argue that people, not just regions or firms, need to be at the forefront of any policy discussion. This necessarily emphasises questions of distribution, welfare and exclusion as leading policy topics in any discussion of promoting local growth.

Taken together, the three sections that comprise this volume contribute both to the new versus old debate on promoting growth, and positions these issues within contemporary critiques in economic geography as it enjoys a bout of introspection (Amin and Thrift, 2000; Martin and Sunley, 2001). The perspectives presented here offer a wide-angled view on the topic both in terms of disciplinary affiliation (contributions from Geography, Planning, Economics, Engineering and Public Policy) and national experiences from different continents.

We are eternally grateful to our families for putting up with the extended overtime that goes with editing a book comprising 20 chapters and demanding co-ordination between 26 authors from 15 countries. Increasing returns would seem to have contributed to the growth of this volume! In our efforts to cope with these returns to scale, we were ably assisted by Michal Stern who converted all manuscripts to camera ready copy and by Michal Kidron from the Cartographic Laboratory at Hebrew University who assisted in transforming maps and figures into standard format.

References

Amin, A. and Thrift, N. (2000), 'What Sort of Economics for What Sort of Economic Geography', *Antipode*, vol. 32, pp. 4-9.

Armstrong H. and Vickerman R. (eds) (1995), *Convergence and Divergence Among European Regions*, Pion, London.

Banks, M. and MacKain, S. (2000), 'Jump In! The Water's Warm: A Comment on Peck's 'Grey Geography'', *Transactions of the Institute of British Geographers, New Series*, vol. 25, pp. 249-254.

Barro R. and Sala-i-Martin X. (1991), 'Convergence Across States and Regions', *Brookings Papers on Economic Activity*, vol. 2, pp.107-158.

Borts G.H. and Stein J.L. (1964), *Economic Growth in a Free Market*, Columbia University Press, New York.

Cohen, S.S. and Fields G. (2000), 'Social Capital and Capital Gains: An Examination of Social Capital in Silicon Valley', in M. Kenney (ed.), *Understanding Silicon Valley: The Anatomy of an Entrepreneurial Region*, Stanford University Press, Stanford, CA, pp. 190-217.

Lovering J. (1999), 'Theory Led by Policy: The Inadequacies of the New Regionalism (Illustrated from the Case of Wales)', *International Journal of Urban and Regional Research*, vol.23, no. 2, pp. 379-395.

Markusen, A. (1996), 'Sticky Places and Slippery Spaces: A Typology of Industrial Districts', *Economic Geography*, vol. 72, pp. 293-313.

Markusen, A. (1999), 'Fuzzy Concepts, Scanty Evidence, Policy Distance: The Case for Rigor and Policy Relevance in Critical Regional Studies', *Regional Studies*, vol. 33, no. 9, pp. 869-994.

Martin R. and Sunley P. (2001), 'Rethinking the 'Economic' in Economic Geography: Broadening our Vision or Losing our Focus?', *Antipode*, vol. 33, pp. 148-161.

Peck, J. (1999), 'Grey Geography?', *Transactions of the Institute of British Geographers, New Series*, vol. 24, pp. 131-135.

Pollard, J., Henry, N., Bryson, J. and Daniels, P. (2000), 'Shades of Grey? Geographers and Policy', *Transactions of the Institute of British Geographers, New Series*, vol. 25, pp. 243-248.

Siebert H. (1969), *Regional Economic Growth: Theory and Policy*, International Textbooks, Scranton PA.

Staber, U. (1996), 'Accounting for Differences in the Performance of Industrial Districts', *International Journal of Urban and Regional Research*, vol. 20, no. 2, pp. 299-316.

PART I
PROCESS

Introduction to Part I

This section presents a broad critique of some of the concepts pervading the discussion of local growth. It attempts to challenge some of the stylized facts and vague definitions all too often associated with this topic. In this spirit, the chapter by Taylor (*Enterprise, Embeddedness and Local Growth: Inclusion, Exclusion and Social Capital*) challenges the cosy view of 'embedded' local growth based in social inclusion, trust and reciprocity. This view ostensibly ignores power relations between firms, regions and countries. The notion of embeddedness is then reformulated into three variants in which processes of inclusion and exclusion operate at differing intensities. Taylor's conclusions point to the partial nature of the common model of embeddedness, its pre-occupation with 'inclusion', the benign role it assigns to the operations of trans-national corporations in the local economy and the misleading causality that can be read from embeddedness to social capital.

In similar vein, Felsenstein (*Analysing Local Growth Promotion: Looking Beyond Employment and Income Counts*) re-examines the concepts of 'growth' and 'development'. He argues for the need for adopting an explicitly stated value system when looking at the outcomes of growth promotion and suggests a local welfare focus. This means moving beyond counting jobs and income as indices of the 'success' of growth promotion and looking instead at outcomes in terms of their distributional effects. Adopting this perspective also means that the distinction between 'growth' and 'development' is artificial and can serve to obscure the incorporation of welfare and distribution issues into analyses of local growth.

Knowledge and learning capabilities are considered key ingredients in any technology-based path towards regional growth. Park (*Knowledge-Based Industry for Promoting Growth*) points out both the opportunities and pitfalls inherent in pursuing knowledge-based local growth. In this view the firm is the leading agent of change. Its ability to consolidate a non-price competitive advantage from the creative use of knowledge and information gives it an edge that is then consolidated through increasing returns. The ability of firms to act this way is a function of the human resources at their disposal and thus the inclusion or exclusion of firms in this process is really a reflection of human capital endowments. Linking firms to regions is the key to regional and local growth. Oinas and van Gils (*Identifying Contexts of Learning in Firms and Regions*) address this issue pointing out that the firm's learning capacity is contingent on its interaction with other firms, sectors and networks in the region. In this way 'regional learning' takes place. The result is a typology of regional growth configurations determined by the resource base available to the firm. An implicit assumption here is that competitive firms lead to competitive regions, a form of expansionism that needs still to be established in measurable, empirical terms.

The final chapter in this section analyses the meaning of 'local governance' in the context of the growth of industrial districts. This is a particularly important issue that is starting to be addressed in contexts as diverse as those of Emilia Romagna and the Santa Clara Valley. For some, governance is the key to regional growth clusters and Eraydin's contribution (*New Forms of Local Governance in the Emergence of Industrial Districts*) outlines the different forms of governance frameworks that have emerged in the most celebrated industrial districts. She further discusses the way they have adapted to changing global economic conditions highlighting local responses to challenges that threaten to erode the advantages of embedded, place-based growth (for example, cross-border strategic alliance between firms).

2 Enterprise, Embeddedness and Local Growth: Inclusion, Exclusion and Social Capital

Michael Taylor

Introduction

In recent years, an increasingly pervasive model of local economic growth has been developed and elaborated in economics, geography and management science that sees growth as being created by complex local processes of 'embedding'. It builds on views of economies as being socially constructed, supported by structures of institutions, operating in a Marshallian 'industrial atmosphere' and driven by entrepreneur-based processes of Schumpeterian creative destruction. The processes of the model are held to underpin successful local economic growth in developed and developing countries alike. At its heart are mechanisms of economic and social *inclusion* based on trust, reciprocity and loyalty amongst SMEs in a place. Those same mechanisms have more recently been extended to include TNCs who, it is maintained, need access to local tacit knowledge to remain competitive both locally and globally and to sustain their role as global information arbitrageurs.

Going against the flow of this stream of research, the argument developed here is that the embeddedness model inappropriately universalises processes of local growth (Bianchi, 1998) and, in so doing, conceals significant *exclusionary* tendencies in embedded inter-firm relationships - especially when those relationships are orchestrated by TNCs. These exclusionary tendencies are implicit in, for example, Grabher's (1993) treatment of power inequalities in embedded networks.

They are implicit in much empirical work on TNCs (Taylor and Thrift, 1986; Phelps, 1997), but they are thrown into stark relief in the developing country context. By neglecting the processes of exclusion and forces that might impair or even destroy local growth mechanisms, it is argued that the model is essentially functionalist as too are the mechanisms of social capital formation that are derived from it. As such, it is contended that policies to promote local growth based on embeddedness and social capital notions and the establishment through interventionist policies of apparently appropriate institutions, must be treated with extreme care.

The argument of the paper is developed in sections. In the first section, the broad parameters of the model are developed and the mounting criticism of it is explored. That critique is extended to recognise the neglected interplay of processes of inclusion and exclusion in local enterprise networks. In the second section, three variants of the embeddedness model are identified: a developed country model; a developing country model; and TNC embeddedness model. In each variant, the interplay of processes of inclusion and exclusion are examined. In section three, Yeung's (1998b) proposed research agenda on enterprise embeddedness is reviewed and extended to build on these processes of inclusion and exclusion and also to question the local growth potential fostered by local institutions. The policy implications of the discussion are drawn together in the conclusion.

The Enterprise Embeddedness Model

During the past decade, a powerful model of local economic growth has developed which draws on a range of complementary literatures on new industrial spaces, learning regions, innovative milieu, regional innovation systems and clustering (Braczyk *et al.*, 1998; Maskell *et al.*, 1998; Porter, 1998; Simmie, 1997; Storper, 1997, Taylor and Conti, 1997). At the heart of the model are complex processes of 'embedding' that recognise the enmeshing of firms' economic relationships within the broader social structures and social relationships of a place. Business enterprises are interpreted as networks of economic, social and cultural relations to the extent that, "[trust], mutual forbearance and reputation may supplement or replace the price mechanism or administrative fiat" (Powell and Smith Doerr, 1994, p. 370). Successful local economies are recognised as islands of sustained local accumulation built on superior local productivity (Porter, 1995, 1998), and integrated into a global mosaic of production (Storper and Scott, 1992; Storper, 1997). Through processes involving place-based trust, reciprocity, loyalty, collaboration and cooperation information, ideas, knowledge and innovation are generated and exchanged via mechanisms of quasi-integration that involve collaboration and sharing rather than more brutal appropriation (Leborgne and Lipietz, 1992). The whole is an inclusionary process of collectivisation and corporatisation that creates 'institutional thickness' and further bolsters these local economic processes. (Powell, 1990; Amin and Thrift, 1994, 1995, 1997) The global mosaic of these successful industrial spaces is, in turn, orchestrated by TNCs, global capital and global political institutions (e.g. IMF, WTO etc). Enmeshed in webs of global coordination and value transfer, local economies are interpreted as nodes of untraded dependencies involving knowledge accumulation, innovation and learning, which are tapped into by internationalised and globalised corporations. By this interpretation, corporations act as global information and knowledge arbitrageurs (The Economist, 1995) in a 'soft capitalism' of networked knowledge (Thrift, 1998, 1999). They too are part of this inclusive, collaborative system needing to be as embedded in their host countries' local economies as they are in those of their home countries (Dicken *et al.,* 1994; Yeung, 1998a, 1998b).

This model of local economic growth is now being rapidly universalised and promoted as a prescription for policy formulation in developing and developed economies alike. It is being proposed as an approach to offset and counteract the threats to local economies that have arisen from the internationalisation and globalisation firms and the forces of uneven development inherent in capitalism. The emphasis the model places on co-operation, collaboration, and the creation of social capital and 'institutional thickness' gives the model obvious policy appeal. This range of processes, with strong overtones of inclusiveness and localism, are now being used increasingly to inform local economic development policies and practices. Indeed, 'social capital' is now being used as a catch-all phrase in political economy to cover the non-economic and non-political relations which underpin successful development and sustainable democracies (Putnam, 1993).

But, the model is not without its critics and the criticisms are getting louder. Bianchi (1998) considers the model to be time and space specific and applicable only to the Third Italy, where it originated, and then only in a very narrow time frame. To him it has been inappropriately universalised to the point where it is now being ideologised. For Baker (1996) the model romanticises small firms and underestimates their weaknesses and insecurities, failing to appreciate their lack of power. That romanticism comes, in part, from the model being based only on analyses of successful regions: what Staber (1996) describes as selecting case studies "on the dependent variable". It has also been contended that the embeddedness and reciprocity that lie at the core of the model, are at "... best temporary ... never total" (Pratt, 1997, p. 132), and might even be illusory (Bresnen, 1996). For Hudson (1999) the model disregards the processes of spatial uneven development that are basic to capitalist development, universalising as policies processes that by definition can not be universalised. Lovering (1999), however, goes much farther and claims that the model is policy driven theory that, "... confuses development *in* a region with development *of* that region ...[and]... fetishises particular kinds of economic and organisational activity" (p. 5). This has the effect of prioritising some and devaluing others. In short, the model is rhetoric at odds with reality.

This paper seeks to extend the critique. It argues that universalising processes of embeddedness as networks of inter-firm collaboration and co-operation based on trust, reciprocity and loyalty highlights stabilising *inclusionary* tendencies that bind enterprises within local economies creating long-term growth. At the same time, however, it excludes equally significant destabilising and *exclusionary* tendencies, especially those orchestrated by TNCs, which may have detrimental long-term local growth consequences. The argument is an extension of the stability/instability dual in the processes shaping geographies of economies proposed by Ekinsmyth *et al.*, (1995). Exclusionary tendencies freezing individual firms and sets of firms out of local economic systems are implicit in the treatment of power inequalities in embedded networks (Grabher, 1993; Taylor, 1995, 2000), but they are thrown into stark relief in the developing country context. In the next section, the notions of inclusion and exclusion are developed further. A number of contexts of embeddedness are recognised in which theorised inclusiveness is matched by empirically identifiable exclusiveness.

Embeddedness in Context

In the extensive literature on small enterprise spatial systems (new industrial spaces, learning regions, economic milieu, regional innovation systems and clustering), three quite distinct contexts of embeddedness can be recognised which are essentially variants of the embeddedness model of local economic growth:

- the developed country context in which 'inclusive' embeddedness is seen as a mechanism for (re)generating international competitiveness in mature industrial economies;

- the developing country context in which 'inclusive' embeddedness is seen as a mechanism and strategy for coping locally with the pressures and problems of economic globalisation; and
- the TNC context in which 'inclusive' embeddedness is a globalisation strategy for tapping into new economic spaces.

In each of these contexts the emphasis in current theoretical literature is on the incorporation of individual firms into networks of reciprocal exchange in local production chains virtually as a prerequisite for entry into the global economy. The route to international competitiveness is through the dynamic inclusiveness of weak ties. Here is a process of local enterprise generation that is said to transcend the individual firm to create a community of commercial interest with the economic virtue of generating productivity (Porter, 1998) sufficient to ensure and maintain international competitiveness. Some sort of 'virtuous capitalism' is conjured up here based on individual wealth creation through co-operative action of a hard-to-recognise breed of entrepreneur. What have been neglected, if not ignored, however, have been the countervailing processes of exclusion created not just by the weakness of strong ties but also by the strength of strong ties.

Embeddedness and Mature, Developed Economies

In the context of local growth in mature industrial economies there is a massive literature on clustering, learning, milieu and untraded dependencies that highlights Schumpeterian innovation, knowledge transfer and the creation of 'atmosphere' as the positive consequences of structural embeddedness. And, when these supply chain mechanisms are coupled with enhanced productivity they enhance or maintain the global competitiveness of a place. It is widely contended in this literature that, in a new era of flexible specialisation and flexible accumulation (Piore and Sabel, 1984; Sabel 1989), local economic growth and the local survival of business enterprises is dependent upon the incorporation of firms into socially embedded networks of collaborative production (Cooke, 1998). In turn, those networks are, "... buttressed by a supportive tissue of local institutions" (Powell and Smith-Doerr, 1994, p. 370). The benefits of these collaborative and co-operative relationships, that are as much social as economic, are seen in terms of heightened place-based capacities for learning, information and knowledge exchange, technological change and innovation amongst network members. Industrial districts are, by this interpretation, regional innovation systems and engines of local economic growth and competitive strength fuelled by the intensified and localised processes of Schumpeterian creative destruction. In this way, local economies remain nimble in the face of processes of 'ubiquification' which leave firms constantly faced by the need to cope with shifting factor mixes (Maskell *et al.*, 1998). Breathing Marshallian 'industrial atmosphere', these firms are incorporated as dynamic nodes of innovation and learning into an emerging global mosaic of regions (Scott and Storper, 1992; Storper, 1997).

But, what are ignored in these essentially functionalist interpretations of dynamic localities in mature industrial economies, are the countervailing processes of exclusion. Already it has been recognised that embeddedness can not replace market mechanisms entirely (see for example Uzzi, 1996, 1997). Co-operation, it is argued, must be tempered by competition. Some inter-firm transactions in local economies might best be undertaken collaboratively and co-operatively while others might best be left to price determined competition (Cooke, 1998; Enright, 1995). A balance is needed to avoid lock-ins, to avoid paternalism in labour markets, to stimulate the knowledge economy and to prevent 'institutional overload' obscuring economic imperatives (Brusco, 1996).

This necessary balance of competition and co-operation is a fundamental issue in appreciating the processes shaping small enterprise economic systems (SESSs). The corollary of the need for such a balance of forces is that an excess of *either* can jeopardise the local system of production and wealth creation – that too much stability is as bad as too much instability (Ekinsmyth *et al.*, 1995). And yet, in the currently developing social capital interpretation of SESSs (Maskell, 1999), co-operation, collaboration and the processes of SME clustering are reckoned to lead to the local accumulation of social capital as a place-specific asset or set of assets, just like money capital. The emergence and support of local institutions then reinforces the processes of local asset formation. But, treating social capital as an asset to be accumulated like money capital is to deny the two-way nature of social processes and their ability to destroy as well as create. The accumulation of money capital in comparison is a one-way process bringing only benefits. In other words, by the use of inappropriate language, the two-way social process of embedding is translated into a one-way process of accumulation that can only ever be beneficial. A social process is converted into a functionalist local growth paradigm.

The case can be argued, moreover, that economic relationships involve more than a dynamic tension between collaboration and competition, they also involve the more brutal exercise of power, through the control of resources, the manipulation of relationships or the exercise of discipline (Taylor, 1995; Allen, 1997). Power inequalities can be seen leading to exclusion in at least three ways. First, they restrict firms' freedom of action. Power inequalities limit the forms of transaction and buyer-supplier relationships open to firms according to their positions in inter-firm networks (Taylor, 1999). This restricts the ways they are able to do business and, thus, their possibilities and potential to accumulate capital. Second, power inequalities can create lock-ins and the ossification of transaction relationships - a process that has been well detailed by Amin and Robins (1990), Amin (1993), Grabher (1993b) and Glasmeier (1991). Indeed, the creation of institutions in a place can be seen just as much as a way of protecting the status quo of doing business in a place (and so promoting lock-in) as it a mechanism for the generation of dynamic development. Third, power inequalities lead to uneven spatial development - a well-recognised inherent characteristic of capitalist accumulation (Hudson, 1999; Massey, 1984). Just as some local economic systems are 'winners', others are 'losers'. Finally, the power inequalities of class lead to exclusion in business communities through its shaping of boards of directors and the strategic decision-making role they have (McNulty and Pettigrew, 1999). However, the arguments that counter the embedded local growth model are as yet only weakly developed in the developed country context.

Embeddedness and Developing Countries

While the local embeddedness model has been accused of being policy-driven theory in the context of the industrial west (Lovering, 1999), it is very clearly seen as a way of promoting socially and culturally sensitive economic growth in the developing country context. A range of analysts have suggested that the embeddedness processes at the heart of the model can be translated into policies whereby regions and localities attempt to accrete social capital and build institutional thickness (Cooke, 1998; Amin and Thrift, 1997; Putnam, 1993). Here, there is a major implicit shift in the focus of the embeddedness model. Though recognised first as a competitive strategy for the incorporation of places in a global economy, here it is subtly translated into a coping mechanism to counter the underdevelopment of places by global economic pressures. However, while analysts generally report examples where institutions have been part of economic success, they are less certain how to build these social relations from scratch. The problem is that these partial analyses reify certain institutional ensembles, elevating them to the status of universal mechanisms, while at the same time cementing key stories in the intellectual-political consciousness (Staber, 1996; Bianchi, 1998). There is now the great danger that the mere existence of certain institutions is sufficient proof of their economic effectiveness without actually demonstrating the causal linkage between their actions and economic outcomes.

In the developing country context the focus is on the creation of 'social capital' and the creation of seemingly appropriate non-economic and non-political institutions as the underpinnings of successful economic growth (Putnam, 1993). Such institutionally driven research is relatively new (Schmitz, 1993; Honig, 1998). The major lending institutions, however, have recently been pushing for 'good governance' to create an 'enabling environment' for economic growth in the developing world, so it can be assumed that such studies will be of growing importance as these policy initiatives mature (Mohan, 1996).

Tentative conclusions from empirical research show, however, that in developing countries where scale economies are lacking, there is a greater need for inter-firm collaboration to foster local economic growth (Rasmussen *et al.*, 1992). However, a study by Schmitz (1993) on Brazil showed that collaboration and reciprocity based on a shared cultural heritage of migrants broke down when competition from other industrial districts - in this case Korean manufacturers - forced firms to interact largely on a cost/price basis. It also showed that social ties between manufacturers actually permitted collusion and the concentration of power within a small number of large firms. The results for their subcontractors were far less autonomy and increased hardship for their workers. Hence, the success of this Brazilian industrial district was predicated only weakly upon processes of social and institutional cooperation that took them far from the 'high road' of economic development envisioned by the regional boosters (Ohmae, 1995; Cooke, 1998). What we do not know is how extensive such *exclusionary* processes are in the global periphery, and we certainly have no strong evidence to assume they do not exist in the developed country context as well. Indeed, there is an urgent need to unlock the complex and specific local relations between individuals, firms and institutions under different conditions of economic development.

Recent research in the small island state of Fiji (Taylor, 1986, 1999) suggests strongly that large foreign firms work 'to get close' to their major host country customers, exhibiting in their buyer/supplier relationships all the characteristics said to indicate local embedding. But, those firms 'get close' only to their largest customers. Small, local, ethnic Indian-owned firms are kept as arms-length on cash-only transactions. In short, embeddedness for some is coupled with exclusion for others. Qualo (1996) explored other dimensions of this commercial divisiveness in Fiji and has shown it to be much more complex. In particular, he has documented the social bias that excludes ethnic Fijians from participation in the enterprise structures of the formal economy – exclusion exacerbated by their own subsistence mindset. And yet, at the same time, ethnic Fijians dominate the political arena, setting the frameworks of real regulation within which the TNCs and ethnic Indian businesses must operate. Here then is a highly complex set of contingent relationships that brings a subtle point and counter-point to processes of enterprise 'inclusion' and 'exclusion'. It denies as naively functionalist 'social capital' approaches to the planning and creation from scratch of local economic growth, especially in the developing country context. What is needed from research, it can be argued, is more evidence and less assertion.

Embeddedness and Transnational Corporations (TNCs)

In the context of TNCs moving into new markets and new economic environments, 'embeddedness' is again a strategy to enhance international competitiveness built on inclusion and incorporation into specific local contexts. In the process of becoming transnational, TNCs grow out of particular local contexts and acquire characteristics of those places through the complex processes of networking and embedding that create them (Dicken and Thrift, 1992; Dicken 1994; Dicken *et al.*, 1994). They carry that local flavour with them as they move into the international arena (Yeung, 1998a, p. 116). However, what might be successful in their home place might not be successful in another where there are significant place-specific differences in, for example:

- gentlemen's agreements on the way business is done;
- the manner and mechanisms of negotiating with other businesses, governments and political elites;
- the understandings developed through personal contacts;
- the accumulation of intelligence, information and knowledge;
- processes of informal recommendation and the building of reputations; and
- the nature of local buyer-supplier relationships and the foibles of local markets (Crew, 1996; Yeung 1998b).

Such place-specific knowledge, it is argued, is only gained from local experience, so foreign firms have to make themselves less 'foreign'. They have to be embedded in the networks of the host countries they operate in so that simultaneously they can be globally integrated and yet locally responsive. Indeed, Yeung (1998a) draws on the work of, amongst others Pryke and Lee (1995), Thrift (1994), Clark (1997) and Amin and Graham (1997) to argue that the external and territorial economies of financial production in centres such as the City of London, New York and

Tokyo that derive from complex networks of social and personal relationships, can only be exploited by transnational financial institutions when they are locally embedded in those places (p. 301). Jonas (1996) has suggested the same local embedding of US TNCs in Mexico's northern border towns, and Fujita and Hill (1995) have outlined Japanese TNCs' locally embedded relationships with firms and institutions in the USA. In a very different context, Yeung (1997a, 1997b) has considered the success of Chinese business organisations to lie in their embeddedness in distinctively local business networks. These networks involve a "perpetual tendency to cultivate complex networks of personal and business relationships that are rooted locally among Overseas Chinese Diaspora" (Yeung, 1998b, p. 119). These essentially spatial networks also offer participants formidable first-mover competitive advantages. Now, as a result of patterns of Chinese migration and settlement, Overseas Chinese business networks are described as borderless, linking "... most of East Asia's economies together in a seamless web of connections" (DFAT, 1995, p. 6).

The message is clear, the inclusiveness of the embeddedness model embraces large corporations as much as it does SMEs, and by being reciprocating network members TNCs can prosper locally and be internationally competitive as part of a corporate globalisation strategy. This is a very different story to the long-standing discourse that highlights the exploitative, exclusionary view of TNCs. That view has been expressed strongly in Dunning's (1979) OLI model, the New International Division of Labour thesis (Fröbel *et al.*, 1980), radical perspectives on the geographical transfer of value (Forbes and Rimmer, 1984), or the extensive critique of US transnational corporations developed by, amongst others, Barnett and Müller (1978) and Bergsten *et al.*, (1978). More recently, Phelps (1997) has demonstrated the lack of local embeddedness of TNCs in South Wales in the UK, and here in a locality where policy has been built explicitly on the embeddedness thesis (Lovering, 1999). It is also far too easy to see the Chinese family networks of the 'gift economy' (guangxi) as cosy, co-operative arrangements offering only benefits to their participants. However, as Yang (1987) has pointed out in discussing the workings of the 'gift economy', it is potentially as *exclusionary* as it is *inclusionary*. 'Face' in this system can be lost as easily as gained, and sanctions when applied can be much deeper and pervasive than mere financial penalties. Behind inclusion lies the threat of exclusion. It would seem important, therefore, to begin to counterbalance the inclusionary tendencies inherent in ideas on the local embeddedness of TNCs with the exclusionary tendencies that stem from their exercise of power in those same places (see Allen and Pryke, 1994) – a power that can be both social and economic.

Extending the Research Agenda

The background of these three contexts of embeddedness and their different tendencies towards the local inclusion and exclusion of business enterprises, reinforces the relevance of Yeung's (1998b) proposed agenda for research on enterprise embeddedness, but at the same time extends it. Yeung's first agenda item is, "to untangle the complex ways in which business organizations are locally embedded" (p. 120). But, to examine the social and institutional contexts of embeddedness and the ways "... business organisations are constituted *in situ*" (p. 121) is to go only part way. Research also needs to address not just the mechanisms, processes and circumstances of *inclusion* in local social networks, but also the other side of the dual, the mechanisms, processes and circumstances of *exclusion*. It is just as important to know why firms and enterprises are not embedded in local systems of reciprocal relations and yet continue to prosper. For the second agenda item, exploring the "... elements that contribute to the 'institutional thickness' arising from local agglomerations" (p. 121) the same point hold true. 'Institutional thickness' is by itself a profoundly unhelpful term, but it is just as important to know how it can operate to *exclude* firms as it is to know how it can *include* them and enhance their growth. Certainly the rules of meaning and membership institutionalised in places (Callon, 1986) are just as exclusionary as they are facilitative (Taylor, 1995). Yeung's final agenda item is to develop an understanding of the impact of enterprise power on local embeddedness. This, it can be argued is the single most important task facing embeddedness research (Taylor, 1995, 1999). As the control of resources, the manipulation of relationships or the exercise as discipline (Clegg, 1989,1990), power within and between enterprises and institutions again has the ability to include or exclude individual actors (Pfeffer, 1981; Wrong, 1995; also see the extensive research of the Aston Group).

To fully understand the significance of local processes of embedding requires empirical assessments of the nature of the embeddedness of firms and enterprises and the associated tissue of institutions (supportive or otherwise) in a range of national and cultural settings. Only in this way will it be possible to assess whether the model as it currently stands and the policy initiatives it has spawned can properly be universalised.

Conclusion

The purpose of this chapter has been to develop a critique of the increasingly prominent enterprise embeddedness model of local economic growth. The aim was to assess the adequacy and appropriateness of the model as an appreciation of the social and economic mechanisms involved in local growth. The need for such a critique is urgent. The embeddedness model is increasingly being used as an intellectual foundation for local policy formulation in both developed and developing countries, and its interpretation as a source of local social capital is gaining prominence. However, in the euphoria of identifying potential (and some would argue potent) mechanisms for advancing local growth, the significant limitations of the embeddedness model have been all but ignored.

The model, however can be recognised as having three variants each with a different though complementary focus:

- a *developed country variant* with a focus on local learning and innovation systems and a policy orientation aimed at the economic regeneration of declining areas and emulating growth processes in buoyant industrial districts;
- a *developing country variant* that seeks to promote local economic growth where none has occurred before, and that emphasises ambiguous processes of local social capital formation; and
- a *TNC variant* that promotes TNCs as sources of local growth through co-operative information exchange rather than viewing them as exploiters of local circumstances.

From the discussion of these three variants of the model four critical flaws can be identified in broad embeddedness thesis. First, it is fundamentally *partial*. It neglects or underplays issues such as the impact of unequal power relationships within enterprise networks and the self-centred nature of wealth creation that lies at the heart of the capitalist system. Greed is not necessarily good but it is still a motivation. It is equally partial in being supported by evidence drawn from specific and possibly quite aberrant case studies (selected on the dependent variable). Second, building on this issue of partiality, the model deals only with *'inclusion' and not 'exclusion'* in inter-firm networks. It highlights co-operation between firms but sees no selectivity in this co-operation. It certainly does not recognise that co-operation among some might be at the expense of the exclusion of others. The fact that network relationships can involve the imposition of sanctions, as in Overseas Chinese business networks, also receives no attention. The embeddedness model of local economic growth needs to be more sensitive to the dualities of process operating within local economies and local enterprise networks - for example, powerfulness versus powerlessness, flexibility versus lock-in, stability versus instability, and coping versus competitiveness – all quite distinctive aspects of the inclusion versus exclusion issue. Third, by emphasising inclusion, the model begins to *rehabilitate TNCs* as inclusionary, co-operative, local growth generators. Here TNCs are co-operative participants in local systems of information exchange: good neighbours rather than extractors of surplus value. They are equal rather than dominant participants in place-based mechanisms of learning and clustering that would enhance a locality's international competitiveness. This interpretation flies in the face of decades of research on TNC investment. It arises from the overwhelming partiality of the model and its ignoring of issues of power. Fourth, the compound of these flaws makes the model *functionalist*. Treating all networked relationships, involving individuals, enterprises and institutions, as co-operative and inclusionary, leads inevitably to the interpretation that embeddedness creates social and economic assets that instigate and sustain local growth – that it creates local social capital. It is argued here that going beyond the model's partiality, recognising exclusion and being more circumspect about TNC behaviour raises the possibility that local enterprise embeddedness has all the potential to destroy as well as create local social capital.

To build local development policies on the embeddedness model as it is currently formulated is, it can be concluded, problematic. The outcomes may not match intentions (Eraydin, this volume). Clearly, there is an urgent need to more fully unpack embeddedness as a concept, to appreciate its social and cultural specificities, and to come to grips with the far from benign influence of power within networked enterprise relationships.

Acknowledgement

The author would like to acknowledge the financial help of the Nuffield Foundation in carrying out this work as part of a project on industrial embeddedness in Fiji.

References

Allen, J. (1997), 'Economies of Power and Space', in R. Lee and J. Wills (eds) *Geographies of Economies*, Arnold, London, pp. 59-70.

Allen, J. and Pryke, M. (1994), 'The Production of Service Space', *Environment and Planning D: Society and Space*, vol. 12, pp. 453-475.

Amin, A. (1993), 'The Globalization of the Economy: an Erosion or Regional Networks', in G. Grabher (ed.), *The Embedded Firm: On the Socioeconomics of Industrial Networks*, Routledge, London, pp. 278-295.

Amin, A. and Graham, S. (1997), 'The Ordinary City', *Transactions of the Institute of British Geographers*, vol. 22, pp. 411-429.

Amin, A. and Robins, K. (1990), 'Industrial Districts and Regional Development: Limits and Possibilities', in F. Pyke, W. Sengenberger and G. Becattini (eds), *Industrial Districts and Inter-Firm Cooperation in Italy*, International Institute of Labour Studies, Geneva, pp. 185-220.

Amin, A. and Thrift, N. (1994), 'Living in the Global', in A. Amin and N. Thrift (eds), *Globalization, Institutions and Regional Development in Europe*, Oxford, Oxford University Press, pp. 1-22.

Amin, A. and Thrift, N. (1997), 'Globalization, Socio-economics, Territoriality', in R. Lee and J. Wills (eds), *Geographies of Economies*, Arnold, London, pp. 147-157.

Baker, P. (1996), 'Spatial Outcomes of Capital Restructuring: 'New Industrial Spaces' as a Symptom of Crisis, Not Solution', *Review of Political Economy*, vol. 8, no. 3, pp. 263-278.

Barnett, R. and Müller, R. (1978), *Global Reach: The Power of the Multinational*, Jonathan Cape, London.

Bergsten, C., Horst, T. and Moran, T. (1978), *American Multinationals and American Interests*, The Brookings Institute, Washington, DC.

Bianchi, G. (1998), 'Requiem for the Third Italy? Rise and Fall of a Too Successful Concept', *Entrepreneurship and Regional Development*, vol. 10, pp. 93-116.

Braczyk, H-J., Cooke, P. and Heidenreich, M. (eds) (1998), *Regional Innovation Systems*, UCL Press, London and Bristol PA.

Bresnan, M. (1996), 'An Organizational Perspective on Changing Buyer-Supplier Relations: a Critical Review of Evidence', *Organization*, vol. 3, no. 1, pp. 121-146.

Brusco, S. (1996), 'Trust, Social Capital and Local Development: Some Lessons from the Experience of the Italian Districts', in OECD (ed.), *Networks of Enterprises and Local Development: Competing and Co-operating in Local Productive Systems*, LEED for OECD, Paris, pp. 115-119.

Callon, M. (1986), 'Some Elements of a Sociology of Translation: Domestication of the Scallops and the Fishermen of St Brieue Bay', in J. Law (ed.), *Power Action and Belief: A New Sociology of Knowledge*, Routledge, London.

Clark, G. (1997), 'Rogues and Regulation in Global Finance: Maxwell, Leeson and the City of London', *Regional Studies*, vol. 31, pp. 221-236.

Clegg, S. (1989), *Frameworks of Power*, London, Sage.

Cooke, P. (1998), 'Introduction: Origins of the Concept', in H-J. Braczyk, P. Cooke and M. Heidenreich (eds), *Regional Innovation Systems*, UCL Press, London and Bristol PA, pp. 2-25.

Crew, L. (1996), 'Material Culture: Embedded Firms, Organizational Networks and the Local Economic Development of a Fashion Quarter', *Regional Studies*, vol. 30, pp. 257-272.

Department of Foreign Affairs and Trade (Australia) (DFAT), East Asia Analytical Unit, (1995), *Overseas Chinese Business Networks in Asia*, AGPS Press, Canberra.

Dicken, P. (1994), 'Global-Local Tensions: Firms and States in the Global Space Economy', *Economic Geography*, vol. 70, no. 1, pp. 101-128.

Dicken, P. and Thrift, N, (1992), 'The Organisation of Production and the Production of Organisation: Why Business Enterprises Matter in the Study if Geographical Industrialisation', *Transactions of the Institute of British Geographers, New Series*, vol. 17, pp. 279-281.

Dicken, P., Forsgren, M. and Malmberg, A. (1994), 'The Local Embeddedness of Transnational Corporations', in A. Amin and N. Thrift (eds), *Globalization, Institutions and Regional Development in Europe*, Oxford University Press, Oxford.

Dunning, J. (1979), 'Explaining Changing Patterns of International Production: in Defence of the Eclectic Theory', *Oxford Bulletin of Economics and Statistics*, vol. 41, pp. 249-265.

Ekinsmyth, C., Hallsworth, A.G., Leonard, S. and Taylor, M. (1995), 'Stability and Instability: the Uncertainty of Economic Geography', *Area*, vol. 27, no. 4, pp. 289-299.

Enright, M. (1995), 'Regional Clusters and Economic Development: a Research Agenda', in U. Staber, B. Schaefer and B. Sharma (eds), *Business Networks: Prospects for Regional Development*, de Gruyter, Berlin, pp. 190-214.

Forbes, D. and Rimmer, P. (eds) (1984), *The Geographical Transfer of Value*, Monography HG17, Department of Human Geography, Australian National University, Canberra.

Fröbel, F., Heinrichs, J. and Kreye, O. (1980), *The New International Division of Labor*, Cambridge University Press, Cambridge.

Fujita, K. and Hill R. (1995), 'Global Toyotaism and Local Development', *International Journal of Urban and Regional Research*, vol. 19, pp. 7-22.

Glasmeier, A. (1991), 'Technological Discontinuities and Flexible Production Networks: the Case of the World Watch Industry', *Research Policy*, vol. 20, pp. 469-485.

Grabher, G. (ed.) (1993a), *The Embedded Firm: On the Socioeconomics of Industrial Networks*, Routledge, London.

Grabher, G. (1993b), 'Rediscovering the Social in the Economics of Interfirm Relations', in G. Grabher (ed.), *The Embedded Firm: On the Socioeconomics of Industrial Networks*, Routledge, London, pp. 1-31.

Honig, B. (1998), 'Who Gets the Goodies? An Examination of Microenterprise Credit in Jamaica', *Entrepreneurship and Regional Development*, vol. 10, no. 4, pp. 313-334.

Hudson, R. (1999), 'The Learning Economy, the Learning Firm and the Learning Region: a Sympathetic Critique of the Limits to Learning', *European Urban and Regional Studies*, vol. 6, no. 1, pp. 59-72.

Leborgne, D. and Lipietz, A. (1992), 'Conceptual Fallacies and Open Questions on Post-Fordism', in M. Storper and A. Scott (eds), *Pathways to Industrialization and Regional Development,* Routledge, London, pp. 332-348.

Lovering, J. (1999), 'Theory Led by Policy: the Inadequacies of the New Regionalism', *International Journal of Urban and Regional Research*, vol. 23, no. 2, pp. 379-395.

McNulty, T. and Pettrigrew, A. (1999), 'Strategists on the Board', *Organization Studies*, vol. 20, no. 1, pp. 47-74.

Maskell, P. (1999), 'Social Capital, Innovation and Competitiveness', unpublished manuscript.

Maskell, P., Eskilinen, H., Hannibalsson, I., Malmberg, A. and Vatne, E. (1998), *Competitiveness, Localized Learning and Regional Development - Specialization and Prosperity in Small Open Economies*, Routledge, London.

Massey, D. (1984), *Spatial divisions of Labour*, Routledge, London.

Mohan, G. (1996), 'Neoliberalism and Decentralised Planning Development', *Third World Planning Review*, vol. 18, no. 4, pp. 433-454.

Ohmae, K. (1995), *The End of the Nation State: The Rise of Regional Economies*, Harper Collins, London.

Pfeffer, J. (1981), *Power in Organizations*, Marshfield MA, Pitman.

Phelps, N. (1997), *Multinationals and European Integration: Trade, Investment and Regional Development*, Jessica Kingsley Publishers, London.

Piore, M. and Sabel, C. (1984), *The Second Industrial Divide: Possibilities for Prosperity*, Basic Books, New York.

Polanyi, K. (1957), 'The Economy as Instituted Process' in K. Polanyi, C. Arensberg and H. Pearson (eds), *Trade and Market in Early Empires*, Free Press, Glencoe ILL, pp. 243-270.

Porter, M.E. (1998), *On Competition*, Macmillan, London.

Powell, W.W. (1990), 'Neither Markets Nor Hierarchies: Network Forms of Organization', *Research in Organizational Behavior*, vol.12, pp. 295-336.

Powell, W.W. and Smith-Doerr, L. (1994), 'Networks and Economic Life', in N. Smelser and R. Swedberg (eds), *The Handbook of Economic Sociology*, Princeton University Press, Princeton NJ, pp. 368-402.

Pratt, A. (1997), 'The Emerging Shape and Form of Innovation Networks and Institutions', in J. Simmie (ed.), *Innovation, Networks and Learning Regions?*, Regional Policy and Development Series, Jessica Kingsley Publishers, London and Bristol PA and Regional Studies Association, London, pp. 124-136.

Pryke, M. and Lee, R. (1995), 'Place Your Bets: Towards an Understanding of Globalisation, Socio-Financial Engineering and Competition Within a Financial Centre', *Urban Studies*, vol. 32, pp. 329-344.

Putnam, R.D. (1993), *Making Democracy Work: Civic Traditions in Modern Italy,* Princeton University Press, Princeton NJ.

Qalo, R. (1996), *Small Business, A Study of a Fijian Family: The Mucanabitu Iron Works Contractor Cooperative Limited*, University of the South Pacific, Suva, Fiji.

Rasmussen, J. *et al.* (1992), 'Introduction: Exploring a New Approach to Small-Scale Industry', *IDS Bulletin*, vol. 23, no. 3, pp. 2-6.

Sabel, C. (1989), 'Flexible Specialization and the Re-emergence of Regional Economies', in P. Hirst and J. Zeitlin (eds), *Reversing Industrial Decline? Industrial Structure and Policy in Britain and Her Competitors,* Berg, Oxford, pp. 17-70.

Schmitz, H. (1993), 'Small Shoemakers and Fordist Giants: Tales of a Supercluster', *IDS Discussion Paper 331*, IDS, Sussex.

Scott, A.J. and Storper, M. (1992), 'Industrialization and Regional Development', in M. Storper and A.J. Scott (eds), *Pathways to Industrialization and Regional Development*, London, Routledge, pp. 3-17.

Simmie, J. (ed.), *Innovation, Networks and Learning Regions?*, Regional Policy and Development Series, Jessica Kingsley Publishers, London and Bristol PA and Regional Studies Association, London.

Staber, U. (1996), 'Accounting for Differences in the Performance of Industrial Districts', *International Journal of Urban and Regional Research*, vol. 20, no. 2, pp. 299-316.

Storper, M. (1997), *The Regional World: Territorial Development in a Global Economy*, New York and London, The Guilford Press.

Taylor, M. (1995), 'The Business Enterprise, Power and Patterns of Geographical Industrialisation', in S. Conti, E. Malecki and P. Oinas (eds), *The Industrial Enterprise and Its Environment: Spatial Perspectives*, Aldershot, Avebury, pp. 99-122.

Taylor, M. (1999) 'Enterprise, Embeddedness and Exclusion: Buyer-Supplier Relations in a Small Developing Country Economy', paper presented at the meeting of the IGU Commission on the Organisation of Industrial Space, Haifa, Israel.

Taylor, M. (2000), 'Enterprise, Power and Embeddedness: an Empirical Exploration', in E. Vatne and M. Taylor (eds), *The Networked Firm in A Global World*, Ashgate, Aldershot, pp. 199-234.

Taylor, M. and Conti, S. (eds) (1997), *Interdependent and Uneven Development: Global-Local Perspectives*, Ashgate, Aldershot.

Taylor M. and Thrift, N. (eds) (1986), *Multinationals and the Restructuring of the World Economy*, Croom Helm, London.

The Economist (1995), *A Survey of Multinationals: Big Is Back*, vol. 335, no. 7920, June, pp. 24-30.

Thrift, N. (1998), 'The Rise of Soft Capitalism', in A. Herod, S. Roberts and G. Ó. Tuathail (eds), *An Unruly World: Globalization, Governance and Geography*, Routledge, London and New York, pp. 25-71.

Thrift, N. (1999), 'The Globalisation of the System of Business Knowledge', in K. Olds, P. Dicken, P. Kelley, L. Kong and H. Yeung (eds), *Globalisation and the Asia Pacific: Contested Territories*, Routledge, London and New York, pp. 57-71.

Uzzi, B. (1996), 'The Sources and Consequences of Embeddedness for the Economic Performance of Organizations: the Network Effect', *American Sociological Review*, vol. 61, pp. 674-698.

Uzzi, B. (1997), 'Social Structure and Competition in Interfirm Networks: the Paradox of Embeddedness', *Administrative Science Quarterly*, vol. 42, pp. 35-67.

Wrong, D.H. (1995), *Power: Its Forms, Bases and Uses*, New Brunswick and London, Transaction Publishers.

Yang, M. (1987), 'The Gift Economy and State Power in China', *Comparative Studies in Society and History*, vol. 31, pp. 25-54.

Yeung, H. (1997a), 'Cooperative Strategies and Chinese Business Networks: a Study of Hong Kong Transnational Corporations in the ASEAN Region', in P. Beamish and J. Killing (eds), *Cooperative Strategies: Asia-Pacific Perspectives*, The New Lexington Press, San Francisco CA.

Yeung, H. (1997b), 'Business Networks and Transnational Corporations: a Study of Hong Kong Firms in the ASEAN Region', *Economic Geography*, vol. 73, no. 1, pp. 1-25.

Yeung, H. (1998a), 'Capital, State and Space: Contesting the Borderless World', *Transactions of the Institute of British Geographers, New Series*, vol. 23, pp. 291-309.

Yeung, H. (1998b), 'The Social-Spatial Constitution of Business Organizations: a Geographical Perspective', *Organizations*, vol. 5, no. 1, pp. 101-128.

3 Analysing Local Growth Promotion: Looking Beyond Employment and Income Counts

Daniel Felsenstein

Introduction

This chapter argues that the analysis of local growth promotion means looking beyond increases in employment, income or the tax base. While these are the tangible expressions of growth promotion at the local level, local growth implies much more than is captured in these simple aggregate measures. An integral part of local growth relates to welfare and distributional outcomes. At stake here is not just the identification and measurement of how much growth but also the question of how that growth is distributed. If extra jobs and income simply accrue to those segments of the population that are least in need, then not much has been achieved. At the local or regional level, this is a particularly acute issue. Labour mobility and the structure of labour markets are such that even if growth is promoted at a certain point in space, there is no guarantee that the beneficiaries of that growth will be the local population. In-migration, commuting and reverse labour flows all ensure that promoting growth locally is no safeguard that the outcomes will be felt locally.

'Growth' is often juxtaposed with 'development'. The former is said to express the objective, measurable, short term and aggregate components of economic progress while the latter captures the structural, qualitative, long-run and distributional outcomes of that progress (Malizia and Feser, 1999). The present chapter argues that such a distinction may be unhelpful and may even serve to smokescreen the important issue of goals: 'growth for who?' and 'development for who?' In the absence of any clearly defined goals, we are left with two further ambiguous and 'fuzzy' concepts (Markusen, 1999). The goal suggested here is that of welfare improvement. Couched in these terms, the distinction between growth and development becomes artificial. It perhaps boils down to time scale (i.e. growth means progress over the short time while development relates to the medium or long term) or ease of measurement, but no more than that. In the absence of a pre-defined goal for growth and development, all attempts to drive a wedge between the two terms, result simply in competing lists of measurement indices. Without an explicitly stated value system for promoting growth and / or development, we are left with definitions bereft of context.

The context suggested here is that of local welfare improvement. While the concern for local welfare and equity issues is hardly new in the economic development debate (see Richardson, 1972; Smith, 1970), this chapter goes on to suggest that the current preoccupation with revisiting past ideas on regional growth (viz. the 'new' economic geography, 'new' regionalism, 'new' endogenous growth theory, 'new' industrial districts and so on) runs the risk of loosing sight of the distributional effects of growth. These tend to get lost amongst the new sectors, regions and institutions that are taken as the motors of growth. The emphasis on new industries and places also means that people get forgotten. This is ironic as much of the celebrated success of the new industries and regions is due to the human capital, community-based networks and personal initiative of individuals. This chapter therefore champions the cause for putting people and their welfare back at the centre of the debate on promoting local growth.

Finally the chapter identifies some of the key welfare and distributional issues that need to be addressed when discussing growth promotion. These include first and foremost the quality and stability of jobs and not just their gross number, the distribution of income and not just its aggregate change and the translation of improved personal welfare into an improved local or community quality of life. While the normative identification of the issues that growth promotion needs to address is fairly easily achieved, the incorporation of these ideas into a framework for evaluation is more difficult. The methodological implications of incorporating these concerns into analyses of growth promotion are considerable. But therein lies the crux. In the absence of suitable methods of analysis we will still be unaware as to whether promoting growth at the local level is really making a difference. This chapter sketches out a first-cut approach to this issue.

'Growth' versus 'Development': Measuring Everything, Valuing Nothing

As noted above, in the absence of a particular value system, the growth versus development debate is devoid of context. The context suggested here is that of welfare improvement. Without it, the debate over definitions does not progress much beyond different ways for counting jobs and income. Early discussions of this issue went to great pains to make the case for separating the two concepts and painstakingly outlined distinctions grounded in theoretical roots, time-scale, origins, geography and the like (Flammang, 1979; Arndt, 1981).

The regional growth literature has similarly expended much energy in finding suitable indices for measuring each of the terms and thereby juxtaposing them. The result is a pot-pourri of defining variables that purport to make the distinction between the two concepts: quantitative progress versus qualitative progress, expansion versus distribution, efficiency versus equity, increase in aggregate output versus structural change, income change versus income distribution, number of jobs versus quality of jobs, static efficiency versus dynamic efficiency, 'hard' versus 'soft' economic structures and so on (Flammang, 1990; Malizia, 1990; Lipshitz, 1993). Some studies do not even make their definitions of growth or development explicit and assume that the context provides the definition. For example, a recent symposium on 'generating growth' via small firms (O'Neill, MacNabb and Ram, 1999), implies that jobs, output, entrepreneurship and innovation all make for economic progress although who exactly gains is never specified. Thus, while some of the above indices do relate directly to measuring improvements in welfare, the implication of the search for dichotomous defining variables is that broader systemic change in a locality or region is only possible through 'development'.

Reese and Fasenfest (1997) suggest a way out of this conceptual cul-de-sac by suggesting a 'social economy perspective' for analysing growth promotion. This puts less stress on dichotomous definitions and more on assessing indices in terms of their contribution to societal welfare (i.e. their social contribution), This can only be assessed on the basis of weighing up alternatives and foregone opportunities to society on the basis of the resources allocated to a particular growth promotion programme. In other words, the important issue is the opportunity cost of the variable being suggested and not the narrow definitional issue. Only on this basis do we have an idea of the value of promoting growth. Without some idea of opportunity costs we are left we attempts at measuring everything but valuing nothing. It is for this reason that the judicious identification of different indices for making the distinction between 'growth' and 'development', may be missing the point.

The search for distinction between the terms has also led to an attempt to anchor 'growth' and 'development' in different theoretical models of regional progress. Malizia and Feser (1999) suggest that economic 'growth' is treated in the stock models of regional development such as economic base theory, neoclassical growth theory, and inter-regional trade theory. Economic 'development' on the other hand, is dealt with in the product cycle model, growth pole theory (despite the name!), entrepreneurship theory and flexible production theory.

Interestingly, this latter group comprises most of the 'soft' models of local economic growth recently identified and empirically tested by Plummer and Taylor (2001). In a path-breaking analysis, they have attempted to add some empirical flesh to the bones of the 'soft' models currently fashionable in economic geography. Using these alternative frameworks, they attempt to explain regional job growth in Australia over the period 1984-1992. A leading finding emerging from this work is that promoting local economic growth is more heavily contingent on the local human resource base than on other popularly touted factors (institutions, information etc.) This finding leads full-circle to our plea for a greater welfare focus in analysing the promotion of regional growth.

Bringing the People Back In

A welfare perspective on promoting growth means a people-centred focus on growth and its outcomes. An irony of the present interest in re-examining old sagas in regional growth is that people and their welfare are no longer of central interest. In the search for new motors of regional growth across new industries, new regions and new institutions, people seem to have been over looked. This is not to say that the down-side of the celebrated success stories of regional growth has gone un-noticed. High growth regions do not necessarily promote a high quality of life for all their citizens. The 'overheating' of Silicon Valley and its attendant problems of degraded physical environment, stressful life-style and regional governance problems are well documented (Saxenian, 1984, 1994; Henton, 1999). Labour conditions and problems of marginality and stability in some of the acclaimed industrial districts are also noted (Bianchi, 1998). The surprising thing however, is that while acclaiming the central role of human capital, personal networks and individual initiative in much of the 'new' interest in regional growth, little attempt is then made to estimate whether in fact all segments of the population are better off, for all this activity. Their welfare seems to have got lost amongst the new sectors, regions and institutions. In some respects, the emphasis suggested here, echoes an earlier call issued by Malecki (1989) in the context of research on innovative industries, asking 'what about people in high technology'?

A case in point is that of 'social capital'. The notion reflects an accumulation of past experiences that affect present behaviour (Bolton, 1998). Regional growth researchers are of course interested in the case where this accumulation occurs within a given territory. The spirit of Putnam (1993) pervades much discussion of the informal, non-institutionalised arrangements that exist between firms for information exchange and dissemination, inter-firm trust and co-operation and 'learning' (i.e. industrial adjustment). These are the rather intangible factors that act as the glue preserving the coherence of regional agglomerations.

Social capital also underlies the interest in networks. This is of particular significance in the context of promoting growth, due to the many network-based 'success stories' (e.g. Jutland, Baden Wurttemburg, Emilia Romagna). The 'network paradigm' outlines the key ingredients of this regional growth: reciprocity, trust, learning, partnership and empowerment (Cooke and Morgan, 1993; Cooke, 1996). This ostensibly puts people at the centre of the universe and emphasises the role of in inter-personal relationships moulded in the past and affecting present behaviour. In a regional context, a cosy view of civic engagement completes the picture, presumably animated by the key ingredients of partnership and empowerment.

The result is an efficient system. Networks, sharing, reciprocity and the formation of clubs make for dynamic scale economies and cost reduction lowering the cost of membership to all involved (Hansen and Echeverri-Carroll, 1997). Little however is mentioned with respect to distributional outcomes. Do networks favour one segment of the population over another? Does social capital formation have greater implications for high wage labour than for low wage labour? Presumably 'partnership' and 'empowerment' indicate some form of welfare outcome although this is hardly made specific and often overlooked in discussions of reciprocity, trust and learning. In fact, the ramifications of the social capital /networks model for effecting welfare change have not been specifically addressed. (For an exception, although bereft of any local growth implications, see Thurow, 1975).

Furthermore, a recent reassessment of the role of social capital in the promotion of the ultimate network-based, regional growth model (Silicon Valley), finds little evidence of the dense networks of civil engagement, cumulative experiences, deep historical roots, familial ties and other people-based indicators that so inform the network paradigm of regional growth (Cohen and Fields, 2000). In this work, Silicon Valley is a lonely, atomistic world with rapid population turnover; an antithesis to regional growth grounded in time-honoured traditions and business practices. While social capital and trust do accumulate in this environment, they are of a very different kind. They are grounded in performance rather than community, commercial reputation rather than civic engagement. Paradoxically therefore, people and their welfare are not on the top of the agenda in the model of local growth grounded in 'social capital'.

An Approach to Analysing the Welfare Outcomes of Promoting Growth

We now suggest an analytic framework for analysing local growth promotion. The point of departure (and arrival) for most analyses in this area is jobs. While political expediency often dictates the equation of job creation with 'growth', there is no assurance that this growth will in fact occur locally. And, even if the effects are felt at the local level there is no guarantee that they will be targeting the 'right' sections of the population and improving their welfare. Often growth of this kind brings aggregate benefits to those least in need: employment to high skilled workers who already hold jobs or income increases to those who have little claim on the local community such as in-migrants. The issue here is one of distribution and economic welfare.

However, in much analysis of the effects of growth, outcomes are invariably overstated. Jobs, income or taxes are attributed to growth promotion programmes, as if no alternatives ever existed. The result is that welfare gains are inflated. In the area of job generation for example, the failure to recognise that workers often have alternatives to the jobs created, means that impacts are over-ascribed. Opportunity cost theory teaches us that the welfare gain to the individual worker from a new job will not in general be equal to that worker's wage. Rather it will be a much smaller amount, equal to the difference between wages on the new job and the worker's wage in his or her next best alternative (Jones, 1989). In the extreme case of smooth and perfectly functioning labour markets, these alternative wages will be close to the wage level on a new job. They indicate the value of alternative production given up when a worker shifts to a new enterprise. In such a situation, simply counting additional employment may not teach us very much about the effectiveness of growth-promoting measures (Courant, 1994).

Conversely, in depressed or high unemployment areas, measures to promote growth through job creation cannot be assessed on the same basis.

These kinds of regions are likely to be characterised by lower opportunity costs, at all levels of employment. While labour markets for high level employment are likely to clear at wage levels close to national levels, for low skilled labour in depressed regions, labour markets are unlikely to clear. Few alternatives exist short of welfare benefits and unemployment pay.

Bearing this in mind, we can suggest a three-stage approach to analysing the welfare effects of promoting growth (Figure 3.1). The framework suggested is purely indicative of the types of issues that need to be tackled when analysing the outcomes of growth promotion. What we are suggesting is a simple accounting approach that could be estimated in a cost-benefit framework. We do not present here a detailed enumeration of all costs and benefits. Nor do we attempt to affix values to the parameters suggested, or even to suggest how some of the concepts such as 'job quality' may be measured. Rather, a framework for analysis is presented and more work would be needed in order to make it fully operational.

The first stage in the analysis involves assessing the employment outcomes of the growth-promoting policy. This however does not just involve a simple count of jobs. Rather, an attempt should be made to assess job quality (q) and stability (s) and not simply aggregate employment increase. Only jobs that meet these criteria are considered in the assessment. This invariably involves a downward adjustment of any reported employment figures. In most job-generation estimates resulting from growth promotion programmes (attracting new firms to a region, assisting local entrepreneurs etc.), the guiding principle is that 'a job is a job'. In line with our welfare perspective however, here we only count jobs that are likely to last and those that improve the quality of work offered to the individual. The result is a net employment estimate (E^*).

The next stage involves translating this employment increase (E^*) into wage gains (W). Again, this step needs to be further refined. Going beyond a simple wage count and to looking at the welfare effects of this wage increase, calls for an opportunity cost adjustment to be made. To do this, wage gains needs to be converted into some simple form of wage distribution across various wage classes. The wage gains for each class then need to be adjusted to account for opportunity costs ($\delta_{1 \ldots n}$) for n wage classes. Higher wage classes will have higher opportunity costs and therefore a growth-promoting programme will be given less credit in creating their wage gain. Low wage classes will have low opportunity costs: most of their wage gain can be credited to the growth promotion programme. This adjustment can also be effected across different regions ($\delta_{1 \ldots k}$). High growth regions will have lower unemployment and thus higher opportunity costs (many viable employment alternatives) than low growth areas. This would give a welfare-based analysis of the benefits of any local growth-promoting programme. The costs side of the balance sheet, which is not outlined here, would simply be involved in the judicious counting of all public outlays involved in the growth-promoting programme. Calculating the difference between local benefits and costs (π), gives us the local welfare effect of a given programme, once we have accounted for opportunity costs. The future stream of this balance, assuming for example that the programme has a 5 or 10 year time horizon, is denoted as NPV(π) (Figure 3.1).

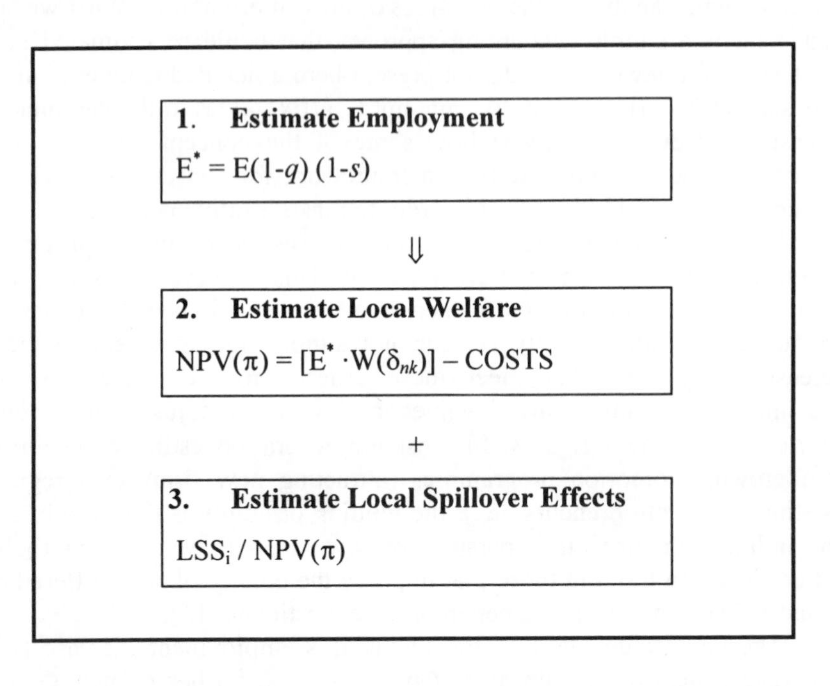

Figure 3.1 A welfare-based framework for analysing local growth promotion

Variable Definitions:
E = total employment; E^* = net employment, W = monthly wage;
π = balance of benefits and costs; LSS = local spillover score

Parameter Definitions:
q = employment quality; s = employment stability;
δ_{nk} = opportunity cost of labour for n different skill levels and k regions

To complete the picture, we need to look at growth-promotion in the wider context of not simply improving the welfare of the individual but also contributing to the community or area. In this third step, we therefore attempt to account for all the positive and negative 'areal spillover' effects connected to the growth promotion programme. For example, if the analysis was centred on the promotion of a local science park, after accounting for all (stable and quality) employment attributable to the park and all (opportunity cost-discounted) wage gains, the next step would be to attempt to capture all the non-price spillovers accruing to the community from proximity to the park, for example improved investor confidence in the area, increased traffic congestion, increased local prestige etc. These are subjective outcomes and not as readily quantifiable as jobs, wages and taxes. They have to be scored as an index if they are to be incorporated into a consistent analytic framework.

The sum of these scores would represent a 'local spillover' score for the *i*th programme (LSS_i). While these qualitative (and subjective) index scores cannot be readily translated into a monetary figure and incorporated into a cost-benefit framework, total scores can be compared with the present value of the monetary balance between costs and benefits (π) (Figure 3.1). Dividing the local spillover scores by the present value of this balance ($NPV(\pi)$) will give an indication in the change of local quality of life effected by a change in personal welfare. The success of the programme in generating local spillover effects will therefore be expressed as the spillover score per unit cost (or benefit) of the programme. This will yield a relative measure of quality of life outcomes and can be useful for comparing across different growth-generating programmes.

Another issue that could perhaps be incorporated into analysing the effect of a welfare improvement on the local community, relates to the fiscal effects of this improvement. Does promoting growth lead to increased fiscal performance? While this issue is not dealt with in the framework for analysis presented above, the traditional view is that localities and regions that engage in growth-promotion often sacrifice fiscal stability in the short term for the prospect of economic growth in the long term. Communities that offer tax breaks and subsidise infrastructure in order to attract outside firms might run fiscal deficits and declining public services in the hope that more jobs and rising incomes will result in the future.

The opposite result however can also occur: local fiscal stability in the short term but with little prospect for promoting growth over the medium term. This happens when ostensibly growth promoting projects (especially in the tourism and entertainment industries) fund local services and infrastructure in return for being granted an operating license. These industries then proceed to establish low-paying, service-type jobs that serve local markets. The result is growth redistribution rather than growth promotion and expansion. It can thus be seen that 'promoting local growth' is not synonymous with 'promoting fiscal growth'. Each can be achieved without the other.

Conclusions

The heart of the welfare-based approach to analysing growth promotion presented above is the differential weighting system used for assessing the opportunity cost of labour at different wage levels. Any such weighting scheme has to be soundly justified. Opportunity cost theory provides a justification for assigning greater credit to those jobs at lower wage levels and in more distressed regions. Adopting such as approach means merging efficiency and equity interests. On the one hand, the distribution of jobs affects the increase in welfare attributable to a growth-promoting programme (i.e. programme efficiency). On the other hand, discounting programme impacts on the basis of opportunity costs adds a welfare-based perspective that moves beyond simple job and income counts.

The argument presented here has been based on the analysis of measures to promote growth. Little has been said about the necessity of local growth promotion policies in the first place. The regional growth literature takes them as a sine qua non. However, the openness of local economies can call this assumption into question. At each stage in the 'new jobs lead to rising income that leads to rising demand that leads to greater fiscal stability that leads to a rising quality of life' scenario, some form of 'leak-out' can occur that breaks the chain of growth. New jobs can be taken by non-locals; rising incomes may not translate into rising local demand; fiscal stability may not transform smoothly into a rising quality of life and so on. As Courant (1994) has succinctly put it:

> Building a local development strategy around subsidizing the next innovation that will spawn the next Silicon Valley, is likely to be about as successful as building a strategy around the possibility that there is a lot of gold buried in your backyard. It is possible, but not likely, and not worth it on the average. (Courant, 1995, p. 874)

If so, is there any justification for local growth-promotion in the first place? Following Bolton (1992) and in line with the welfare emphasis presented here, we can suggest that there might be a market failure in the process of generating local growth. The existence of this market imperfection may provide the justification for growth promotion policies.

This failure relates to the provision of local public goods, not in terms of local infrastructure such as schools and education, but in terms of promoting a local 'sense of place' or 'community'. A 'sense of place' represents local social capital of the type that is said to form the glue that binds together 'clusters', 'new innovations spaces' and 'industrial districts'. Trust, reciprocity and non-traded relationships breed off this sense of place. While it is an intangible, place-specific asset, it provides returns to the locality or the region. It can also be considered a local public good as it is characterized by non-rivalry and non-exclusion in consumption and by returns to investment that accrue to the public at large and not just to the original investors.

As with other public goods, private markets may fail to allocate the necessary resources to promote a sense of place. To the extent that promoting a sense of place has a wider spillover effect on a locality or region and creates a valuable social asset and demonstration effect for other localities, then growth promotion would seem justified. In fact some of the stock policy prescriptions for promoting growth especially in depressed areas can be re-interpreted as promoting a sense of place. For example, attracting external investment in order to provide local jobs, income and taxes in a declining region can also be viewed as promoting a sense of place and preservation of social capital. In its absence, community fabric would be destroyed by selective out-migration and valuable, intangible local assets would be lost. In this respect promoting local growth would also be promoting local welfare.

References

Arndt, H.W. (1981), 'Economic Development: A Semantic History', *Economic Development and Cultural Change*, vol. 29, pp. 457-466.

Bianchi, G. (1998), 'Requiem for the Third Italy? Rise and Fall of a Too Successful Concept', *Entrepreneurship and Regional Development*, vol. 10, pp. 93-116.

Bolton, R. (1992), 'Place Prosperity vs. People Prosperity: An Old Issue with a New Angle', *Urban Studies*, vol. 29, no. 2, pp. 185-203.

Bolton, R. (1998), *A Critical Examination of the Concept of Social Capital*, Paper presented at the Annual Conference of the AAG, Boston, March.

Cohen, S.S. and Fields G. (2000), 'Social Capital and Capital Gains: An Examination of Social Capital in Silicon Valley', in M. Kenney (ed.), *Understanding Silicon Valley: The Anatomy of an Entrepreneurial Region*, Stanford University Press, Stanford, CA, pp. 190-217.

Cooke, P. (1996), 'Building a Twenty-First Century Regional Economy in Emilia Romagna', *European Planning Studies*, vol. 4, no. 1, pp. 53-62.

Cooke, P. and Morgan K. (1993), 'The Network Paradigm; New Departures in Corporate and Regional Development', *Society and Space*, vol. 11, pp. 543-564.

Courant, P.N. (1994), 'How Would You Know a Good Economic Development Policy If You Tripped Over One? Hint: Don't Just Count Jobs?', *National Tax Journal*, vol. 47, no. 4, pp. 863-881.

Flammang, R.A. (1979), 'Economic Growth and Economic Development: Counterparts or Competitors?', *Economic Development and Cultural Change*, vol. 28, pp. 47-61.

Flammang, R.A. (1990), 'Development and Growth Revisited', *Review of Regional Studies*, vol. 20, pp. 49-55.

Hansen, N. and Echeverri-Carroll E. (1997), 'The Nature and Significance of Network Interactions for Business Performance and Exporting to Mexico: An Analysis of High Technology Firms in Texas', *Review of Regional Studies*, vol. 27, no. 1, pp, 85-99.

Henton, D. (1999), *An Innovative Region Takes its Next Step: Governance in the Global City Region - Lessons from Silicon Valley*, paper presented at the UCLA Global City-Regions Conference.

Jones, S. (1989), 'Reservation Wages and the Cost of Unemployment', *Economica*, vol. 56, pp. 225-246.

Lipshitz, G. (1993), 'The Main Approaches to Measuring Regional Development and Welfare', *Social Indicators Research*, vol. 29, pp. 161-181.

Malecki, E.J. (1989), 'What About People in High Technology? Some Research and Policy Considerations', *Growth and Change*, vol. 20, no. 1, pp. 67-79.

Malizia, E.E. (1990), 'Economic Growth and Economic Development: Concepts and Measures', *Review of Regional Studies*, vol. 20, pp. 30-36.

Malizia E.E. and Feser E.J. (1999), *Understanding Local Economic Development*, Center for Urban Policy Research, New Brunswick, NJ.

Markusen, A. (1999), 'Fuzzy Concepts, Scanty Evidence, Policy Distance: The Case for Rigor and Policy Relevance in Critical Regional Studies', *Regional Studies*, vol. 33, no. 9, pp. 869-994.

O'Neill, K., MacNabb, A. and Ram, M. (1999), 'Generating Growth: Guest Editorial', *Environment and Planning C, Government and Policy*, vol. 17, pp. 508-510.

Plummer, P. and Taylor, M. (2001), 'Theories of Local Economic Growth (Part 2): Model Specification and Empirical Validation', *Environment and Planning A*, vol. 33, pp. 385-398.

Putnam, R.D. (1993), *Making Democracy Work: Civic Traditions in Modern Italy*, Princeton University Press, Princeton, N.J.

Reese, L.A. and Fasenfest, D. (1997), 'What Works Best? Value and the Evaluation of Local Economic Development Policy', *Economic Development Quarterly*, vol. 11, no. 3, pp. 195-207.

Richardson, H. (1972), *Regional Economics: Location Theory, Urban Structure and Regional Change*, Weidenfeld and Nicholson, London.

Saxenian, A. (1984), 'The Urban Contradictions of Silicon Valley: Regional Growth and the Restructuring of the Semiconductor Industry', in L. Sawers and W.K. Tabb (eds), *Sunbelt Snowbelt: Urban Development and Regional Restructuring*, Oxford University Press, Oxford, pp. 163-197.

Saxenian, A. (1994), *Regional Advantage: Culture and Competition in Silicon Valley and Route 128*, Harvard University Press, Cambridge MA.

Smith, D.M. *Human Geography: A Welfare Approach*, Edward Arnold, London.

Thurow, L.C. (1975), *Generating Inequality: Mechanisms of Distribution in the US Economy*, Basic Books, New York.

4 Knowledge-Based Industry for Promoting Growth

Sam Ock Park

Introduction

One of the most important but less well understood phenomena at the beginning of the 21st century has been a shift toward knowledge-based economic activity as the foundation of comparative advantage in modern industrialised countries. There have been two major changes in the global economy since the 1980s. First, the output of the world's science and technology system has grown rapidly, and second, the nature of investment has changed, with intangible investment having grown more rapidly than physical investment (Miller, 1996). There is now greater complementarity between physical and intangible investments, with high technology being more important in both kinds of investment (Miller, 1996). Even in the newly industrialised countries, the growth of technology intensive industries, the increase of R&D activities and the growth of the knowledge intensive producer services have become common features in recent years. In this changed structure of productive assets, knowledge has become recognised as a fundamental resource (OECD, 1996; World Bank, 1998). The development of information and communication technology (ICT) and globalisation have promoted this shift toward a knowledge-based economy.

Recognition of the economic and social significance of knowledge is not new. In Western society, Francis Bacon asserted long ago that 'knowledge is power', and similar views have long been held in oriental societies. Indeed, "Adam Smith refers to new layers of specialists who are men of speculation and who make important contributions to the production of economically useful knowledge" (OECD, 1996, p. 11). The role of knowledge has become increasingly important through history and has periodically come to the fore as implied in Kondratiev's long waves.

However, if the importance of knowledge in the economy is not a new idea and was recognised several centuries ago, then why should the shift towards a knowledge-based economy be emphasised now?

There would appear to be three main reasons for emphasising the most recent shift towards a knowledge-based economy. First, new job creation is predominantly occurring in knowledge-intensive economic activities, which include both the knowledge industry and knowledge-based industry. Knowledge industries are those sectors whose output is knowledge, in forms such as patents, inventions, and new products, as well as services that are mainly knowledge. Knowledge-based industries are those sectors whose main product or service is dependent on technology or knowledge. Knowledge industries and knowledge-based industries are interdependent since the output from the former and input into the latter. Together they make up the knowledge-based economy.

Second, a significant part of the workforce in modern industrial countries comprises knowledge workers such as information system designers, managers, professionals, educators, scientists, skilled manufacturing teams, and the like. It is expected that the traditional 'job' will fade and knowledge entrepreneurs will replace traditional service and factory workers in the more flexible workplace of tomorrow (Halal, 1996). In the most of industrialised countries, the principal contribution to employment growth has been the advanced service sector which creates and uses knowledge products in exactly the way that manufacturing transforms raw materials into physical products (Berry *et al.*, 1997). Concurrently, new technology has been adopted in manufacturing and, as a result, there has been a dramatic growth of high-tech industrial production involving substantial R&D inputs. These same trends are also evident in newly industrialised countries in recent years (Clark and Kim, 1996). In these places, too, knowledge industries and knowledge-based industries are driving forces in the knowledge-based economy.

Third, the rapid development of ICT has promoted the shift toward the knowledge-based economy. Many characteristics of the knowledge-based economy, such as the new dynamics of tacit and codified knowledge, the growing importance of networked knowledge, and the acceleration of process of interactive learning, are related with the increasing use of ICT (Freeman and Soete, 1993). The rapid diffusion of ICT to developing countries in recent years has reduced the information gap between industrialised and developing countries, making the knowledge-based economy more important in the global society as a whole.

The widespread emergence of the knowledge-based economy and its impact of economic activity posses a challenge for regional development theory and for regional and industrial policy to understand and come to grips with the dynamics of economic space in the 21st century. The emergence of the region and the locality as the main arenas of growth in a globalising world has been well recognised over the last two decades. Much growth at the regional and local level results from the growth of new industries, which include cultural, health, high-tech and knowledge industries. Basically, these new types of industries are knowledge-based activities. The new motors of growth driving the emergence of new growth regions have received less attention in economic geography. The purposes of this chapter is to begin to explore the role of knowledge-based industries in promoting regional growth and to derive some directions for regional and industrial policy to promote growth in these new circumstances.

Knowledge, Development, and ICT

Knowledge is a crucial and well-recognised factor for advancing economic and social development. The difference between poor countries and rich ones can be ascribed not only to less capital but also to less knowledge (World Bank, 1998). The results of empirical studies by the OECD reveal that the overall economic performance of the OECD countries is increasingly based on their stocks of knowledge and their learning capabilities (OECD, 1996). Developing countries have less knowledge about technology than industrialised countries, creating knowledge gaps. Even within countries, knowledge gaps in acquiring, absorbing, and communicating knowledge exist among regions. But, the knowledge gaps between industrialised and developing countries are much greater because knowledge is often much more costly to create in the developing country context.

The model of national or sector scale knowledge production functions, formalised by Griliches (1979), also supports the positive relationship between knowledge and development. In the knowledge production function model, firms exist exogenously and then engaged in the pursuit of new economic knowledge as an input into the process of generating innovative activity. Considerable empirical evidences suggest that there is a strong relationship between R&D and innovative output, when R&D is measured as either patents or new product innovation (Audretsch, 1999). Some empirical analysis also shows that investment in knowledge is characterised by increasing returns to scale (OECD, 1996). However, the model of knowledge production functions is particularly weak when small firms are included in analysis. This result at the firm level is consistent with the results of some empirical studies which suggest that formal R&D is concentrated among the largest corporations and small firms account for a disproportionate share of new product innovation, given their low R&D expenditures (Acs and Audretsch, 1990).

In industrial engineering and industrial production, knowledge and technology are closely correlated, since knowledge of technology is emphasised in these fields. In the knowledge-based economy, however, not only knowledge of technology but also knowledge of advanced services, such as management, marketing, financing and so on, is also important. In the knowledge-based economy product innovation is not limited to tangible, manufactured products, but it also includes intangible products, such as new financial products and new management systems which are mainly based on new knowledge and ideas. However, some high technology sectors require high levels of knowledge, while others require relatively lower levels of new knowledge. Equally, some low technology sectors and activities require high levels of knowledge, while others require lower levels, as would be expected. Owing to the inconsistent relationship between knowledge and technology in the knowledge-based economy, it is possible for low technology industries to join the knowledge-based industry through the application of new knowledge. These two dimensions of knowledge and technology will be elaborated more fully in the next section of the chapter.

The ICT revolution has generated knowledge and technology gaps between countries, regions and firms in the global economy. The development of ICT has accelerated radical change in the way knowledge is produced, stored, and diffused. With communication costs plummeting in the last two decades, especially optical fibre transmission costs, the cost of sending information and knowledge is cheaper than ever (World Bank, 1998). These reduced costs and rapidly narrowing knowledge gaps have the potential to raise levels of growth and well-being in developing economies. In the long run, the development and accelerated use of ICT offers three development trajectories:

- the faster and cheaper generation of tools and instruments for basic research and R&D;
- the enhanced ability to create technological options; and
- the growing power of electronic networks as research tools (OECD, 1996).

On offer are new productivity gains through the generation, distribution and exploitation of knowledge in the knowledge-based industries.

In reality, however, while ICT is pervasive in all sectors of industry and economic life, it does not seem to have produced the expected rise in productivity in all economic sectors, creating a productivity paradox. This paradox - the apparent contradiction between high levels of investment on ICT and low levels of industrial productivity gain - can easily be explained in less developed regions compared to advanced regions because of their relatively lower capacity to absorb ICT investment aimed at improving productivity growth. Investment on ICT may not have a direct impact on productivity in some regions because it is not always easy to apply ICT to a region's specialist, traditional industries. In addition, ICT may be associated with sectors, such as high-tech manufacturing and the information industries, that are not well developed in a region.

However, even though ICT can reduce knowledge gaps, developing countries are still at a disadvantage because of their lack of knowledge about such things as product quality and the creditworthiness of firms (World Bank, 1998). For example, the financial crises in East Asian countries in 1997 were related to knowledge limitations. Moreover, tacit knowledge, which has been identified as the foundation of innovation and local growth, is locationally 'sticky' and can not easily be transferred from place to place, even though ICT has speeded the diffusion of information and codified knowledge between countries (von Hippel, 1998; Malmberg *et al.*, 1996). However, in the long term, investment in and the spread of ICT will improve productivity even in traditional industries and peripheral areas.

Because of the uneven distribution of knowledge in the space and uneven increase of productivity by sectors and regions, appropriate regional policy with regard to knowledge-based economy is urgently needed, especially in the less favoured regions. Considering the complexity in creating, acquiring, and using knowledge, more detail examination of the knowledge-based industries and regions is necessary to understand the process of promoting regional growth.

Knowledge, Knowledge-Based Industry and Regions

The Knowledge Production Function and Knowledge Spillover

Based on empirical studies, the most important source of new economic knowledge is R&D. A high level of human capital, a skilled labour force, and a strong local presence of scientists and engineers are also important factors generating new knowledge. Because of this close link between knowledge inputs and innovative output, empirical analyses using aggregated data also supports this model of knowledge production (Audretsch, 1999). However, knowledge can be generated through a diversity of learning modes. In some activities, knowledge is accumulated by informal mechanisms, such as learning-by-doing and learning-by-interacting with customers and suppliers. In others, it is generated through the more formalised activities of R&D. Knowledge is generated not only from a firm's internal R&D activities but also through university research and scientific advances. Because of this diversity of learning modes, when the knowledge production function linking knowledge inputs and innovative outputs is examined at the level of the firm, the relationships are weak or non-existent (Audretsch, 1999). These weak relationships arise because many firms, especially small firms, have considerable innovative output but little or even no R&D expenditure.

The question is then, where do innovative small firms or new firms with little or no R&D expenditure get their knowledge inputs from? Recent studies suggest that knowledge spillover from other firms conducting R&D or from research institutions such as universities is the major source of knowledge inputs for those innovative small firms (Audretsch, 1995; Baptista, 1997; Cohen and Levinthal, 1989).

There are three major mechanisms that produce knowledge spillover. First, firms can develop the capacity to adapt new technologies and innovations developed in other firms. In this way, firms can appropriate some of the returns accruing to investments in innovative outputs made by other firms or research institutions (Cohen and Levinthal, 1989). Most of leading the large firms in the newly industrialising economies (NIEs) may belong to this category. Indeed, most NIEs benefited considerably from this form of knowledge spillover during periods of rapid industrialisation. Second, inter-firm networks, collaboration with public institutions and universities can provide synergy effects for participating firms, generating greater innovative outputs than would be expected from their knowledge inputs. Firms can develop new ideas and knowledge through interactions with customers and suppliers, workshops or forums provided by public institutions or universities. They can appropriate some of the returns accruing to these externally developed new ideas or knowledge. Third, knowledge workers may leave the firms or universities where the knowledge has been created in order to establish a new firm. If knowledge workers can pursue a new idea within the organisational structure of the firm in which he or she works, developing knowledge and appropriating the expected value of that knowledge, then they can continue to work at the firm. However, if the scientist or engineer places a greater value on ideas than on decision-making bureaucracy of the incumbent firm, he or she might decide to start a new firm to appropriate the value of those ideas (Audretsch, 1999). In this mechanism of knowledge spillovers, the knowledge production function is reversed. Knowledge is exogenous to the firm and embodied in a knowledge worker.

There is some agreement that knowledge spillover effects are spatially limited. According to Malmberg *et al.* (1996), three elements are involved in the local creation and accumulation of knowledge and in the sustainability of competitiveness. First, innovation processes are locally confined because of the nature of trial and error process of problem solving, the need for repeated interaction between related firms, and the need to exchange knowledge through face-to-face contacts. Second, there are barriers to the spatial diffusion of locally embedded knowledge. The ability to gain access to local informal and formal networks of knowledge exchange and accumulation is mostly limited to insiders within milieus. Third, both knowledge resources brought in by outsiders and initiatives taken by incumbents to tap knowledge resources from outside can enhance the process of knowledge accumulation within the local milieu. Because of these elements, knowledge spillover effects have clear spatial dimensions.

Empirical evidence suggests that proximity is important in the exploitation of knowledge spillover (Audretsch and Feldman, 1996; Feldman, 1994; Jaffe, 1989) both from university research laboratories as well as from firms' R&D laboratories. Also, the geographical clustering of innovative activity tends to be greater in industries where new economic knowledge plays a particularly important role, such as the bio-technology industry (Audretsch and Feldman, 1996). This evidence suggests that the industry life cycle of knowledge-based industries has a relationship with spatial location.

Knowledge-Based Industry and Regions

Product life cycle and global commodity chain models suggest that key location factors differ from one stage of the cycle to another as the importance of new knowledge inputs changes. In the early stages of the product life cycle, when new knowledge is critical for development, industries tend to cluster geographically in a few locations. In older industries, however, when new knowledge is less important, mass production and product standardisation are more likely to be associated with geographical dispersal. For example, firms in bio-technology, which is an industry based mainly on new knowledge, tend to cluster in just a handful of locations (Prevenzer, 1997). And, in the Newly Industrialised Economies (NIEs), relatively new and knowledge intensive industries, making computers and medicines for example, are mostly concentrated in a few clusters (Park, 1995; Park and Nahm, 1998).

Location patterns also differ according to the positions of industries in commodity chains (Gereffi and Korzeniewiez, 1994). Even in old industries, like textiles production and apparel manufacturing, the early stage of the commodity chain may require considerable knowledge resources, with the result that firms in those stages tend to cluster. For example, design and the development of new materials in the textile and apparel industries is a knowledge intensive process and is usually concentrated in major metropolitan areas. Producer services in commodity chains, which are mainly regarded as knowledge intensive activities, also tend to concentrate to major metropolitan areas.

Knowledge inputs, therefore, significantly shape patterns of industrial location, and when knowledge is a critical factor clustering is essential. This is the case in the 'knowledge-based industries', where 'new knowledge' is more important than land, labour or capital. These industries are knowledge intensive in terms of needing new techno-scientific knowledge. They need high R&D expenditures, scientists and engineers, and highly skilled labour. When these characteristics are taken into account, industries can be classified according to their use of knowledge and technology. Such a classification is attempted in Figure 4.1. Knowledge intensive industries fall into two cells in Figure 4.1. Industries in Cell I are highly knowledge-intensive and technology-based industries, the typical knowledge-based industries. Innovative activity is characteristic of these industries, which include biotechnology and ICT, for example. Strong local enterprise networks, research institutions, and university research laboratories are important sources of new knowledge, both tacit and codified. Innovative ideas and knowledge are, in general, sticky in local milieux, and clustering of knowledge-based industries is natural. The development of these clusters depends on local social capital, learning mechanisms and external economies of scale made available through networking: Silicon Valley is an example of a place where these conditions come together.

Industries in Cell II of Figure 4.1 are low-tech but require intensive knowledge inputs. High value added craft industries and advanced producer services such as product design, finance, market research, firm management, coordination, etc. belong to this category. Even in low-tech or traditional sectors like textiles or apparel, the design parts of those industries fall into this category reflecting the need to continuously develop new products. Knowledge is an important input for product design, firm management, coordination, market research, finance and other advanced services, and accordingly the industries in this cell can be regarded as knowledge-based. In fact some aspect of firm management, such as corporate and regional headquarter functions, market research, finance, and

	Intensity of Knowledge Inputs	
	High	***Low***
High **Level of Technology**	I. Innovative New Products	III. Scale-Based Products
Level of Technology ***Low***	II. Product Design and Advanced Services	IV. Scale-Based Low Skill Products

Figure 4.1 Intensity of knowledge and level of technology in industries

design are more knowledge intensive than R&D (Enright, 2000). This type of the knowledge-based industry can be clustered in specific parts of the large metropolitan areas or in industrial districts owing to the need for face-to-face contacts for the exchange of information and tacit knowledge, the division of labour, local skill formation, and collaboration and cooperation among firms. In large metropolitan areas, clusters of these industries and activities have emerged newly industrialising countries. Associated with globalisation, Hong Kong for example has specialised in knowledge intensive activities even though its R&D activities are not strong compared to other NIEs (Enright, 2000). Hong Kong can be regarded as typical of a place with activities that fall into Cell II.

Industries in Cell III of Figure 4.1 are high-tech sectors, but belong to industries that mass produce and are mature in the industry life cycle. Generally these industries have decentralized to peripheral areas in industrialised countries and to NIEs. The production of electrical appliances, such as TVs, refrigerators, air conditioners etc., are good examples of this type of industry. Relocation has been to reduce production costs in the face of severe competition and product standardisation. The development of ICT and the globalisation of MNEs' production have contributed to the decentralisation of these industries. Even though new technologies are used in production in these industries, their mass production of standardised products does not demand new knowledge. Accordingly, these industries cannot be regarded as knowledge-based.

However, in recent years, the application of new knowledge to these mature industries has created a new story. The application of ICT has raised the possibility of product innovation or improvement and process innovation. Product improvement and new product development in these sectors in recent years have contributed to growth regional agglomerations, especially in the NIEs. Technology catch-up and leapfrogging in these high-tech, mature industrial sectors has been successful in some of the NIEs such as Korea (Lee and Lim, 1998), creating new industrial districts (Park and Markusen, 1995; Fromhold-Elisebeth, 1998; Wang, 1999). Also, in traditional agglomerations of these industries, creative destruction unleashed by industrial restructuring has enabled regions to regain competitive advantage (Florida, 1996). High-tech parks in NIEs, such as Kumi in Korea, and Silicon Glen in UK are good examples of places with industries that fall into Cell III.

Industries in Cell IV of Figure 4.1 are low-tech and scale-based producers of standardised products. These industries are typical, traditional, mature industries in which production technology is stabilised and new knowledge inputs are not important. The mass production of textiles, apparel and shoes in low cost locations are examples of these industries. These are not the high value-added textile and apparel products into which new design knowledge is incorporated. Here the emphasis is on reducing labour costs, and these industries have decentralised even to the peripheral areas of developing countries outside specific agglomerations.

In sum, the four types of industries outlined above have developed distinctive locational characteristics in recent years. First, the industries characteristic of Cells II and III in Figure 4.1 are now more likely to cluster owing to the renewed importance of new knowledge creation and accumulation, and the growing significance of knowledge spillover in local milieu (though these are weaker than for industries in Cell I). ICT is transforming these industries into knowledge-based activities. Second, in some maturing and restructuring industries, the creation and application of new knowledge is causing clustering to re-emerge. In contemporary metropolitan areas, high-tech activities can be found alongside traditional specialised activities in all four types of activity recognised in Figure 4.1. In some large metropolitan areas, all the four types of activity can co-exist, even though a dominant type will emerge over time. At the same time, new innovation centres can emerge in new places owing to the intellectual lock-in that can occur among maturing industries in traditional innovation centres, a form of intellectual 'congestion' that contrasts with traditional notions of congestion associated with rents, commuting and pollution (Audretsch, 1999).

The Knowledge-Based Economy and New Industry

The evolution of the knowledge-based economy in recent years has contributed to the development of new industries and new forms of industrialisation. New regional motors of growth seem to have emerged in modern industrialised countries relating, first, to the creation and expansion of knowledge-intensive industries, and second, to the emergence of knowledge-intensive service-oriented industries.

The importance of knowledge inputs and knowledge-based industries is as important in NIEs as in modern industrialised, owing to the processes of globalisation and the expanding activities of MNEs. The development of ICT has promoted globalisation and the expansion of the knowledge-based industries (Dunning, 1998; Solvell and Zander, 1998), together with the conspicuous clustering of these new activities. In Europe, these clusters have been labelled as 'islands of innovation', and contribute to regional growth. Significantly, it has bee estimated that 90% of the 'islands' are in old or traditional industrial regions (Hilpert, 1992; 1996). In the NIEs, similar local clusters of these industries can be identified forming new industrial districts and new centres of growth (Markusen *et al.*, 1999; Park and Markusen, 1995). The industries are not actually new, but rather introduced to the NIEs.

Knowledge inputs are also important in new services such as health, tourism and leisure, and the cultural industries. With increasing per capita incomes, people become more interested in health products and good health services which are for the most part knowledge-based and high value adding. Health service centres enhance regions' environments and contribute to their development, and the local manufacture of new health products can reinforce that development. Rapidly growing culture and image industries can have similar impacts, as in Los Angeles, for example (Scott and Soja, 1996). Modern tourism and leisure industries are also becoming intensive users of knowledge with the development of new tourism goods such as theme tours and ecological tours, which are different from traditional sightseeing. The clustering of these new knowledge intensive industries is not confined to the industrialised world but is also found in the NIEs and even in the developing countries. These knowledge-intensive industries are expected to be the motor of regional growth in some regions in the coming century.

Knowledge is equally critical for producer services such as market research, product design, engineering, computer software, finance, and corporate management and co-ordination. In recent years, many new products have been developed based on new knowledge, even in the finance sector. The rapid growth of producer services in the industrialised countries has been obvious during the last two decades. Rapid employment growth in R&D, engineering services and computer software industries is also well recognised in the NIEs as well as in modern industrialised countries (Meyer-Krahmer, 1996; Fromhold-Eisebith, 1998; Wang, 1998). These activities are, in the main, clustering in a few regions and promote regional economic growth and change (Malecki and Oinas, 1999).

Policy Directions for Promoting Regional Growth

In the knowledge-based economy, linkage between research and industry is a key component. It is argued that 'learning firms' in the knowledge-based economy seek to build non-price competitive advantages from the creative use of information whose cost is rapidly decreasing with the ongoing ICT revolution. These firms contrast with 'neo-classical firms' which are more concerned with issues of scale and cost and production of standardised and homogeneous goods (Guinet, 1996). The competitive advantages of the learning firms in the knowledge-based economy is not in low-cost production based on low-wage employment, but in the production of non-standardised, non-routinised, tradable goods that is most usually associated with high-wage employment. Thus, the learning firm can be seen as a source of employment and growth in the knowledge-based economy (Guinet, 1996). The knowledge-based economy, however, does not mean that all employment will be in the 'learning firms' of knowledge-based industries. It coexists in the knowledge economy with low wage, but declining, employment in 'neo-classical' firms. These changes necessitate the formulation of new policies and strategies to promote growth in the knowledge-based economy. In this chapter it is argued that human resources and institutions are of most importance in framing policies to achieve this structural shift.

The successful creation, acquisition, absorption, and accumulation of knowledge in learning firms are fundamental for the creation of knowledge-based industries and the knowledge-based economy. Accordingly, government policy needs to focus on the development of the human resources. R&D activities, inter-firm networks of buyers and suppliers, innovative manufacturing and knowledge-intensive advanced services increase the stock of knowledge that knowledge-based regional growth requires.

Government policies in the knowledge-based economy can not replace the firm as a learning organisation, but they can create a climate or soft infrastructure that is favourable for knowledge-based development (Hilpert, 1996). Even with ICT, the spatial distribution of technology and knowledge is uneven. Government policy needs to enhance access to knowledge and technology in the less favoured regions, and to develop a qualified and flexible labour force in those areas through educational provision and vocational training. Learning must be recognised as a life-long process with training directed at technology-related fields.

But, the same human resources policies will not be appropriate in all countries and regions. In industrialised countries, the greatest emphasis needs to be placed on product innovation and the development of promising new technologies to enhance profits. Strategies of strong networking among firms, universities, and research institutions to foster collaborative R&D will be important for the regional clustering of highly innovative industries in Cell I of Figure 4.1. In old industrial regions or newly industrialising countries, however, promoting the application of new technology to mature industries, the diffusion of knowledge, and knowledge spillover are important for regional development. These mechanisms assist local skill formation and the development of incremental knowledge in these locations. Strategies of labour training with the collaboration of universities, firms, trade unions, and local governments are promising policy directs for newly industrialising countries, along with the provision of ICT.

Recent studies have emphasized the importance of institutional factors for the formation of industrial clusters and innovation in the knowledge-based economy (Guinet, 1996; Storper, 1996, 1999; Storper and Salais, 1997). Institutions are "persistent and connected sets of rules, formal and informal, that prescribe behavioural roles, constrain activity and shape expectations" and overlap with conventions (Storper, 1996, p. 265). To promote regional growth, in the knowledge-based economy, institutional reform is required. According to Guinet (1996) the main objectives of institutional reform are:

- to improve coordination between training and research institutions, and between training and research functions within any given institution;
- to reinforce the institutional mechanisms for the assessment of labour and firm competency, to increase workers' incentives to learn and financial institutions' propensities to invest in innovation;
- to promote the development of those networks which enable producer/user and basic research/development interactions;
- to improve the physical (e.g. information infrastructure) and legal conditions of access to existing stocks of knowledge;

- to stimulate demand for innovative services and products, through the removal of trade and regulatory obstacles to market access and development; and
- to remove structural impediments to technological entrepreneurship.

These directions of institutional reform emphasise the production system rather than the firm as the target of policy. Given this orientation, a number of policy directions can be identified, especially for newly industrialised or industrialising countries. First, networking between large and small firms, among large firms and among small firms needs to be encouraged and supported (Park, 1996). Second, industry specific or region-specific labour training should be provided by public institutions to help workers secure jobs in the face of flexibility in specific, regionally-concentrated sectors (Storper, 1996). Third, to promote knowledge spillover effects, support for new venture businesses, technology incubators, and spin-offs is needed. Fourth, technopoles or science parks in the newly industrialised countries needs to be promoted, emphasising relationships between research and industry, collaborative R&D activities, and skill formation. Lastly, centres providing producer services such fields as design, computer software, legal, managerial and marketing services need to be developed to promote knowledge-based industries.

In this chapter the intention has not been to explore all policy directions that might be followed to promote regional growth in the knowledge-based economy. Rather, the focus has been on those human resource and institutional reforms that appear most promising for the achievement of such growth – an inter-related set of reforms compatible with the local promotion of the knowledge-based economy.

Acknowledgement

This paper was supported by the Development Fund of Seoul National University.

References

Acs, Z. and Audretsch, D. (1990), *Innovation and Small firms*, MIT Press, Cambridge, MA.

Audretsch, D. (1995), *Innovation and Industry Evolution*, MIT Press, Cambridge, MA.

Audretsch, D. (1999), 'Knowledge, Globalization and Regions', in J.H. Dunning (ed.), *Regions, Globalization and the Knowledge-Based Economy*, Oxford University Press, Oxford.

Audretsch, D. and Feldman, M. (1996), 'R&D Spillovers and the Geography of Innovation and Production', *American Economic Review*, vol. 86, no. 4, pp. 253-273.

Baptista, R. (1997), *An Empirical Study of Innovation, Entry and Diffusion in Industrial Clusters*, Ph.D. Dissertation, London Business School, University of London.

Berry, B.J.R., Conkling, E.C., and Ray, D.M. (1997), *The Global Economy in Transition*, Second Edition, Prentice Hall, Upper Saddle River, NJ.

Castells, M. (1996), *The Rise of Network Society. The Information Age: Economy, Society and Culture*, vol. 1, Blackwell, Oxford.

Clark, G.L. and Kim, W.B. (1996), *Asian NIEs & the Global Economy,* The Johns Hopkins University Press, Baltimore and London.

Cohen, W. and Levinthal, D. (1989), 'Innovation and Learning: the Two Faces of R&D', *Economic Journal*, vol. 99, no. 3, pp. 569-596.

Dunning, J.D. (1998), 'Globalization, Technological Change and the Spatial Organization of Economic Activity', in A.D. Chandler, P. Hagstrom and O. Solvell (eds), *The Dynamic Firm: The Role of Technology, Strategy, Organization, and Regions*, Oxford University Press, Oxford and New York, pp. 289-314.

Enright, M.J. (2000), 'Globalization, Regionalization, and the Knowledge-Based Economy in Hong Kong', in J.H. Dunning (ed.), *Regions, Globalization, and the Knowledge-Based Economy,* Oxford University Press, Oxford, pp. 381-406.

Florida, R. (1996), 'Regional Creative Restruction: Production Organization, Globalization, and the Economic Transformation of the Midwest', *Economic Geography*, vol. 72, pp. 314-334.

Freeman, C. and Soete, L. (1993), *Information Technology and Employment*, Universitaire Pers, Maastricht.

Fromhold-Eisebith, M. (1999), 'Bangalore: A Network Model for Innovation-Oriented Regional Development in NICs', in E.J. Malecki. and P. Oinas (eds), 1999, *Making Connections – Technological Learning and Regional Economic Change,* Ashgate, Aldershot, pp. 231-260.

Gereffi, G. and Korzeniewiez, M. (1994), *Commodity Chains and Global Capitalism*, Praeger, Westport and London.

Griliches, Z. (1979), 'Issues in Assessing the contribution of R&D to Productivity Growth', *Bell Journal Of Economics*, vol. 10, pp. 92-116.

Guinet, J. (1996), 'Institutional Framework Conditions for the Development of the Knowledge-Based Economy', in OECD, *Employment and Growth in the Knowledge-based Economy,* OECD, Paris, pp. 207-211.

Halal, W. (1996), 'The Rise of the Knowledge Entrepreneur', *The Futurist*, November-December, pp. 13-16.

Hilpert, U. (1992), *Archipelago Europe – Islands of Innovation*, FAST, Commission of European Communities, Brussels.

Hilpert, U. (1996), 'The Role of Social Partners in Designing Learning Organizations', in OECD, *Employment and Growth in the Knowledge-based Economy,* OECD, Paris, pp. 143-156.

von Hippel, E. (1998), '"Sticky Information" and the Locus of Problem Solving: Implications for Innovation', in A.D. Chandler, P. Hagstrom and O. Solvell (eds), *The Dynamic Firm: The Role of Technology, Strategy, Organization, and Regions*, Oxford University Press, Oxford and New York, pp. 60-77.

Jaffe, A. (1989), 'Real Effects of Academic Research', *American Economic Review*, vol. 79, pp. 957-970.

Lee, K. and Lim, C. (1998), *Technological Regimes, Catch-Up and Leapfrogging: Findings from Korean Industries*, A paper presented at the 1998 Annual Convention of the Korean Economic Association, Seoul, Korea.

Malecki, E.J. and Oinas, P. (eds) (1999), *Making Connections – Technological Learning and Regional economic change,* Ashgate, Aldershot.

Malmberg, A., Solvell, O. and Zander, I. (1996), 'Spatial Clustering, Local Accumulation of Knowledge and Firm Competitiveness', *Geografiska Annaler*, vol. 78 B, no. 2, pp. 85-97.

Markusen A.R., Lee Y.S. and DiGionanna (eds) (1999), *Second Tier Cities*, University of Minnesota Press, Minneapolis, MN.

Meyer-Krahmer, F. (1996), 'Dynamics of R&D-Intensive Sectors and Science and Technology Policy', in OECD, *Employment and Growth in the Knowledge-Based Economy*, OECD, Paris, pp. 213-236.

Miller, R. (1996), 'Towards the Knowledge Economy: New Institution for Human Capital Accounting', in OECD, *Employment and Growth in the Knowledge-based Economy*, OECD, Paris, pp. 69-80.

OECD (1996), *Employment and Growth in the Knowledge-Based Economy*, OECD, Paris.

Park, S.O. (1995), 'Seoul, Korea: City and Suburb', in G.L. Clark and W.B. Kim (eds), *Asian NIEs and Global Economy*, Johns Hopkins University Press, pp. 147-167.

Park, S.O. (1996), 'Networks and Embeddedness in the Dynamic Types of New Industrial Districts', *Progress in Human Geography*, vol. 20, no. 4, pp. 476-493.

Park, S.O. and Markusen, A. (1995), 'Generalizing New Industrial Districts: A Theoretical Agenda and an Application from a Non-Western Economy', *Environment and Planning A*, vol. 27, pp. 81-104.

Park, S.O. and Nahm, K-B. (1998), 'Spatial Structure and Inter-Firm Networks of Technical and Information Producer Services in Seoul, Korea', *Asia Pacific Viewpoint*, vol. 39, no. 2, pp. 209-219.

Prevenzer, M. (1997), 'The Dynamics of Industrial Clustering in Biotechnology', *Small Business Economics*, vol. 9, no. 3, pp. 255-271.

Scott, A.J. and Soja, E.W. (eds) (1996), *The City, Los Angeles and Urban Theory at the End of the Twentieth Century*, University of California Press, Berkeley, Los Angeles, and London.

Solvell, O. and Birkinshaw, J. (1999), 'Multinational Enterprises in the Knowledge Economy: Leveraging Global Practices', in J.H. Dunning (ed.), *Regions, Globalization and the Knowledge-Based Economy*, Oxford University Press, Oxford.

Solvell, O. and Zander, I. (1998), 'International Diffusion of Knowledge: Isolating Mechanisms and the role of MNE', in A.D. Chandler, P. Hagstrom and O. Solvell (eds), *The Dynamic Firm: The Role of Technology, Strategy, Organization, and Regions*, Oxford University Press, Oxford and New York, pp. 402-416.

Storper, M. (1996), 'Institution of the Knowledge-Based Economy', in OECD, *Employment and Growth in the Knowledge-based Economy*, OECD, Paris, pp. 255- 283.

Storper, M. (1999), 'Rethinking the Economics of Globalization: The Role of Ideas and Conventions', in J.H. Dunning (ed.), *Regions, Globalization and the Knowledge-Based Economy*, Oxford University Press, Oxford and New York.

Storper, M. and Salais, R. (1997), *Worlds of Production: the Action Frameworks of the Economy*, Harvard University Press, Cambridge MA.

The World Bank (1998), *Knowledge for Development, World Development Report, 1998/1999,* The World Bank.

Wang, Jici (1999), 'In Search of Innovativeness: The Case of Zhong'guancun', in E.J. Malecki and P. Oinas (eds), *Making Connections – Technological Learning and Regional Economic Change,* Ashgate, Aldershot, pp. 205-230.

5 Identifying Contexts of Learning in Firms and Regions

Päivi Oinas and Hein van Gils

Introduction

In the course of the 1990s, a relatively broadly shared view has emerged concerning learning as a key process leading to local growth. The reasoning behind this view builds on a growing literature (summarised in Oinas 1999, 2000). The contemporary tendency towards 'globalisation' makes inter-firm competition increasingly intense. As one group of firms develops competitive advantages others immediately either imitate or innovate to erase that competitive edge (Maskell, 1999). Since firms are often not able to own, create or internally organise the resources they need for this competitive race, it pays if they have specialist 'core competences' (Prahalad and Hamel, 1990), outsource parts of their activities and share resources with other firms within flexibly organised networks (see Piore and Sabel, 1984, and the literature on flexible specialisation). It is argued that collaborative networks are most effectively established in localities where actors are embedded in 'institutionally thick' social relations (Amin and Thrift, 1993). In these social relations, interaction is governed by mutually shared conventions in specific action domains (Storper, 1992, 1997; Storper and Salais, 1992, 1997). These conventions develop against shared backgrounds in local communities. The resulting geographically and socio-culturally proximate and trustful relationships facilitate the exchange of tacit knowledge that is fundamental to processes of collective learning (Asheim and Cooke, 1999; Kirat and Lung, 1999; Lawson, 1999; Storper, 1997). These learning processes are supported by a rich history of local interaction between users and producers of specific technologies and a favourable institutional environment (Lawson and Lorenz, 1999). The resulting knowledge accumulation enables the creation of unique, or 'noncosmopolitan' (Storper, 1997), assets that are needed for competitiveness (Maskell and Malmberg, 1999a, 1999b).

This thinking is in line with the so-called 'resource-based' view of the firm. We build on this view and suggest extending it to include both firms and broader entities (networks, industries, sectors and regions) where learning takes place. This extension is suggested to recognise that the upgrading of firm-specific resources involves collective learning that does not happen in isolation but is related to other actors' resources and learning processes through external relations in specific interaction contexts. Thus, the availability and quality of resources in various contexts forms the basis for, or imposes constraints on, firms' learning capabilities.

We aim to clarify the discussion in economic geography on learning and the creation of regional competitiveness by explicitly considering the different contexts within which collective learning processes take place. This means that we do not assume the existence of 'learning regions' generally but try to outline an approach that is helpful in analysing regional learning by identifying learning in different organisational and regional contexts. In these contexts, learning is understood as the *actualisation of potential resource relatedness* between (or within) actors.

The expression 'learning region' is now commonly used in economic geography (Florida, 1995; Morgan, 1997; Simmie, 1997). Here, however, in place of 'learning region', we use the expression '*regional learning*'. We suggest that analysing regional learning cannot be separated from identifying the relevant actors and understanding their interrelationships. Thus, regional learning is understood here as *inter-organisational learning* that is tied to the location where it takes place because it would be impossible or difficult to create the same circumstances for learning elsewhere (Oinas and Virkkala, 1997).

The discussion in this chapter is developed in five sections. First, the resource-based view of the firm is outlined. Based on this approach, a typology of key organisational resources (termed *collective competences)* is proposed, incorporating the notion of *resource (competence) relatedness*. The chapter goes on to discuss the role of resource relatedness in collective learning. This discussion provides the conceptual means for outlining a typology of organisationally and regionally distinct learning contexts based on the opportunities provided by various types of resource relatedness. Finally, it is argued that the proposed framework might be useful for understanding actual processes of learning in real world situations for analysts and policy makers alike.

Resources and Resource Relatedness in Firms and Regions

The Resource-Based Approach

The resource-based view of the firm has been developed in strategic management research, organisational economics and industrial organisation (Mahoney and Pandian, 1992). It draws on the seminal work of Penrose (1959) who saw firms, fundamentally, as collections of productive resources. Wernerfelt (1984) reintroduced the Penrosian ideas to the research community by arguing that "[f]or the firm, resources and products are two sides of the same coin" (Wernerfelt, 1984, p. 171). Thus, instead of focusing on the product-market aspects of competition, this approach focuses on resources as the fundamental bases for creating and sustaining competitiveness within product markets.

The resource-based view stems from a simple insight. When a firm possesses resources that are specific to it and not to its competitors it is able to develop unique products. If those products create more value for their customers than those of their competitors, such firm-specific resources create competitive advantage (Barney, 1986; Dierickx and Cool, 1989; Conner, 1991; Peteraf, 1993). In the most recent literature, this view has evolved into accounts where the *competences* or *capabilities* of the actors involved are highlighted rather than resources in general (see, Prahalad and Hamel, 1990; Teece and Pisano, 1994; Teece *et al.*, 1997). This approach has been introduced into economic geography through the work of Maskell and Malmberg (e.g. Maskell, 1999; Maskell and Malmberg, 1999a). In economic geography the importance of the *local context* is emphasised for the creation of competitiveness among business firms.

In this chapter we extend the idea that resources external to the firm are involved in creating competitive advantage. Yet, we emphasise that it is not just the *local* context that provides additional resources. Instead, the argument is developed that individual firms or the separate units of corporations derive resources from several *entities* in which they are a part; *corporations*, *networks*, *industries, sectors*, *regions, nations*. They participate in a *coevolution* process that incorporates these entities. It is contended that evolution as part of a larger whole leads firms and other actors to adopt certain aspects of those larger entities, and to share common characteristics with other actors forming a part of that entity. Along with this sharing of certain characteristics, a degree of homogenisation emerges among the actors involved. The acquired characteristics *affect their (trans)formation, perception, use of and access to resources*. These propositions are elaborated in the discussion that follows.

Resources

We conceptualise firms as bundles of activity-specific resources. These sets of resources coevolve through their integrated use and other interrelationships: used in the same or a related process; used by a particular group of people; or used within the same knowledge system. Similarly, resources in the broader entities of which firms are a part – networks, industries and so on – coevolve as their use is connected through the use of other resources in the same 'system'.

But what are these resources? In his seminal paper, Wernerfelt (1984) viewed resources broadly: "By a resource is meant anything which could be thought of as a strength or weakness of a given firm" (p. 172). Traditionally, resources have been classified as land, labour and capital, or as physical, human, and financial capital. Some add technology, others organisational capital (Barney, 1997). For what we call here 'resources', different terms are used in the contemporary literature, such as 'assets', 'capital', 'competences', and 'capabilities'.

In the context of 'learning', we focus on *firm-specific combinations of coevolved individual competences* (human capital) and label them *collective competences*. Individuals are bearers of collective competences, but collective competences do not reduce to individual competences. Collective competences enable the collective work of carrying out certain tasks, functions or projects within organisations or different types of network. Different activities or functions within firms develop bundles of resources, which typically combine different types of collective competence. It seems possible to identify at least four fundamental collective competences in economic activities, 'technical', 'economic', organisational and 'institutional.

Technical competences refer to the ability to develop and implement technological solutions, design products and processes, and to operate machines and facilities effectively (Teece *et al.,* 1994).

Economic competences refer to the ability to find economically efficient solutions to activities carried out in business organisations. Teece *et al.* (1994) subdivide these competences into:

- *allocative competences* ("deciding what to produce and how to price it");
- *transactional competences* ("deciding whether to make or buy, and whether to do so alone or in partnership"); and
- *administrative competences* ("how to design organizational structures and policies to enable efficient performance").

Teece *et al.* (1994) call these competences 'economic' or 'organisational'. We call them simply economic competences, and make a sharper distinction between economic and organisational competences. For us, 'economic competences' are more focused on strategic decision-making capabilities in the face of competitive pressures to achieve (relative) economic efficiency.

Organisational competences, we argue, enable the functioning of an organisation as a co-ordinated, yet evolving, entity. When equipped with organisational competences at the collective level, individuals in organisations are able to communicate the meanings they give to their activities, to motivate joint action, to agree on given goals, and to handle the internal politics associated with reaching those goals. Collective entities have to set up certain fixed procedures but they also have to maintain a sufficient degree of flexibility. This is successfully done by finding a balance between *stabilising* and *dynamising routines* (this view resembles Teece and Pisano's (1994) idea of dynamic capabilities).

Stabilising routines create relative stability in organisational structures, social practices and resource utilisation, and thus allow continuity in organisational life. They also help to maintain an organisation's identity within a collectivity. In times of change, they prevent an organisation from falling into chaos. Stabilising routines involve such procedures as, for example, the systematisation of simple tasks, the establishment of organisation-specific ways of dealing with certain types of situation, and the creation of a strong organisational culture and commitment among employees.

Dynamising routines facilitate change in the structure and content of social relations within organisations and thus provide the opportunity to alter organisational procedures. In essence, they help firms to learn better or new economic and technical competences and to put them into practice. Examples of such dynamising routines are continuous training of the labour force, task rotation, or incentive systems to develop new ideas from the shop-floor.

Both of these aspects of organisational competence are reflected in organisational cultures. Organisational competences can result from successful leadership and/or stem from the adoption of appropriate management styles which allow for individuals and groups to self-organise and coordinate themselves. They are pronounced in coordinating collective action not only in firms or multi-unit corporations but also in networks. To the extent that firms operate within relatively tightly integrated networks, organisational competences are needed in the coordination of networked collaboration. In other words, like economic competences, these competences are used in coordinating relationships that extend beyond organisational boundaries.

Institutional competences refer to the understanding of and relative compliance with prevailing social practices, values, and perceptions concerning resource utilisation in specific *institutional environments.* Institutional environments, through their *cultural* (practices based on values, norms and beliefs) and *political* (laws, rules and regulations) aspects, *condition* economic action. Thus, Oinas (1995) points out that institutional environments can be seen as '*institutional resources*' for economic actors. For institutions to serve as resources, actors need to understand how the environment functions and how to deal with it. When they are able to make use of institutional resources, they possess *institutional competences*. Such institutional competences enable economic actors to appropriate external resources in their institutional environments so as to function effectively within them.

These are the kinds of competences that units of multinational corporations, for example, need to develop when they set up activities in new national and regional environments in foreign countries. They need to understand the socio-politico-cultural circumstances of those places to be able to act according to local formal and informal rules and, thus, interact effectively with local employees, partners, customers and administrators. Being able to deal with a specific institutional environments is reflected in issues as varied as relationships with public sector actors, inter-firm relations, setting up effective governance structures, building reputations, marketing skills and customer relations.

Organisational competences and institutional competences can be interpreted as 'meta-capacities'. They 'function' at different levels to integrate and coordinate the use of other resources – those resources that are actually employed in the design and production of goods and services. Inside the firm/corporation/network, 'organisational competences' function to coordinate individuals and groups with their specific business-related competences. Outside, 'institutional competences' serve to integrate internal resources with the external institutional environment.

The gradual processes through which all these competences coevolve as a consequence of use makes them complex, firm-specific, and partially tacit (Dosi and Malerba, 1996). Thus, they are difficult to transfer from firm to firm (Dierickx and Cool, 1989; Prahalad and Hamel, 1990). Sometimes they are integrated or shared within broader contexts than just firms or corporations, for example, within industries and regions. This may be essential for integrated action between different firms or corporate units emerging from these different contexts. It is suggested in what follows that this integration is especially important for collective learning.

Resource Relatedness

In addition to the theoretical-conceptual toolkit developed so far – building on the resource-based view of the firm and invoking 'resources' and their 'coevolution' in different 'contexts' - we still need a conceptual tool that helps us understand how and why relationships between actors emerge and persist in different contexts, and why they emerge the way they do. In other words, what *conditions* and *directs* the dynamic process of resource (co)evolution? In answering this question we build on the idea of the path-dependent evolution of resources. Firms learn by building on their existing resource bases and upon the opportunities to enhance them with resources available both internally and externally. In creating new resources, individuals or groups within firms join their existing resources, or firms join their resources with those in other firms. Yet, as Storper points out, there are limited ways in which economic actors can coordinate their joint action:

> Though in principle there are unlimited ways to coordinate economic action, there is in practice a limited number of practically coherent combinations of actions for each kind of material good or service produced in the economy. (Storper, 1997, p. 46)

In the following, we try to explain the emergence of 'practically coherent combinations of actions' at the level of resources. These emerging combinations of actions are 'steps' on the 'path' of resource evolution. The combinations are conditioned by a kind of 'fit' between the resources of the actors involved. We call this 'fit' *relatedness*. Relatedness between actors' resources creates an opportunity for actors to coordinate their action and develop them in new directions.

We identify two types of relatedness, 'similarity' and 'complementarity'. Relatedness in terms of resource *similarity* refers to the fact that resources are in effect interchangeable. Similarity can be identified in physical capital and in human and collective competences. Two resources are similar when they can be replaced one by the other, and they can fulfil the same task equally well. Thus, similarity can refer to identicality or substitutability. Resources are *identical* when they are exactly the same and thus can replace one another (like two identical computers) or are 'practically' the same, for example, when one laboratory assistant, with basic skills in chemistry, can be replaced by another. Resources are *substitutable* when, for a certain task, two dissimilar resources can replace one another. Substitutability can occur, for example, between different types of computer that are equally capable of word processing; between a metal component and a plastic version; between a middle manager with a degree in engineering and a manager with a degree in economics when they both perform the same tasks.

Relatedness in terms of *complementarity* refers to *dissimilar* resources functioning together in an economic activity. By complementing each other, resources provide the basis for interaction both *within* and *between* entities. Simple examples include: the combining of different chemicals or materials in equipment; the interaction within organisations between engineers (whose professional training provides them with general technical competences which become more specialised with work experience) and economists (economic competences) or between computer hardware and software developers (different types of technical competences). Essentially, all socially organised action is based on various complementarities. In capitalist economic systems, physical, human, and privately owned financial capital, as well as the collective competences outlined above, complement each other to form business organisations. When an individual or a firm is not able to make or acquire the needed resources of its own, it resorts, if possible, to external sources. Sometimes this leads to the creation of new resources and new knowledge when learning takes place based on complementarities.

Collective Learning

Building on this discussion, the analysis of the present section attempts to build an understanding of the opportunities for learning in the various contexts identified above. It is argued that the emergence of processes of collective learning will differ depending on the degree of relatedness of the resources in the possession of the actors involved in learning. It is not our intention to enter into an elaborate discussion of learning but to look at learning from the point of view of relatedness. The literature on learning is as yet rather unclear on the very process of learning (Hudson, 1999).

Following discussion in the literature (Lundvall, 1992; Asheim and Cooke, 1999), learning is understood as a collective interactive and iterative process between actors which builds on organisation-specific and region-specific resources. Learning processes build actors' stocks of knowledge, and they enhance collective competences. Learning does not move in random directions, however, but tends to follow specific paths (David, 1986; Arthur, 1989) and can be characterised as 'relatedness over time'. Thus, successive and path-dependent steps use the resources that have been developed earlier and combine them to create new ones. *Learning, therefore, is a bridge between potential and actual relatedness*. It is the process that integrates potentially related sets of resources within or between firms and sectors, and within regions or inter-regionally.

Each 'step' on a learning path introduces new elements to an actor's path that have both 'depth' and 'scope' (McKee, 1992). Moving in the *depth* direction, actors' knowledge is deepened, intensifying their pre-existing resources. Moving in the *scope* direction, actors integrate new types of competence into their resource bundles (as with the integration of biotechnology into the pharmaceuticals industry, or semiconductors into the machine tool industries). "The merging of disparate skills, spanning different industries, is facilitated by the networking resulting from interorganizational relationships" (Pennings and Harianto, 1992, p. 30). Usually learning involves both depth and scope. Together, these dimensions give the learning path its direction, as illustrated in Figure 5.1.

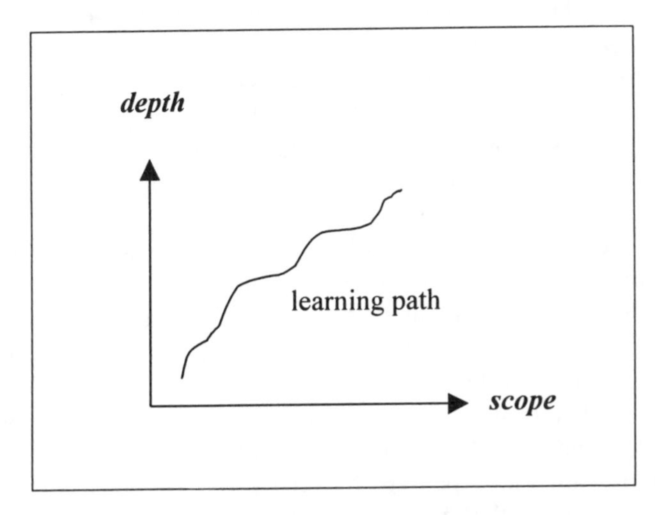

Figure 5.1 Learning path, combining depth and scope

The learning actor can be seen either as an individual or as a collective actor (such as a firm, a corporation or even a network). Learning processes can be internal to an actor, or they can take place in an inter-actor context. In the latter case, new elements are integrated into actors' learning processes from paths followed by other actors. Relatedness – similarity and complementarity – between the new elements and the pre-existing resource-bundles condition the directions (depth or scope) that learning can take.

When a learning path moves only in the depth direction, an actor basically improves the skills or the resources, s/he has already acquired (McKee, 1992). Usually, however, scope is also added to the learning path. As a skill is developed, new skills can be added to complement the old ones. Also, because we view learning as a collective process – and thus, the discussion is focussed not on individuals learning alone but on individuals learning together – this also means that the resources of several actors are brought together. Under these circumstances, the opportunity arises to identify complementarities between potentially related resources, especially when there is resource similarity. Resource *similarity* provides 'commonality' (Cohen and Levinthal, 1990) or what Nooteboom (1999) calls 'communicability' (or 'cognitive proximity'). In sum, in most cases, both similarity and complementarity are needed for learning. *Similarity* creates the shared basis for creating novel combinations for discovering and actualising *potential complementarities*.

Nooteboom (1999) suggests that learning processes should be viewed as cycles of discovery. Put simply, learning cycles consist of alternating periods of *exploration* (emergence of novel combinations, or increasing scope in terms of Figure 5.1) and *exploitation* (increasing depth in terms of Figure 5.1). Exploitation involves the application of existing resources in new contexts which is likely to add some scope as well. Thus, various combinations of depth and scope may be involved at different points in actors' learning paths (Figure 5.1). We regard 'novelty' in new combinations as a matter of degree (amount of scope). Thus, instead of accepting a dichotomy between radicality (during the exploration stage) and incrementality (during the exploitation stage) in learning, we regard incrementality and radicality as ends on a continuum. Often what is called a 'radical innovation', may perhaps be seen as taking place through '*rapid but incremental* steps' that move in the direction of increasing scope on the learning path. Even if a learning actor intuitively identifies a potentially radical complementarity, s/he may have to take a number of small steps to understand the task at hand before being able to actualise that potential.

Conceptualising learning as the actualisation of potential relatedness makes us focus on the opportunities and constraints for learning that actors face. Both are grounded in the existing resource bases possessed by actors. In the next section, we move from the abstract discussion of similarities and complementarities and speculate on the kinds of relatedness that are likely to exist in different configurations of actor-specific and context-specific resource bundles.

Resource Relatedness and Learning in Different Contexts

The discussion in this section seeks to identify the learning contexts that may emerge as actors make connections with each other under various circumstances, based on their specific resource bundles. Our goal is to develop a typology of learning contexts based on the distinctions that have been discussed in the earlier sections of the chapter:

- different *entities* – firms, corporations, networks, regions and industries – as *contexts* where actor's resources coevolve and where learning takes place; and
- *resource relatedness* in terms of similarity and complementarity.

The typology of learning contexts is presented in Table 5.1. It builds on these two distinctions. The *columns* distinguish between networks and corporations as contexts. These contexts are further subdivided on the basis of:

Resource Similarity:

- *intra-industry networks;*
- corporations with related businesses; and

Resource Complementarity:

- *inter-industry networks, i.e., networks with participants from several industries, forming sectors or clusters (local or non-local);* and
- diversified corporations (unrelated businesses).

The *rows* distinguish between '*milieux*' and 'cities' (Camagni, 1995). *Milieux* are characterised by relatively specialised resources (one or a limited number of key industries – often supporting a dominant one – i.e., many firms related to each other via technical and other *similarities*). *Cities* are characterised by relatively diversified (dissimilar but *potentially complementary*) resources (several industries, potentially interconnected through relationships based on technical and other complementarities that form sectors or local clusters). Building on these basic distinctions, Table 5.1 identifies eight kinds of learning context. Four are learning contexts involving networks and four are learning contexts involving corporations in different types of region. Each of these contexts is elaborated in turn.

Table 5.1 Contexts of learning

Industrial organisation / **Type of region**	***Networks* with resource similarity**	***Networks* with resource complementarity**	***Corporations* with resource similarity**	***Corporations* with resource complementarity**
Milieux	1 Traditional Industrial District	2 Modern Industrial District	5 Traditional Company Town	6 Modern Company Town
Cities	3 Fragmented City	4 Integrated City	7 Fragmented Corporate City	8 Integrated Corporate City

Traditional industrial district In these types of milieux, a majority of firms operate in one dominating industry together with auxiliary industries that form a 'sector' or 'cluster'. Firms form horizontal and vertical relations within networks with other firms. Thus, these milieux consist of technically related firms operating in intra-sector networks. This is essentially how Becattini (1990) characterises Italian industrial districts. "The firms of the district belong mainly to the same industrial branch, but the term industrial branch must be defined in an especially broad sense. For instance, in studies on industrial districts the term 'textile branch' also includes the machines and the chemical products used in the textile industry, and the various services required by it" (p. 40).

We call this type of region a 'traditional' industrial district because it is predominantly self-contained in its key relationships associated with learning. Also, there is relatively little variety in the products and processes of each industry branch (both the dominant industry and its 'auxiliaries'). Internally, these districts are characterised by relatedness in terms of similarity. Furthermore, economic actors tend to be strongly embedded in the region's social relations, i.e. they share the same institutional competences. Under these circumstances, this type of milieu is culturally and economically stable. Also, relations between industries remain unchanged, and the directions that potential learning processes take tend to be narrow in their scope and are concentrated principally at the intra-industry level. Learning is likely to be incremental, especially if the competitive environment is also relatively stable (Asheim and Cooke, 1999). Owing to relatively weak external connections and meagre stimuli emerging from internal variety, lock-in is a potential threat in these industrial communities.

Modern industrial district As actors in 'traditional' industrial districts develop external collaborative relations, they can be interpreted as evolving into 'modern' industrial districts. Within these milieux, too, firms with complementary resources operate in sector-specific networks. Lock-in is avoided through connections with other firms in other areas. Owing to the limited number of industrial branches and connections between them, non-local relations serve to bring variety into network relationships. These may give rise to the identification of potentially novel complementarities and thus provide the possibility to increase the scope of learning. As is the case in collective learning, a 'commonality' among some resources is needed before such learning relationships can be established. Besides having to share some similar technical or economic competences, in non-local relationships actors may especially need to broaden their institutional competences by becoming sufficiently embedded in extra-local institutional environments in order to tap potential complementarities with more distant partners. The nature of those social relations may differ considerably from those in traditional districts and, on occasions, can considerably broaden the scope of learning among actors in this type of milieu. The level of dynamism in these milieux is notably greater than in traditional industrial districts. They are able to create and maintain 'cultures of change' which stimulate an innovative attitude among members.

Fragmented city In these types of (usually urban) areas, technically related firms operate within several intra-industry networks. This type of region incorporates several unrelated industries that are co-located for historically contingent reasons or for reasons related to natural endowments. They remain relatively isolated from each other. There is a lack of potential relatednesses between industries – or an inability to identify them. There are no strong incentives, or needed competences, to engage in more radical forms of innovation. Although actors share the same regional institutional competences this does not trigger their use as common ground for increasing the scope of learning in inter-industry networks. This might be due to relative backwardness, lack of competitive pressures, and/or strong external control. Gradual steps that deepen existing technical, economic and organisational competences (the 'depth' dimension in Figure 5.1) are the dominant form of collective learning leading to incremental innovation. This may lead to high quality but seldom to competences, processes, and products which are genuinely novel on the world scale.

Integrated (innovative) city These types of city may develop as networks in 'fragmented cities' extending industrial boundaries and establish technical and other complementary relationships between previously unrelated industries. The institutional environment in this type of city facilitates the emergence of an atmosphere where heterogeneity in the region (Grabher and Stark, 1997) results in the generation and dissemination of new ideas (Amin and Thrift, 1992) and the *identification* of potential complementarities between industries. It stimulates processes of learning within inter-industry networks which broadens the scope of learning, the *actualisation* of potential complementarities, and potentially gives rise to the emergence of new sectors and clusters. The scope of learning depends on the variety of industries present in the city and on the actual links that are built between industries. Such constellations of interrelated activities crossing industrial boundaries may also extend regional boundaries and involve extra-regional relations which link up with and add to the processes of learning.

Traditional company town These urban areas are characterised by actors with similar competences. These areas tend to be dominated either by a locally headquartered one-business firm, or an externally controlled branch plant specialised in one or related lines of production. Mature products – possibly of top quality – are produced with dependable but unoriginal and sometimes old-fashioned methods. In the case of a multi-divisional and multi-locational corporate organisation, contacts to other units within the corporate organisation are not frequent and do not stimulate learning. Also this type of learning context is very limited in terms of the 'genetic pool' of new ideas, offering only incremental innovation possibilities. Lock-in may be an even greater threat than in 'traditional industrial districts'.

Modern company town In contrast to the 'traditional' company town, the 'modern' company town is dominated by a diversified corporation with one or several related businesses operating in the milieu and others elsewhere. These businesses are coupled under the 'organisational umbrella'. In the case of large diversified corporations, intra-corporate relationships provide the common ground for identifying potential complementarities which broaden the scope of learning between the business units located in the town and other units elsewhere. The degree to which this happens, depends largely on the organisational competences in the corporation - on the degree to which inter-unit exchanges in various forms are encouraged so as to result in activities that actualise complementarities.

Fragmented corporate city These are urban environments with firms operating in several unrelated industries. The firms may be units of possibly locally headquartered, one-business firms, locally headquartered corporations with highly related businesses, or branch plants of corporations with related or unrelated businesses elsewhere. The characteristic feature of these places is that there is very little local integration between competences in the various industries. The learning context in fragmented corporate cities resembles that of fragmented cities, yet the scope of learning may be widened in those multi-locational corporations where more distant units bring novel ideas to local incremental learning processes. Some formerly unrelated competences may become perceived as potentially related (complementary) in the course of intra-corporate interactions.

Integrated (innovative) corporate city Integrated corporate cities may host a number of locally important diversified corporations encompassing activities in different industries. These intra-organisational connections between units with complementary competences increase the potential scope of learning. The firms are also able to create complementary network relationships with other local firms. These processes are enhanced because they operate within a shared institutional environment. Additionally, the scope of learning may be widened through intra-corporate or network-based interactions with actors in other firms or units elsewhere.

The typology of learning contexts outlined above is not intended to cover all possible real world configurations of corporations and networks in regional contexts. The multiple variations of the basic types described here and the details related to different industries can be discovered as empirical research is done on different organisational configurations in different regions. Networks and corporations were discussed as separate cases to maintain clarity in the discussion. Clearly, corporate units and network members, for example, are often co-located, and corporate units often perform the role of network partners as well.

Conclusion

This chapter suggests that recent literature on learning regions is far less systematic than it might be in its approach to the actual variety of contexts within which learning takes place. The term 'learning regions' is questioned because it is difficult to regard regions in their entirety as actors engaged in collective learning. The alternative term 'regional learning' is preferred. The chapter develops a framework that enables the identification of different contexts where regional learning may take place. The analysis builds on the resource-based view of the firm, and extends it by distinguishing between different types of resource (collective competences) and elaborating the role of resource relatedness in processes of competence building (learning). It is argued that this approach provides us with better tools for understanding real world situations than the highly general discussion on 'learning regions' which has prevailed in the literature in recent years.

There are a several kinds of resource-activity bundles in firms and regions, which can stimulate learning in diverse, but not entirely random, constellations of actors. That learning can proceed in a number of path-dependent directions, constrained by the kinds of competence possessed by the actors at any particular time.

Recognising these features is useful for understanding learning in real world situations by analysts and policy makers alike. For scientific analysis, the approach suggests that it is important to identify the actors involved, the kinds of regional environments within which they operate, and the kinds of connections they maintain with local and more distant partners, in order to understand the kinds of learning that might be possible in different learning contexts. For policy makers, the approach seeks to highlight the point that no one kind of regional learning exists, and that there is no one way of boosting it.

Acknowledgements

The authors wish to thank Mike Taylor for useful comments. Päivi Oinas gratefully acknowledges the support of the Foundation for Economic Education and the Jenny and Antti Wihuri Foundation in Finland, and the Foundation *Vereniging Trustfonds Erasmus Universiteit Rotterdam* in the Netherlands.

References

Amin, A. and Thrift, N. (1992), 'Neo-Marshallian Nodes in Global Networks', *International Journal of Urban and Regional Research*, vol. 16, no. 4, pp. 571-587.

Amin A. and Thrift, N. (1993), 'Globalization, Institutional Thickness and Local Prospects', *Revue d'Economie Regionale et Urbaine*, vol. 3, pp. 405-427.

Arthur, W. B. (1989), 'Competing Technologies, Increasing Returns, and Lock-in by Historical Events', *Economic Journal*, vol. 99, no. 394, pp. 116-131.

Asheim, B. and Cooke, P. (1999), 'Local Learning and Interactive Innovation Networks in a Global Economy', in E.J. Malecki and P. Oinas (eds), *Making Connections: Technological Learning and Regional Economic Change,* Ashgate, Aldershot, pp. 145-178.

Barney, J. (1986), 'Strategic Factor Markets: Expectations, Luck, and Business Strategy', *Management Science*, vol. 32, no. 10, pp. 1231-1241.

Barney, J. (1997), *Gaining and Sustaining Competitive Advantage*, Addison-Wesley, Reading MA.

Becattini, G. (1990), 'The Marshallian Industrial District as a Socio-Economic Notion', in F. Pyke, G. Becattini and W. Sengenberger (eds), *Industrial Districts and Inter-Firm Co-operation in Italy*, ILO, International Institute for Labour Studies, Geneva, pp. 37-51.

Camagni, R. (1995), 'Global Network and Local Milieu: Towards a Theory of Economic Space', in S. Conti, E. Malecki and P. Oinas (eds), *The Industrial Enterprise and Its Environment: Spatial Perspectives*, Avebury, Aldershot, pp. 195-214.

Cohen, W. and Levinthal, D. (1990), 'Absorptive Capacity: a New Perspective on Learning and Innovation', *Administrative Science Quarterly*, vol. 35, pp. 128-152.

Conner, K. (1991), 'A Historical Comparison of Resource-Based Theory and Five Schools of Thought Within Industrial Organization Economics: Do We Have a New Theory of the Firm?', *Journal of Management*, vol. 17, no. 1, pp. 121-154.

David, P. (1986), 'Understanding the Economics of QWERTY: the Necessity of History', in W. Parker (ed.), *Economic History and the Modern Economist*, Basil Blackwell, Oxford, pp. 30-49.

Dierickx, I. and Cool, K. (1989), 'Asset Stock Accumulation and Sustainability of Competitive Advantage', *Management Science*, vol. 35, no. 12, pp. 1504-1511.

Dosi G. and Malerba, F. (1996), 'Organizational Learning and Institutional Embeddedness. An Introduction to the Diverse Evolutionary Paths of Modern Corporations', in G. Dosi and F. Malerba (eds), *Organization and Strategy in the Evolution of the Enterprise*, Macmillan, London, pp. 1-24.

Florida, R. (1995), 'Towards the Learning Region', *Futures*, vol. 27, no. 5, pp. 527-536.

Grabher, G. and Stark, D. (1997), 'Organizing Diversity: Evolutionary Theory, Network Analysis and Post-Socialism', in G. Grabher and D. Stark (eds), *Restructuring Networks: Legacies, Linkages, and Localities in Postsocialism*, Oxford University Press, London and New York, pp. 1-32.

Hudson, R. (1999), 'The Learning Economy, the Learning Firm and the Learning Region: a Sympathetic Critique of the Limits to Learning', *European Urban and Regional Studies*, vol. 6, no. 1, pp. 59-72.

Kirat, T. and Lung, Y. (1999), 'Innovation and Proximity. Territories as Loci of Collective Learning Processes', *European Urban and Regional Studies*, vol. 6, no. 1, pp. 27-38.

Lawson, C. (1999), 'Towards a Competence Theory of the Region', *Cambridge Journal of Economics*, vol. 23, no. 2, pp. 151-166.

Lawson, C. and Lorenz, E. (1999), 'Collective Learning, Tacit Knowledge and Regional Innovative Capacity', *Regional Studies*, vol. 33, no. 4, pp. 305-317.

Lundvall, B. (ed.) (1992), *National Systems of Innovation: Towards a Theory of Innovation and Interactive Learning*, Pinter, London.

Mahoney J.T. and Pandian, R. (1992), 'The Resource-Based View Within the Conversation of Strategic Management', *Strategic Management Journal*, vol. 13, pp. 363-380.

Maskell, P. (1999), 'Globalisation and Industrial Competitiveness: the Process and Consequences of Ubiquitification', in E. Malecki and P. Oinas (eds), *Making Connections: Technological Learning and Regional Economic Change*, Ashgate, Aldershot, pp. 35-59.

Maskell, P. and Malmberg, A. (1999a), 'The Competitiveness of Firms and Regions. 'Ubiquitification' and the Importance of Localized Learning', *European Urban and Regional Studies*, vol. 6, no. 1, pp. 9-25.

Maskell, P. and Malmberg, A. (1999a), 'Localised Learning and Industrial Competitiveness', *Cambridge Journal of Economics*, vol. 23, no. 2, pp. 167-186.

McKee, D. (1992), 'An Organizational Learning Approach to Product Innovation', *Journal of Product Innovation Management*, vol. 9, pp. 232-245.

Morgan, K. (1997), 'The Learning Region: Institutions, Innovation and Regional Renewal', *Regional Studies*, vol. 31, no. 5, pp. 491-503.

Nooteboom, B. (1999), 'Innovation, Learning and Industrial Organization', *Cambridge Journal of Economics*, vol. 23, pp. 127-150.

Oinas, P. (1995), 'Organisations and Environments: Linking Industrial Geography and Organisation Theory', in S. Conti, E. Malecki and P. Oinas (eds), *The Industrial Enterprise and its Environment. Spatial Perspectives*, Avebury, Aldershot, pp. 143-167.

Oinas, P. (1999), 'Activity-Specificity in Organizational Learning: Implications for Analysing the Role of Proximity', *GeoJournal,* vol. 49, pp. 363-372.

Oinas, P. (2000), 'Distance and Learning: does Proximity Matter?', in R. Rutten, S. Bakkers, K. Morgan and F. Boekema (eds), *Knowledge, Innovation and Economic Growth: Theory and Practice of the Learning Region*, Edward Elgar, Aldershot, pp. 57-69.

Oinas, P. and Virkkala, S. (1997), 'Learning, Competitiveness and Development. Reflections on the Contemporary Discourse on 'Learning Regions', in H. Eskelinen (ed.), *Regional Specialisation and Local Environment - Learning and Competitiveness*, NordREFO 1997, no. 1, pp. 263-277.

Pennings, J. and Harianto, F. (1992), 'The Diffusion of Technological Innovation in the Commercial Banking Industry', *Strategic Management Journal*, vol. 3, no. 1, pp. 29-46.

Penrose, E. (1959), *The Theory of the Growth of the Firm*, Oxford University Press, Oxford.

Peteraf, M. (1993), 'The Cornerstones of Competitive Advantage: a Resource-based View', *Strategic Management Journal*, vol. 14, pp. 179-19.

Prahalad C. and Hamel, G. (1990), 'The Core Competence of the Corporation', *Harvard Business Review*, vol. 3, pp. 79-91.

Simmie, J. (ed.) (1997), *Innovation, Networks and Learning Regions*, Jessica Kingsley, London.

Storper M. (1992), 'The Limits to Globalization: Technology Districts and International Trade', *Economic Geography*, vol. 68, no. 1, pp. 60-93.

Storper, M. (1997), *The Regional World: Territorial Development in a Global Economy*, Gilford, New York and London.

Storper, M. and Salais, R. (1992), 'The Four 'Worlds' of Contemporary Industry', *Cambridge Journal of Economics*, vol. 16, pp. 169-193.

Storper, M. and Salais, R. (1997), *Worlds of Production*, Harvard University Press, Cambridge MA.

Teece D. and Pisano, G. (1994), 'The Dynamic Capabilities of Firms: an Introduction', *Industrial and Corporate Change*, vol. 3, no. 3, pp. 537-556.

Teece, D., Pisano, G. and Shuen, A. (1997), 'Dynamic Capabilities and Strategic Management', *Strategic Management Journal*, vol. 18, no. 7, pp. 509-533.

Teece, D., Rumelt, R., Dosi, G. and Winter, S. (1994), 'Understanding Corporate Coherence: Theory and Evidence', *Journal of Economic Behavior and Organization*, vol. 23, pp. 1-30.

Wernerfelt, B. (1984), 'A Resource-based View of the Firm', *Strategic Management Journal*, vol. 5, no. 2, pp. 171-180.

Stoppard, [illegible] (1977) [illegible] [illegible] [illegible] and [illegible] [illegible] [illegible] [illegible], [illegible]
[illegible] (1977) [illegible] [illegible] [illegible] [illegible], London.
[illegible] and [illegible] (1993) [illegible] [illegible] [illegible] [illegible] [illegible] [illegible] [illegible]
Morgan, [illegible] and [illegible] [illegible] [illegible] [illegible] [illegible]
[illegible] [illegible] [illegible] [illegible] [illegible]
[illegible] [illegible] [illegible] [illegible] [illegible] [illegible]
[illegible] [illegible] [illegible] [illegible] [illegible]
[illegible] [illegible] [illegible] [illegible] [illegible] [illegible]
[illegible] [illegible] [illegible] [illegible] [illegible]
[illegible] [illegible] [illegible] [illegible] [illegible]

6 New Forms of Local Governance in the Emergence of Industrial Districts

Ayda Eraydin

Introduction

Recent research on spatial development has emphasised the importance of flexible production and specialisation within 'industrial districts' and 'clusters' as a way of achieving and maintaining local economic 'success' and internationally competitive industrialisation in developed economies (Brusco, 1982; Sabel, 1989; Piore and Sabel, 1984; Capecchi, 1989; Beccatini, 1991; Storper, 1993). Based on Marshallian ideas and emphasising hi-tech industry, clusters of flexibly organised, networked firms are seen in this research as being able to respond quickly to economic change in the face of globalisation.

In the past decade, a limited number of success stories have begun to emerge from developing countries (Eraydin, 1997): from Latin America, especially from Brazil (Schmitz, 1995; Storper, 1990); South Korea (Park and Markusen, 1995); Mexico (Rabelotti, 1995, 1997); India (Cawthorne, 1995); Pakistan (Nadvi, 1992); Indonesia (Symth, 1992); Turkey (Eraydin, 1998a, 1998b); and South Africa (Rogerson, 1994). But, in those developing countries, the definition of what constitutes an industrial district has been widened to include places with distinctive patterns of economic specialisation, production networks and embeddedness (Park, 1996). All these success stories come from regions that are currently internationally competitive. Now, however, doubt has been expressed about the sustainability of growth in these regions and their future development patterns (Harrison, 1994; Cooke, 1996; Staber, 1996). Indeed, now it is questioned whether textbook interpretations of 'industrial districts' are supported by factual evidence (Rabelotti, 1995). Nevertheless, industrial districts are still in the policy agenda in both developed and developing countries. Clusters based on innovation are now defined as 'learning regions' (Florida, 1995) 'innovative milieux' (Camagni, 1991) or 'regional innovation systems' (Cooke *et al.*, 1998). They are seen as the spin-offs from or advanced forms of industrial district founded information, knowledge and innovation (Hassink, 1997).

Governance and the Success of Industrial Districts

A first question that must be asked is why are some industrial districts more successful than others? The existing literature highlights the importance of the organisation of production and the strength of relationships between workers, entrepreneurs and local institutions (Amin and Malmberg, 1992; Park and Markusen, 1995), and the principal characteristics of the newest forms of industrial district have been identified as:

- specialisation in fields of production that maximise interactive relations (Beccatini, 1989, 1990; Pyke and Sengenberger, 1991);
- local support networks (Harrison, 1992; Schmitz, 1990); and
- quality based competition (Pyke and Sengenberger, 1991; Brusco, 1990).

The information exchange facilitated by various forms of social capital, especially the cultural norms of trust, cooperation and reciprocity (Brusco, 1986; Fukuyama, 1995), worker-entrepreneur relations and common social and cultural characteristics are defined as important in facilitating the emergence of local networks (Rabelotti, 1995). It is also claimed that local networks promote local innovative capacity and the dissemination of technology-based information, which are the end results of interactive learning processes and face-to-face relations in local industrial clusters (Digiovanna, 1997). Local institutions are emphasised as mediators in local development (Schmitz and Musyck, 1994), especially education and training facilities for the supply of skilled, adaptable labour (Pyke and Sengenberger, 1991).

What is clear from this discussion is that successful industrial districts have grown through the internal evolution of relationships. As a result, they are good examples of *governance* as an increasingly important new form of regulation. In fact, governance is the key issue that enables us to understand the dynamics of industrial districts and to discuss the conditions of growth and development within them in an era of globalisation and deregulation.

The academic literature on governance is eclectic and springs from a variety of theoretical roots, with the result that the term itself has a number of definitions (Stoker, 1998). In practice, governance is accepted as '*a code for less government*' and reflects the search for reduced government spending. However, governance has a wider meaning that can be generalised as "different modes of coordination of interdependent activities" (Jessop 1993, p. 29) and the creation of conditions for collective self-organising action. Indeed, Jessop has refined his definition of governance as the "self-organisation of inter-organisational relations" (Jessop, 1998, p. 30).

Obviously, the success of a self-organising and self-evolving system depends on a range of conditions, such as the local mode of coordination, the nature of the objects of governance, and the environment within which relevant actors coordinate their activities. Jessop (1998) identifies three different forms of governance. First, the simplest type stems from the selective formalisation of interpersonal networking among actors who build their relationships on past familiarity. The partnerships formed are targeted and partners share a community interest. In this form of governance, individuals represent themselves, but they may also speak for the institutional order from which they are recruited. Second, governance can refer to the mode of coordination among formally autonomous organisations to secure mutually beneficial joint outcomes which, through, synergy magnifies the powers of individual actors. Third, governance can reflect the mutual understanding and co-evolution of different institutional orders to secure agreed societal objectives. In this situation, relationships are no longer pure, direct, mutual benefits Instead, they are the means to realise an inter-systemic consensus, inter-organisational arrangements and relevant activities to achieve specific objectives and to respond to continuous change in the external world.

The discussion above enables us both to redefine industrial districts and to understand differences in the development trajectories of these areas. Successful industrial districts can be defined as self-organising areas, in which mutual benefits are maximised through cooperation, collaboration and different types of networking. It is obvious that the emergence of distinct forms of governance in industrial districts is context depended and evolution of one or the other depends on the basic socio-cultural and economic characteristics of the industrial district.

The first part of this chapter explores the different aspects of governance in industrial districts that are considered to be important for growth. However, it is clear from the discussion that initial structural conditions and forms of governance in industrial districts are inadequate for responding or adapting to changing global economic conditions. For this reason, the second part of the chapter explores the nature of successful transformation in different industrial districts in the last decade. Finally, the third part of the chapter focuses on the impact of deregulation on industrial districts and emerging trends in the governance of industrial clusters with different development trajectories.

Governance in Industrial Districts

From the literature, it can be suggested that the nature of governance differs between industrial districts. In 'Marshallian' industrial districts, production organisations are embedded in local communities of actors, and in hi-tech industrial districts local production, innovation and information networks define the potential of these clusters. In contrast, the literature on industrial districts in less developed countries emphasises government's role in initiating as 'satellite' districts and clusters centred on lead firms. However, when latter group of districts become more complex, local institutions and authorities become increasingly important (Park, 1996).

A range of studies emphasises different aspects of governance in industrial districts. While the importance of self-governing networks of local actors is widely discussed (Porter, 1990), governance in industrial districts is defined as a change from individual to collective interests (Ottai, 1994). There is also a strong emphasis on the role of institutions: the changing role of central government and the increasing importance of local institutions. Central governments are seen as mediator, enabler and 'catalytic agent' (Capecchi, 1989), whereas local government led partnerships are accepted as the most important agents of governance promoting industrial district performance (Piore and Sabel, 1984). In fact, the declining role of central government and the increasing role of public-private partnerships are at the heart of the shift from government to governance. The next section of this chapter explores the different forms that governance can take and the roles of actors within them.

The Importance of the Self-Governing Networks as the Basis of Governance

Self-governance is the core concept in industrial district formation. This type of development is obviously different than central-local government promoted growth, since the growth that is generated is principally based on local networked capacities. Individual firms are the main actors in these networks, but other formal and informal institutions act as catalytic agents, especially the enterprise associations formed by local entrepreneurs. In Baden-Wurttemberg, these associations have been very influential, whereas in the districts of the other institutions, such as local labour unions, local financial institutions and family ties, as well as business associations, have been significant players (Brusco, 1982; Piore and Sabel, 1984; Garofoli, 1991). These network participants combine to create different governance structures - from relatively egalitarian structure, to loose or even more rigid hierarchies – that can be defined according to their main motivations. At least four sets of motivations can be identified in recent research.

Networks that enable learning-by-interacting The local concentration and networking of specialist activities can bring specific advantages to industrial clusters. They create local pools of knowledge and specialist labour markets. They build local cultures of work flexibility and cooperation based on trust and reciprocity. They reduce transaction costs and improve the local availability of specialist services (Sternberg, 1996). As a consequence of these 'untraded interdependencies', local innovation capacities are said to be enhanced (Storper, 1995).

Innovation is said to stem from local inter-firm interactions that promote learning (Hassink, 1997) coupled with interactions with public research establishments, higher education institutes and technology transfer establishments. From hi-tech industrial districts to 'Marshallian' industrial districts and even in the LDC districts, it is possible to observe the crucial importance of learning-by-interacting, especially when it is dependent upon tacit knowledge. Several studies note that significant numbers of innovations have their origins in 'learning-by-doing' and 'learning-by-using' (Massey *et al.*, 1992; Cooke, 1998).

The self-help activities and informal networks The industrial district literature demonstrates the changing weight of quasi-public interest associations in local development processes. Cooperatives formed by small business firms and local institutions initiated by entrepreneurs, such as the export trade firms and the institutions specialised in promotion, advertising and marketing, have been identified as being especially important in the initial phases of industrial district development (Heidenreich, 1996). In addition, organisations that provide information and connect firms to production technologies have also been identified and being (Ottai, 1994).

Other types of self-help institution have also been effective in generating local development stimuli. In some industrial districts, venture capital providers are important for generating small and medium sized technology based firms that have ideas but not capital. These venture capitalists have played a major role in Silicon Valley where they have accelerated the pace of technological innovation (Saxenian, 1990; Florida and Kenney, 1990). The activities of venture capitalists are information and interaction intensive and create highly personalised, informal and localised networks that can take over some functions of formal financial institutions.

In developing country industrial districts, venture capital is also available to small firms, but here family circles are more important than public and non-public institutions (Eraydin, 1998a; Saracoglu, 1993). In these industrial districts, active self-help organisations, sometimes supported by prominent families, facilitate trust and collaboration between local entrepreneurs (Erendil, 1998; Schmitz, 1995). The same situation has been identified in the districts of northeast Italy where families have been seen as 'communitarian families' (Storper, 1993).

Collaborative networks for problem solving The problems and inadequacy of public institutions force entrepreneurs to get together to solve their problems. Sometimes, collaboration is needed because of acute competition, as occurred in the Swiss watch industry, for example (Glasmeier, 1991). Collaboration may in some circumstances, create self-help institutions, especially in countries where central government institutions are not very efficient. For example, in Gaziantep in Turkey, entrepreneurs recently formed a quality control unit to tackle the problems of low quality production that had begun to undermine the reputation of local industry (Eraydin, 1998b). In the Sinos Valley in Brazil (Schmitz, 1995) and in Guadalajara and Leon in Mexico (Rabelotti, 1997), examples from the footwear industry show that, besides local institutions, voluntary organisations and local enterprise associations are very influential in activating local entrepreneurs and facilitating cooperation between them over particular issues.

But, small firms can also be highly protective of their proprietary knowledge and independence (Semlinger, 1993). For example, in Denizli in Turkey, firms tried to overcome problems in meeting large orders for high quality products by integrating some stages of production within their factories rather than by using local production networks or cooperating among themselves to increase the quality of production (Ozcan, 1995; Erendil, 1998). In these situations, self-help institutions and formal relations can play a vital role. Studies on several industrial districts in Turkey have shown that small scale enterprises mainly depended on their own financial resources or get support from their families, relatives and even from other local entrepreneurs when they need financial help (Eraydin, 1998b). These firms use limited credit from money lending institutions because of its high cost and the problems of finding collateral. In the industrial districts of developed countries, regional and local institutions perform this credit function, and in Baden Wurttemberg, for example, local banks and financial institutions help small-scale industries to raise capital money (Schmitz and Musyck, 1994).

Networks facilitated by local mediators In addition to informal networks local mediators are also important in relatively less developed countries in promoting interaction and solving some local problems. In India, the Trippur example indicates the very important role of local mediators (Cawthorne, 1995), which is also observed in the Sinos Valley in Brazil (Schmitz, 1995). Similarly, in Denizli in Turkey, the production of low quality products for low and medium income groups is organised by local tradesmen (Erendil, 1998). They act as mediators among small and medium size production units and supply raw materials. In fact, a classical putting-out system can be observed in the industrial districts of many less developed countries, as well as some industrial districts in advanced countries, at least at the initial phases of development.

Governance as a Shift from Individual to Collective Interests

In the formation of local networks within clusters emphasis now is increasingly on collective interests, institutional benefits and mutual developments, instead of individual gains. Collective interests are increasingly important for all the participants in an industrial cluster because they facilitate regional collective learning through the synergies of knowledge sharing (Asheim, 1996; Florida, 1995; Hassink, 1997).

Governance as a New Deal Between Network Participants

The struggle between workers and entrepreneurs, the militancy of labour and high wages are said to define the end of the Fordist production system (Harvey, 1989; Leborgne and Lipietz, 1992). Thus, flexible production has flourished in the areas where labour supply and labour conditions are favourable for entrepreneurs. In Marshallian industrial districts, family labour (Capecchi, 1989), the availability of a trained, adaptable and co-operative workforce (Pyke and Sengenberger, 1991), good worker-entrepreneur relations and limited conflict between entrepreneurs and labour unions are have been identified as important features of this type of district (Lazerson, 1990). However, strong labour unions have been identified as an important source of competitiveness and innovations in Baden-Wurttemberg, for example (Digiovanna, 1997). In some studies, the lack of social security systems and anti-union sentiments are seen as the attractions of certain industrial clusters. As Massey (1984) points out, scientific workers at the top of the labour hierarchy in flexible production systems may benefit from anti-unionism, because their autonomy brings higher wages and flexible working conditions, but for shopfloor workers this system bring low wages and insecurity. It is interesting that, notwithstanding these tensions and contradictions in the labour market, most of the regions of flexible specialisation have been said to have smooth worker- entrepreneur relations. These findings suggest that a new deal between workers and entrepreneurs is emerging which is central to the successful governance of industrial districts.

Public-Private Partnerships and the Changing Roles of Governments

Governance as a way of getting things done does not rest solely on the power of government. Government's new role, is as a 'mediator', 'enabler' or 'catalyst'. In this respect, public-private partnerships are essential in the transition from government to governance.

Many studies emphasise local institutions rather than government institutions in the promotion of local growth. However, recent studies also indicate the strong impact of macro economic regulation on administrative systems (Cooke and Morgan, 1994; Staber, 1996). Although government policies appear to play no direct role in promoting industrial districts in the USA and UK, government funded defence subcontracting still plays an important role. This subcontracting, in southern California in the USA and along the M4 Corridor of UK, is not regulated by market norms. Instead, it is shaped by government procurement policies (Castells and Hall, 1994). The public-private partnerships that develop from these arrangements are major stimuli for many hi-tech industrial districts.

Central government and regional growth Central governments' regional policies also support local growth. While local processes shape the economic activities of industrial clusters, even in Emilia Romagna and Baden-Wurttemberg small innovative firms are supported by central, state and local governments through a range of in institutions (Cooke and Morgan, 1994). Special tax provisions are common in many industrial districts and almost all hi-tech regions (Steed, 1987; Fujita, 1988; Masser, 1989; Castells and Hall, 1994; Scott and Storper, 1987; Lin, 1997). Local governments attract firms by providing tax incentives to create 'a good business climate'.

In the industrial districts of less developed countries (LDCs), such as Turkey (Denizli, Gaziantep and Corum), central government schemes are important, despite local entrepreneurs' usual claims that industrial development areas succeeded because of the efforts of local entrepreneurs without any contribution from the state (Eraydin, 1998a, 1998b). Central government is similarly important in Southeast Asia (Park, 1996). For example, in South Korea, central government policies and programmes to provide tax incentives, financial support, import controls, education facilities and infrastructure are important for generating development in industrial districts (Park and Markusen, 1995). Publicly owned firms can also be instrumental in creating initial growth in some industrial districts, although their roles tend to be hidden in the industrial district literature. In fact, empirical studies show the importance of state manufacturing firms that became the leaders in knowledge accumulation and dissemination in Turkey (Pinarcioglu, 1998), and in the initial development of the Emilian industrial districts (Brusco, 1990).

Administrative reform, the transfer of government functions and local development The reform of administrative systems can also foster local economic growth and the emergence of industrial districts. These reforms usually enable local governments to be more active and to use their own resources to initiate activities. For example, the reorganisation of the administration system in Italy increased the rights and responsibilities of local governments and supported local and regional institutions in their development activities (Capecchi, 1989). It built on past experience such as the Artisan's Law that was enacted after the Second World War that helped small enterprises through low tax rates and tax exemptions. Other regulations reduced the current costs of small firms and provided low interest credit to firms as cooperatives (Cooke and Morgan, 1994).

In countries such as the UK, administrative support is less well developed. But, in the industrial districts of less developed countries, central government plays an active role through collaboration with local bodies rather than through command. In Turkey, central government institutions have strengthened their local offices and seek to encourage collaborative partnerships with local industrial enterprises.

Local governments and institutions as active agents in development In recent decades, local authorities have come to play an increasingly active role in development, even in very centralised countries (Castells and Hall, 1994). These authorities are not only public organisations but they are also local public-private co-operative institutions that support local governance systems.

Malmberg (1996, p. 399) emphasises the importance of local institutions in the formation of local industrial agglomerations. He notes that "a localised industrial system may be understood in several different ways: a system of activities directly related; as a system made up by separate value chains linked through competition; or merely by the fact that firms coexist within some shared regulatory and institutional structure". In other words, the new governance system is mainly based on local institutions and cooperation between different types of organisation. The evaluation of the development process of pioneering industrial districts in Emilia-Romagna in Italy indicates the very important role of local institutions and local policies in accelerating development and the dissemination of information and innovative knowledge in these regions (Digiovanna, 1997; Putnam *et al.*, 1993).

The institutions critical for growth vary between industrial clusters, but the literature makes clear that the institutions of finance, education and training are pivotal. Local banking facilities have been vital in the development of the Third Italy because of their support programmes for small industry operated through cooperative banks, banks owned by municipalities and commercial local banks (Beccatini, 1991). These banks provide risk capital for innovative small firms (Schmitz and Musyck, 1994). In Baden-Wurttemberg, local financial institutions provide guarantees for small firms when they apply to commercial banks for credit.

Education and training establishments aid local development through the socialisation of local communities (Scott and Storper, 1987; Lyons, 1995). Local education programmes are directed both to workers and to entrepreneurs. They are designed collaboratively by local and central government institutions to meet local needs. It is claimed that this targeting is more effective and cost efficient when local entrepreneurs finance these local education programmes. In the 1970s, especially in Italy, local education programmes were initiated by the municipalities, business associations and entrepreneurs to overcome the discrepancy between needs and available schemes. They have been judged to have been very useful for upgrading regions' technology bases innovative capacities (Capecchi, 1989). Universities are thought to play a particularly crucial role in the development of hi-tech districts, through the generation of knowledge, the training of labour, the spinning-off of new business ventures and the provision of cultural amenities (Castells and Hall, 1994).

Partnerships between private enterprise, central government and local government have supported the development of medium and large enterprises in some industrial districts through the creation of industrial estates, science parks and technoparks (Eraydin, 1998a; Massey *et al.*, 1992). In Baden-Wurttemberg, they have created 'technology centres' specialising in technology transfer and technological improvement that finance their services through their own commercial activities. According to Schmitz and Musyck (1994), the Baden-Wurttemberg experience shows that the services provided directly by public organisations are not particularly effective, while semi-public institutions that work in close collaboration with entrepreneurs are more influential in innovative activities. The principal role of central government would appear, therefore, to be initiate cooperative action.

Culture and local governance Cultural norms embedded in regions have also been highlighted as affecting local governance. Lazerson (1990), for example, has claimed that the most important source of uniqueness in the Modena knitwear district in Italy is its cooperative culture. A common culture is said to generate shared trust and knowledge exchange within regions and localities, especially those involving small numbers of actors and limited mobility because of the limits these place on opportunistic behaviour (Maskell and Malmberg, 1999a). Even in the context of strong competition, cultural norms limit the exploitation of information asymmetries, the passing of defective and substandard goods or the creation of hold-ups gain benefit at the expense of others.

The Transformation of Industrial Districts and the New Forms of Local Governance

In the 1990s, deregulatory forces and volatile global economic conditions brought major shifts in the governance structures of industrial districts. Deregulation has occurred in capital markets, in international trade and in labour markets, accompanied by a dismantling of the Keynesian welfare state (Dunford, 1990, Jessop, 1990). As a result of these pressures, some districts have prospered while others have declined. From the existing literature, three trajectories of change in industrial districts can be identified.

First, some industrial districts have been negatively affected by the changes in global markets. They have been unable to adapt and have lost their competitive edge (Schmitz, 1998). In some districts, such as Prato in Italy and Jura in Switzerland, the numbers of firms declined after mid 1980s (Bigarelli and Crestallo, 1994; Glasmeier, 1994). Global competition required constant, steady adaptation, but these new conditions disturbed the assets bases of these districts, especially the collaborative networks that had been the foundation of growth in previous decades (Furmagalli and Mussatti, 1993; Staber, 1997). From his studies in the Sinos Valley in Brazil, Schmitz (1998) asserted that the centrifugal forces of globalisation made local cooperation increasingly difficult.

Second, in other industrial clusters new and different approaches have been adopted to sustain and maintain local competitiveness. In both developed and developing countries, mergers among the local firms and increased local networking have been adopted as adaptation mechanisms in some industrial districts. These strategies have been identified in Prato in Italy (Ottai, 1996) and in Italy more generally (Locke, 1995), and in Bangkok in Thailand (Scott, 1994). In other districts, firms have merged to become parts of international production networks. In these districts in LDCs, growth has been achieved through increasing firm size and vertical integration or through local firms becoming strongly linked with large local or multinational firms (Eraydin, 1997). Here, in developed and developing countries alike, powerful lead firms have become increasingly dominant, threatening the collaborative nature of inter-firm relations in industrial clusters (Harrison, 1992).

Third, some industrial districts have remained successful. While some Marshallian industrial districts sustained their competitive power by successfully transforming themselves into innovative learning regions (Cooke, 1996), some hi-tech industrial clusters have become knowledge creation centres. Among them there are clusters where design and innovation have become separated from production, but in others innovation continues through learning-by-interacting.

All these trajectories of change in industrial districts indicate that dynamic innovative capacity is at the core of sustained economic performance rather than static competitive advantage. Now, existing global institutions and systems of regulation appear to be incapable of solving emerging economic problems and countering distortions in the distribution of the benefits of globalisation re-emphasising the importance of these local capacities (Eraydin, 1998c). Deregulation has brought increasing interaction at all scales with the result that "regional fortunes are increasingly intertwined with global events that are largely beyond a single community's control" (Glasmeier, 1994, p. 139). The new institutions and regulatory forms that are currently being produced on both subnational and supranational scales, while at the national scale governance is being refined in response to the current round of specialisation (Brenner, 1999).

The Adaptive Capacities of Successful Industrial Clusters

In the literature on industrial clusters it has been asserted that to be competitive in a global environment firms must now be simultaneously locally embedded and globally integrated (Gordon, 1996; Camagni, 1991; Maskell and Malmberg, 1999b). Because globalisation makes many previously localised capabilities ubiquitous, local *learning and institutional capacities* are increasingly seen as being important in achieving this global-local balance.

An economic agent's learning capacity has been described as "the capacity to create, acquire and transform knowledge and upgrade its skills, expertise and competencies to fulfil its objectives in a fast changing and turbulent economic environment" (Jin and Stough, 1998). These competencies are most readily achieved where knowledge creation is supported through the institutional embodiment of tacit knowledge (Maskell and Malmberg, 1999b) – through a local knowledge system, norms, social conventions, values and institutions (Hassink, 1997).

Four characteristics of industrial clusters are now recognised as fostering learning capacities. First, reduced *spatial transaction costs* in the creation, diffusion and appropriation of tacit knowledge is expected to raise learning capacities in spatial clusters (Scott, 1996). Proximity encourages learning and the creation of relational assets that are essential for the learning process (Amin and Wilkinson, 1999). Second, it is widely accepted that learning capacities are closely related the *social capital* (Coleman, 1988), which includes trust, reciprocity and collaborative action. The term social capital is important since it covers both non-economic and non-political relationships within local communities that are critical in sustaining local development (Putnam *et al.*, 1993). Third, regional institutions, organisations and infrastructure remain central to learning processes and maintaining local competitiveness (Florida, 1995; Asheim, 1996). To sustain the learning capabilities of successful clusters, local governments, communities and voluntary associations have to provide a range of collective goods (Jin and Stough, 1998), because learning also involves interaction with research centres, higher education institutes and technology transfer agencies. Fourth, it is now recognised that learning processes also necessitate '*unlearning*'. Unlearning involves the dismantling and removal of formerly important institutions, which negatively affect further development. In the Sinos Valley in Brazil, for example, buyers who had previously contributed to a process of learning-by-exporting, began to obstruct learning about the potential future benefits of firms moving up the value chain (Schmitz, 1998).

Paralleling this emphasis on learning, interest is still focused on the institutional setting of norms, routines and conventions for the creation and maintenance of regional economic competitiveness and performance (Cooke *et al.*, 1998). However, this appreciation, especially of 'institutional thickness', is now more critical than it was in the past (Raco, 1999; Amin and Thrift, 1995). Several studies now claim that the institutional structures that supported growth in some industries in the earlier decades now increasingly act as barriers to innovation (Raco, 1999) - a situation described as 'institutional overload'. Interestingly, vocational training and financial systems that once were praised as the most important institutions of successful development are recognised in some studies as obstacles to innovation in production. For example, Rabelotti (1997) has identified local institutions as the source of resistance to change in the Brenta footwear industrial district in Italy, and Schmitz (1998) has found institutional thickness unable to resolve conflict in the Sinos Valley footwear industries in Brazil. Similar findings have emerged from studies of the Swiss watch industry (Glasmeier, 1994), and even for Baden-Wurttemberg, where it is claimed that institutional inertia and local resistance is impeding growth (Heidenreich and Krauss, 1998). Although there are successful examples where institutions have turned decline to prosperity, there is always the danger of asset erosion when important institutions a region can no longer stimulate and motivate technological change.

Conclusion

In the post-Fordist era of flexible accumulation, new approaches were needed to sustain economic growth, and in this period, beginning in the 1970s, export orientated industrial agglomeration emerged as a successful system of local development and governance. Empirical studies highlighted the significant of network relationships between firms in agglomerations, but places that were successful in the 1980s began to face problems in the 1990s. Globalisation in this decade radically altered the conditions facing these local networked economies, although some successfully improved their learning and institutional capacities through locally embedded relationships and their local governance structures. However, it would now appear that local capacities and local institutional thickness may no longer be enough to maintain a place's competitive advantage.

As industrial clusters have begun to experience different trajectories of growth in the new globally competitive conditions of the 1990s, factors of local competitiveness that were once lauded a promoting growth are now being questioned. For example, 'learning' is now coupled with 'unlearning' and 'institutional thickness' with 'institutional overload'. Successful industrial clusters are now likely to be imitated, though imitation is limited by *asset mass efficiency* (stocks of R&D, experience-based knowledge and specialised labour), *time compression diseconomies* (the time taken to create capabilities) and the *interconnectedness of asset stocks* (the complex webs of national, regional and local institutions) (Maskell and Malmberg, 1999a). However, even these conditions do not guarantee against the deterioration of localised economic capacities.

It would now appear that the local embeddedness of economic relations alone might not be enough to sustain local competitive strength in a globalised economy. Globalisation is now creating new roles of governments, cross-border strategic alliances among firms (non-local embeddedness), and the new roles for transnational corporations (TNCs).

There is now a reviving interest in the role of national governments in shaping development and guarding places against the unwanted economic, social and cultural consequences of globalisation process (Immerfall *et al.*, 1998). National governments play a major role in providing learning infrastructure and promoting innovation in both developed and developing counties (Asheim and Isaksen, 1997; Amsden, 1989; Park and Markusen, 1995). In many places they are still the most legitimate mediators to resolve conflict (Heidenreich, 1996; Schmitz, 1998), and central government policies and support can be vital for economic growth in underdeveloped areas (Eraydin, 1998a). In the absence of adequate collaborative local initiatives, Gordon (1996, p.124) claims that "government intervention is required for construction of scientific and technological infrastructure, assistance with the establishment of linkages to global networks and, perhaps above all, the articulation of coherent cross-regional strategies in order to avoid destructive competition between regions for new technological activities".

Technological complementarities, increased buyer-supplier interaction and the need for access to global markets have moved firms' strategic alliances beyond the borders of local economies (Gordon, 1996; Maskell and Malmberg, 1999a), making non-local embeddedness increasingly important (Cooke and Morgan, 1994; Park, 1996). Now there is increasing interest in the local embeddedness of TNCs and also multi-locational enterprises as they try to tap into the embedded capacities of industrial clusters and engage with local innovative capacities.

The link between local and global processes in shaping economic growth is, therefore, complex. Thus, in a globalising world, there is a 'persistence of territory' and a 'place dependency of globalisation'. It is still proximity that encourages learning (Nooteboom, 1999), favours interactive learning and institutional embodiment (Maskell and Malmberg, 1999a) and creates competitive power.

Governance is still central to understanding the dynamics of industrial agglomerations and their learning and adaptative capacities. But in the new, complex environment of globalisation it is important to begin to begin to understand processes of *mixed embeddedness* (Kloosterman *et al.*, 1999), the emerging *new networks* that operate beyond the local level and the *power relationships* between local and non-local actors to begin to more fully understand local growth and local competitive capacity building.

References

Amin, A. and Malmberg, A. (1992), 'Competing Structural and Institutional Influences on Geography of Production in Europe', *Environment and Planning A*, vol. 24, pp. 401-416.

Amin, A. and Thrift, N. (1995), 'Globalisation, Institutional Thickness and the Local Economy', in P. Healey, S. Cameron, S. Davoudi, S. Graham and A. Madani-Pour (eds), *Managing Cities: The New Urban Context*, John Wiley and Sons, London.

Amin, A. and Wilkinson, F. (1999), 'Learning Proximity and Industrial Performance: an Introduction', *Cambridge Journal of Economics*, vol. 23, pp. 121-125.

Amsden, A. (1989), *Asia's Next Giant: South Korea and Late Industrialization*, Oxford University Press, New York.

Asheim, B.T. (1996), 'Industrial Districts as Learning Regions: a Condition for Prosperity', *European Planning Studies*, vol. 4, no. 4, pp. 379-397.

Asheim, B.T. and Isaksen, A. (1997), 'Location, Agglomeration and Innovation: Towards Regional Systems in Norway', *European Planning Studies*, vol. 5, no. 3, pp. 299-330.

Beccatini, G. (1989), 'Sectors and/or Districts: Some Remarks on the Conceptual Foundations of Industrial Economics', in E. Goodman and J Bamford (eds), *Small Firms and Industrial Districts in Italy*, Routledge, London.

Beccatini, G. (1990), 'The Marshallian Industrial Districts as a Socio-Economic Notion', in F. Pyke, G. Beccatini and W. Sengenberger (eds), *Industrial Districts and Inter-firm Cooperation in Italy*, Geneva, International Institute for Labour Studies.

Beccatini G. (1991), 'The Industrial District as a Creative Milieu', in G. Benko and M. Dunford (eds), *Industrial Change and Regional Development*, Belhaven, London, pp. 102-113.

Bigarelli, D. and Crestanello, P. (1994), 'An Analysis of the Changes in the Knitwear/Clothing District of Carpi During the 1980s', *Entrepreneurship and Regional Development*, vol. 6, pp. 127-144.

Brenner, N. (1999), 'Globalisation as Reterritorialisation: The Rescaling of Urban Governance in the European Union', *Urban Studies*, vol. 36, pp. 431-451.

Brusco, S. (1982), 'The Emilian Model: Productive Decentralization and Social Integration', *Cambridge Journal of Economics*, vol. 6, pp. 167-184.

Brusco, S. (1986), 'Small Firms and Industrial Districts: The Experience of Italy', in E. Keeble and E. Wever (eds), *New Firms and Regional Development*, Croom Helm, London, pp. 184-202.

Brusco, S. (1990), 'The Idea of Industrial Districts: Its Genesis', in F. Pyke and W. Sengenberger (eds), *Industrial Districts and Local Economic Regeneration*, Geneva, International Institute for Labour Studies.

Camagni, R. (1991), 'Local Milieu, Uncertainty and Innovation Networks: Towards a New Dynamic Theory of Economic Space' in R. Camagni (ed.), *Innovation Networks*, Belhaven, London, pp. 121-144.

Capecchi V. (1989), 'The Informal Economy and the Development of Flexible Specialization in Emilia-Romagna', in A. Portes, M. Castells, L.A. Benton (eds), *The Informal Economy Studies in Advanced and Less Developed Countries*, John Hopkins University Press, Baltimore, pp. 189-215.

Castells, M. and Hall, P. (1994), *Technopoles of the World*, Routledge, London and New York.

Cawthorne, P.M. (1995), 'The Rise of Networks and Markets of a South Indian Town: The Example of Tiruppur's Cotton Knitwear Industry', *World Development*, vol. 23, pp. 43-56.

Coleman, J.S. (1988), 'Social Capital in the Creation of Human Capital', *American Journal of Sociology*, vol. 94, pp. 95-120.

Cooke P. (1996), 'Building a Twenty-first Century Regional Economy in Emilia-Romagna', *European Planning Studies,* vol. 4, no. 1, pp. 53-62.

Cooke, P. (1998), 'Introduction: Origins of the Concept', in H.J. Braczyk, P. Cooke and M. Heidenreich (eds), *Regional Innovation Systems: The Role of Governance in a Globalized World*, UCL Press, London, pp. 2-27.

Cooke, P., Morman, K. (1994), 'Growth Regions Under Duress: Renewal Strategies in Baden-Wurttemberg and Emilia Romagna', in A. Amin and N. Thrift (eds), *Globalisation, Institutions and Regional Development in Europe*, Oxford University Press, Oxford.

Cooke, P., Uranga, M.G. and Etxebarria, G. (1998), 'Regional Systems of Innovation: an Evolutionary Perspective', *Environment and Planning A*, vol. 30, pp. 1563-1584.

Digiovanna, S. (1997), 'Industrial Districts and Regional Economic Development: A Regulation Approach', *Regional Studies*, vol. 30, no. 4, pp. 373-386.

Dunford, M. (1990), 'Theories of Regulation', *Environment and Planning D: Society and Space,* vol. 8, pp. 297-321.

Eraydin, A. (1997), 'LDC Industrial Districts: the Challenge of the Periphery', in K.I. Westeren (ed.), *Cross Border Cooperation and Strategies for Development in Peripheral Regions*, Nord-Trondelags, Forskning, Oslo.

Eraydin, A. (1998a), 'From an Underdeveloped Region to a Locality: The Experience of Corum', Paper prepared for World Bank.

Eraydin, A. (1998b), 'The Role of Regulation Mechanisms and Public Policies at the Emergence of the New Industrial Districts', Paper presented to New nodes of growth in Turkey: Gaziantep and Denizli, September 1998, Ankara.

Eraydin, A. (1998c), *Forming and Bursting Bubbles in Tokyo: Global Cities Under the Pressure of Global Forces and the Local Regulation Systems,* Institute of Developing Economies, Tokyo, Japan.

Erendil, A. (1998), *Using a Critical Realist Approach in Geographical Research: An Attempt to Analyze the Transforming Nature of Production and Reproduction in Denizli*, Unpublished Ph.D Thesis, Middle East Technical University, The Department of City and Regional Planning.

Florida, R.L. (1995), 'Towards the Learning Region', *Futures*, vol. 27, no. 5, pp. 527-536.

Florida, R.L. and Kenney, M. (1990), 'Silicon Valley and Route 128 Won't Save US', *California Management Review*, vol. 33, no. 1, pp. 68-88.

Fujita, K. (1988), 'The Technopolis-High Technology and Regional Development in Japan', *International Journal of Urban and Regional Research,* vol. 12, no. 4, pp. 566-594.

Fukuyama, F. (1995), *Trust: The Social Virtues And The Creation Of Prosperity*, The Free Press, New York.

Furmagalli, A. and Mussatti, G. (1993), 'Italian Dynamics from the 1970s to the 1980s: Some Reflections on Entrepreneurial Activity', *Entrepreneurship and Regional Development*, vol. 5, pp. 27-37.

Garofoli, G. (1991), 'The Italian Model of Spatial Development in the 1970s and 1980s', in G. Benko and M. Dunford (eds), *Industrial Change and Regional Development: The Transformation of New Industrial Spaces*, Belhaven, London, pp. 85-101.

Glasmeier, A. (1991), 'Technological Discontinuities and Flexible Production Networks: The Case of Switzerland and the World Watch Industry', *Research Policy*, vol. 20, pp. 469-485.

Glesmeier, A. (1994), 'Flexible Districts, Flexible Regions? The Institutional and Cultural Limits to Districts in an Era of Globalization and Technological Paradigm Shift', in A. Amin and N. Thrift (eds), *Globalisation, Institutions and Regional Development in Europe*, Oxford University Press, Oxford.

Gordon, R. (1996), 'Industrial Districts and the Globalization of Innovation: Regions and Networks in the New Economic Space', in X. Vence-Deza and J.S. Metcalfe (eds), *Wealth from Diversity*, Kluwer, Rotterdam.

Harrison, B. (1992), 'Industrial Districts: Old Wine in New Bottles?', *Regional Studies*, vol. 26, pp. 469-483.

Harrison, B. (1994), 'The Italian Industrial Districts and the Crisis of Cooperative Form: Part II', *European Planning Studies*, vol 2, pp.159-174.

Harvey, D. (1989), *The Condition of Post Modernity: An Inquiry into the Origins of Cultural Change*, Basil Blackwell.

Hassink, R. (1997), 'What does the Learning Region mean for Economic Geography?', Paper presented at the Regional Studies Association "Regional Frontiers" EURRN European Conference, 20-23 September 1997, Frankfurt (Oder), Germany.

Heidenreich, M. (1996), 'Beyond Flexible Specialization: The Rearrangement of Regional Production Orders in Emilia Romagna and Baden-Wurttemberg' *European Planning Studies*, vol.4, pp. 401-417.

Heidenreich M. and Krauss, G. (1998), 'The Baden-Wurttemberg Production and Innovation Regime: Past Success and New Challenges', in H.J. Braczyk, P. Cooke and M. Heidenreich (eds), *Regional Innovation Systems: The Role of Governance in a Globalized World*, UCL, Press London, pp. 214-244.

Immerfall, S., Conway, P., Crumley, C. and Jarausch, K. (1998), 'Disembeddedness and Localization: The Persistence of Territory', in S. Immerfall (ed.), *Territoriality in the Globalizing Society : One Place or None?*, Springler, Berlin.

Jessop, B. (1990), *State Theory: Putting the Capitalist State in its Place*, Pennsylvania University, University Park.

Jessop, B. (1993), 'Towards a Shumpeterian Workfare State? Preliminary Remarks on Post-Fordist Political Economy', *Studies in Political Economy*, vol. 40, pp. 7-39.

Jessop, B. (1998), 'The Rise of Governance and the Risks of Failure: the Case of Economic Development', *International Social Science Journal*, vol. 3, pp. 29-45.

Jin, D.J and Stough, R. (1998), 'Learning and Learning Capability in the Fordist and Post-Fordist Age: an Integrative Framework', *Environment and Planning A*, vol. 30, pp. 1255-1278.

Kloosterman, R., van der Luen, J. and Rath, J. (1999), 'Mixed Embeddedness: Informal Economic Activities and Immigrant Business in the Netherlands', *International Journal of Urban and Regional Research*, vol. 23, no. 2, pp. 252-266.

Lazerson, M.H. (1990), 'Transactional Calculus and Small Business Strategy', in Z.J. Acz and D.B. Audretsch (eds), *The Economics of Small Firms*, Dordrecht, Kluwer.

Leborgne, D. and A. Lipietz (1988), 'New Technologies, New Modes of Regulation: Some Special Implications', *Environment and Planning D: Society and Space,* vol. 6, pp. 263-280.

Lin, C.Y. (1997), 'Technopolis Development: An Assessment of the Hschchu Experience', *International Planning Studies,* vol. 2, no. 2, pp. 257-272.

Locke, R. (1995), *Remaking the Italian Economy*, Cornell University Press.

Lyons, D. (1995), 'Agglomeration Economies Among High-Technology Firms in Advanced Production Areas: The Case of Denver Boulder', *Regional Studies*, vol. 29, no. 3, pp. 265-278.

Malmberg A. (1996), 'Industrial Geography: Agglomeration and Local Milieu', *Progress in Human Geography*, vol. 20, no. 3, pp. 392-403.

Maskell, P. and Malmberg, A. (1999a), 'Localised Learning and Industrial Competitiveness', *Cambridge Journal of Economics*, vol. 23, pp. 167-185.

Maskell, P. and Malmberg, A. (1999b), 'The Competitiveness of Firms and Regions: Ubiquitification and the Importance of Localized Learning', *European Urban and Regional Studies*, vol. 1, pp. 9-25.

Masser, I. (1989), 'Technology and Regional Development Policy: A Review of Japan's Technopolis Programme', *Regional Studies*, vol. 24, no. 1, pp. 41-53.

Massey, D. (1984), *Spatial Divisions of Labour: Social Structures and the Geography of Production*, Macmillan, London.

Massey, D., Quintas, P. and Wield, D. (1992), *High-Tech Fantasies,* Routledge, London.

Nadvi, K. (1992), *Flexible Specialization, Industrial Districts and Employment in Pakistan*, Geneva, International Labour Office.

Nooteboom, B. (1999), 'Innovation, Learning and Industrial Organisation', *Cambridge Journal of Economics*, vol. 23, pp. 127-150.

Ottai G.D. (1994), 'Cooperation and Competition in the Industrial Districts as an Organisation Model', *European Planning Studies*, vol. 2, no. 4, pp. 463-483.

Ottai, G.D. (1996), 'Economic Changes in the Industrial District of Prato in the 1980s: Towards a more Conscious and Organized Industrial District', *European Planning Studies*, vol. 4, no. 1, pp. 307-330.

Ozcan, G.B. (1995), *Small Firms and Local Economic Development*, Avebury, London.

Park, S.O. (1996), 'Networks and Embeddedness in the Dynamic Types of New Industrial Districts', *Progress in Human Geography*, pp. 476-493.

Park, S.O. and Markusen, A. (1995), 'Generalizing New Industrial Districts - A Theoretical Agenda and an Application from a Nonwestern Economy', *Environment And Planning A*, vol. 27, no. 1, pp. 81-104.

Pinarcioglu, M. (1998), *Industrial Development and Local Change: The Rise of Textiles and Clothing since 1980s and Transformation in the Local Economies of Bursa and Denizli*, Unpublished PhD Thesis, University College, University of London.

Piore, M. and Sabel, C.F. (1984), *The Second Industrial Divide*, Basic Books, New York.

Porter, M.E. (1990), *The Competitive Advantage of Nations*, Macmillan, London.

Putnam, R.D., Leonardi, R. and Nanetti, R.Y. (1993), *Making Democracy Work: Civic Traditions in Modern Italy*, Princeton University Press, Princeton.

Pyke F. and Sengenberger, W. (1991), *Industrial Districts and Local Economic Regeneration*, Geneva, International Institute for Labour Studies.

Rabelotti, R. (1995), 'Is There an Industrial District Model - Footwear Districts in Italy and Mexico Compared', *World Development*, vol. 23, no. 1, pp. 29-41.

Rabelotti, R. (1997), *External Economies and Cooperation in Industrial Districts: A Comparison of Italy and Mexico*, Macmillan, Basingstoke.

Raco, M. (1999), 'Competition, Collaboration and the New Industrial Districts: Examining the Industrial Turn in Local Economic Development', *Urban Studies*, vol. 36, no. 5-6, pp. 951-968.

Rogerson, C.M. (1994), 'Flexible Production in the Developing World: The Case of South Africa', *Geoforum*, vol. 25, no. 1, pp. 2-17.

Sabel, C. (1989), 'Flexible Specialization and Re-emergence of Regional Economies', in P. Hirst and J. Zeitlin (eds), *Reversing Industrial Decline?*, Berg, Oxford.

Saracoglu, Y. (1993), *Local Production Networks: An opportunity for Development*, unpublished MCP Thesis, Middle East Technical University, Ankara.

Saxenian, A.L. (1990), 'Regional Networks and Resurgence of Silicon Valley', *California Management Review*, vol. 33, no. 1, pp. 89-112.

Schmitz, H. (1990), 'Small Firms and Flexible Specialization in Developing Countries', *Labor and Society,* vol. 15, pp. 3.

Schmitz, H. (1995), 'Small Shoemakers and Fordist Giants: Tale of a Supercluster', *World Development*, vol. 23, pp. 9-28.

Schmitz, H. (1998), *Responding to Global Competitive Pressure: Local Co-operation and Upgrading in the Sinos Valley*, IDS Working Paper 82, Institute of Development Studies.

Schmitz, H. and Musyck, B. (1994), 'Industrial Districts in Europe - Policy Lessons for Developing-Countries', *World Development*, vol. 22, no. 6, pp. 889-910.

Scott, A.J. (1994), 'Variations on the Theme of Agglomeration and Growth: the Gem and the Jewellery Industry in Los Angeles and Bangkok', *Geoforum*, vol. 25, no. 3, pp. 249-263.

Scott, A. J. (1996), 'Regional Motors of the Global Economy', *Futures*, vol. 28, pp. 391-411.

Scott, A.J. and Storper, M. (1987), 'High Technology Industry and Regional Development: a Theoretical Critique and Reconstruction', *International Social Science Journal*, vol. 112, pp. 215-232.

Semlinger, K. (1993), 'Economic Eevelopment and Industrial Policy in Baden-Wurttemberg: Small Firms in a Benevolent Environment', *European Planning Studies*, vol. 1, pp. 435-463.

Smyth, I. (1992), 'Collective Efficiency and Selective Benefits: the Growth of Rattan Industry of Tegalwangi', *Bulletin Institute of Development Studies*, vol. 23, no. 3, pp. 51-56.

Staber, U. (1996), 'Accounting for Variations in the Performance of Industrial Districts: The Case of Baden-Wurttemberg', *International Journal of Urban and Regional Research*, vol. 299, pp. 316.

Staber, U. (1997), Specialization in a Declining Industrial District, *Growth and Change*, vol. 28, pp. 475-495.

Steed, G. (1987), 'Policy and Technology Complexes: Ottowa's Silicon Valley North', in I. Hamilton (ed.), *Industrial Change in Advanced Economies*, Croom Helm, Beckenham.

Sternberg, R. (1996), 'Regional Growth Theories and High-Tech Regions', *International Journal of Urban and Regional Research*, vol. 20, no. 3, pp. 518-538.

Stoker, G. (1998) 'Governance As Theory: Five Propositions' *International Social Science Journal*, vol. 3, pp. 17-28.

Storper, M. (1990) 'Industrialization and the Regional Question in the Third World: Lessons of Post-Imperialism, Prospects of post-Fordism' *International Journal of Urban and Regional Research*, vol. 14, pp. 423-444.

Storper, M. (1993), 'Regional Worlds of Production: Learning and Innovation in the Technology Districts of France, Italy and USA', *Regional Studies*, vol. 27, pp. 433-455.

Storper, M. (1995), 'The Resurgence of the Regional Economies. Ten Years Later: The Region as a Nexus of Untraded Interdependencies', *European Urban and Regional Studies*, vol. 2, no. 3, pp. 191-215.

PART II
PRACTICE

Introduction to Part II

This section draws together empirical evidence and case study detail of local growth stories by exploring the realities of change as they are currently impacting on places. The evidence on processes is drawn from a range of scales; from the global processes associated with the development of the Internet and the emergence of B2B e-commerce, to the shifts of hyper-footloose business in the European Union, and to the specific circumstances of change and the processes shaping change in particular places as disparate as Hong Kong, Tel Aviv in Israel, Rjukan in Norway, and Oxford in the UK. In these studies, the spotlight of detailed empirics is thrown not just on 'successful' places – the Achilles heel of much eurocentric institutionalism – but also on 'less successful' places and communities in Norway and Israel, for example. In this way, the chapters can be viewed as a first move to bolster the 'thin empirics' that Martin and Sunley (2001) have complained of, and to move away from the overwhelmingly limiting selection of case studies 'on the dependent variable' that Staber (1996) has so roundly criticised.

Through the nine quite disparate chapters included in this section, we begin to explore the limitations, constraints, pressures and potentials for local growth associated with the shifting economic circumstances of places tied into a global economy. Individually, the chapters are sectoral and place-based case studies, but together they begin to demonstrate the diversity of local experience, both positive and negative, and the diversity of processes that affect places. They also give a flavour of changing processes and pressures, and 'things to come' in terms of the problems and potentials of the Internet, B2B e-commerce, the knowledge economy, institutional constraints and hyper-mobile business.

Tamásy (*Evaluating Innovation Centres in Germany: Issues of Methodology, Empirical Results and International Comparison*) focuses on technology and innovation as engines of local growth, and reviews attempts in Germany to position places within the knowledge economy through the creation of innovation centres. She shows clearly how ineffective such initiatives can be and demonstrates the imperative that they be demand-led. Patchell and Eastham (*Creating University-Industry Collaboration in Hong Kong*) extend this local perspective on innovation and the knowledge economy by exploring the often discussed but all too infrequently analysed collaboration between universities and industry to foster technological change and the creation of new businesses. In the Hong Kong situation, they reflect on the importance of establishing local institutions to create effective collaboration, and the tensions that people experience in balancing their academic and commercial roles in universities. Lawton Smith (*Promoting Local Growth in the Oxfordshire High-tech Economy: Local Institutional Settings*) takes this theme further in the context of institutional support for the small firm high-tech sector in Oxfordshire. She demonstrates the significance of networks of institutions in fostering growth, shows the importance of powerful institutions (in this case Oxford University and local government) in fostering growth, but also highlights the problems of multi-layered bureaucracy and parochialism that can also be created.

The chapters by Vatne and Sofer, Benenson and Schnell cast a different light on local growth and demonstrate the social and ethnic impediments that may impact in different places. Vatne (*Technological Change, Local Capabilities and the Restructuring of a Company Town*) develops a theoretically informed empirical analysis of the circumstances of change in the company town of Rjukan in Norway. He shows very clearly how the accumulated knowledge and work attitudes of a community – built in this instance through 80 years of conformity to the work practices of Norsk Hydro – can stifle initiative and innovation and hinder the very restructuring that a community might urgently need. Sofer, Benenson and Schnell (*Ethnic Minorities' Strategies for Market Formation: The Israeli Arab Case*) demonstrate the equally significant constraints that ethnicity can place on market development and local growth. Their case study concerns Arab Israeli businesses and the difficulties they have accessing Jewish markets. However, they show very clearly that geographical peripherality combines with ethnicity to reinforce exclusion and, by inference, limit local growth.

In separate chapters, Malecki and Langdale reflect on aspects of the Internet as they affect and will continue to affect local growth, and draw somewhat different conclusions. Malecki (*The Internet Age: Not the End of Geography*) explores the creation, extension and elaboration of the network, and the access of places to it. He argues that the Internet that is evolving will foster the persistence of existing urban hierarchies and dominant nodes on the information highway, spawning only very few changes as the system becomes more complex and elaborate. Langdale (*Electronic Commerce: Global-Local Relationships in Financial Services*) focuses on B2B e-commerce, particularly the global twenty-four hour financial market. He identifies new growth opportunities arising from time zone differences in the operation of these twenty-four hour markets especially in services such as design and engineering. In these developments he recognises the potential for small industrialised and developing countries to be locked into new forms of dependency. He sees B2B e-commerce as having the potential to both lower barriers to entry and to reinforce existing dominance-dependency relationships. But, what the outcome will be is hard to predict.

The business services that Langdale and Malecki see linked by the Internet, Kipnis and Borenstein (*Patterns of Suburban Office Development: The Spatial Reach of Office Firms in Metropolitan Tel Aviv*) see being rearranged spatially at the intra-urban scale. At this scale, and using Tel Aviv as an empirical example, they see the growth of office activities in post-industrial society and the Internet age bringing new pressures to bear on urban areas in terms of employment concentration, spatial patterning and pressures on infrastructure, driven by the need for intra-urban accessibility. Alvstam and Jönsson (*Hyper-Footloose Business Services: The Case of Swedish Distance Workers in the Mediterranean Sun Belt*), however, recognise the geographical liberalising potential of the Internet and ICT at the level of the individual business. For them, new personal values, life-styles and attitudes, when coupled with new communications technologies, facilitate spatial dispersal and create the opportunity for people to locate their business for reasons of 'quality of life'. In this case study, certain types of small Scandinavian businesses are identified as shifting to the Mediterranean, needing only accessibility to an international airport to compensate for economic peripherality.

References

Martin R. and Sunley P. (2001), 'Rethinking the 'Economic' in Economic Geography: Broadening our Vision or Losing our Focus?', *Antipode*, vol. 33, pp. 148-161.

Staber, U. (1996), 'Accounting for Differences in the Performance of Industrial Districts', *International Journal of Urban and Regional Research*, vol. 20, no. 2, pp. 299-316.

7 Evaluating Innovation Centres in Germany: Issues of Methodology, Empirical Results and International Comparison

Christine Tamásy

Introduction

Since the beginning of the 1980s, the establishment of innovation centres has become one of the most popular instruments among local and regional technology policymakers in Germany. Following the founding of the first innovation centre – the Berlin Innovation Centre (BIG) – in 1983, around 200 municipalities have established similar facilities (see Figure 7.1). The regional distribution of innovation centres in Western Germany reflects the technology policy of each state. For example, Lower Saxony has only supported such facilities to a limited extent. On the other hand, states like North Rhine-Westphalia have invested heavily in innovation centres from the start. Other states like Hesse have been catching up in development in recent years. In Eastern Germany the establishment of innovation centres (since reunification in 1990) has been publicly funded by the former Federal Department of Research and Development (BMFT) - a novelty for German technology policy. Massive state funding has led to a situation, in which a higher concentration of innovation centres exists today in Eastern than in Western Germany, on a per capita basis (Tamásy, 1996).

Figure 7.1 Innovation centres in Germany

Innovation centres exert a certain power of attraction on German politicians. Why is this the case? One explanation can be found by exploring the roots behind the establishment of innovation centres. At the end of the 1970s and the beginning of the 1980s traditional regional policy underwent a crisis. The usual strategy of government investment to strengthen peripheral regions proved to be rather unsuccessful. Furthermore, the success of Silicon Valley, emerging debates about international competitiveness, as well as the exposure of small and medium-sized firms in the USA pointed out the necessity of new ideas. Regional policy focused on exploiting endogenous potential and promoting entrepreneurs, technologies and the transfer of technology and knowledge. Innovation centres were seen as an opportunity for addressing all new demands in one policy instrument. Technology policy, which was purely a sector for federal or state policymakers, could also be entered and shaped by municipalities. Innovation centres were particularly attractive from the perspective of political rationality. Such facilities easily gain acceptance in a region, since funding from the European Union, the federal government and the state governments can be directed into a region for this purpose. Innovation centres can also be utilised for their media impact, so that political activities achieve visibility.

In Germany innovation centres have primarily been supported and built using public funds. Employing tax revenues in such projects, means the need for public accountability over the use of those funds. A further justification for evaluating innovation centres is that scientific interest exists in the measurement of the efficiency and effectiveness of such centres. This chapter attempts to combine both perspectives. It provides a framework for evaluating the following two questions:

- What level of success do innovation centres achieve in relation to their own goals, i.e. the support of business start-ups, the creation of qualified jobs and the intensification of knowledge and technology transfers?
- What factors determine the success (or failure) of innovation centres?

The chapter begins with a brief list of criteria used to define innovation centres. Various issues and methods for evaluating innovation centres are discussed. Then, the data for the empirical analysis are described. Results are presented and contrasted with experiences from abroad based on available empirical studies. This comparison on an international scale provides further support for interpreting the results.

Defining an Innovation Centre

Various property-based initiatives have been introduced in Germany focusing on the needs of technology-based start-ups. However, not all of these projects can be regarded as innovation centres. Facilities also exist, under the rubric 'innovation centre', but are actually industrial parks. Content and packaging can differ substantially. Thus, in order to analyse innovation centres a working definition is necessary emphasising the conceptual characteristics of the facilities. Taking major political goals, which have hardly changed since the beginning of this development in the 1980s, innovation centres can be defined as business communities, in which young, innovative firms take advantage of a spatially, concentrated offer of rental space, communal facilities, technical services and consulting for a limited period of time (Sternberg *et al.*, 1997). Initiatives under other names, for example, research parks or technology parks, that meet the criteria set out above are also included within this definition.

Supply availability in innovation centres is based on subsidised and, thus, inexpensive rental space and central services, which eases the difficult start-up phase of young, innovative businesses by reducing the fixed costs. The spectrum of central services includes meeting rooms, telecommunication services, copy facilities, secretarial functions and a central telephone operator, which are available in most of the facilities. In addition, many of the innovation centres provide a cafeteria as a meeting place and centre for possible synergy effects. Special offers like computers for technical applications, laboratories and measuring instruments are rarely offered due to their relatively high investment and operational costs. The truly innovative elements in the innovation centre concept, however, are the consulting services. It is assumed that the founders of young, innovative businesses normally possess technical qualifications, yet, at the same time they tend to lack expertise in marketing and management. General consulting services are, therefore, available at no cost to the occupants of practically all innovation centres. Contacts to specialised consultants and capital providers outside the innovation centres are often included as well.

Issues and Methods for Evaluating Innovation Centres

A general set of instruments for analysing the effectiveness of innovation centres has yet to be developed. Nevertheless, various issues and methods exist for evaluating the effectiveness of instruments from technology policy, which have proven themselves in the past and could certainly be applied to innovation centres (see Table 7.1).

Table 7.1 Issues and methods for evaluating the effectiveness of innovation centres

Issues	*Possible Questions (Selection)*	*Feasibility*
Clientele analysis	To what extent have the target groups actually been reached? Does a branch of business or field of technology exist, which is over-proportionally represented?	1
Acceptance	How do the entrepreneurs rate the ICs and their services? How important is the IC for the selection of a business location? How well utilised are the centres over time (with respect to target groups)? How high is the demand for rental space in the ICs? What share of firms desire a location within ICs and how has this figure changed? To what extent were federal and state funds for ICs used by eligible cities and communities?	1
Intended effects	Has the number of successful firms increased due to ICs? How high is the employment effect of ICs? How has the technology transfer from local universities intensified due to ICs?	2
Unintended effects	To what extent are firms in ICs, those that would have been founded without public funding? How many municipalities would have founded ICs without the financial support of the government? Has the multitude of ICs resulted in a loss of image for the overall concept? Has the disparity between the centre of economic activity and the periphery or between Eastern and Western Germany increased due to ICs? Have the time-restricted rental contracts led to a poor utilisation of ICs?	2
Learning curves	Are universities and large firms more willing to financially or technically support ICs after a certain period of time in operation?	2
Check of assumptions	Are business and technical consulting services important for young entrepreneurs? Do entrepreneurs perceive a great advantage in the IC location due to a reduction in fixed costs and the proximity to other firms?	2
Implementation & administration	Which parties are involved in the conception of ICs? What relationships exist between the parties involved?	1
Commercial Exploitation	Does a business location in an IC prove to be advantageous for receiving projects and securing external sources of funding?	2

Methods	*Possible Questions (Selection)*	*Feasi-bility*
Before-and-after comparison	Do significant differences exist in the structure and behaviour of firms before and after entering an IC? How has the economic and technological level of a region changed since the establishment of ICs?	1
Control-group concept	What differences exist between comparable firms inside and outside ICs with respect to business features (e.g. technological intensity)? Have comparable cities with and without ICs developed fundamentally differently from an economic and technological point of view?	2
Case-study approach	This covers all questions with respect to the relationship between an IC and the regional development or the economic development of firms involved.	1
Econometric models	Does technical consulting increase IC productivity?	3

1 = highly feasible; 2 = partially feasible; 3 = not/hardly feasible; IC = innovation centre

Source: adapted by Sternberg *et al.* (1997), following a study by Meyer-Krahmer (1989)

Among the issues covered in various studies, the following are included as central: clientele analysis, programme acceptance by target groups, intended and unintended effects, the realisation of learning curves, an evaluation of basic assumptions for a programme, the implementation and administration as well as the commercialisation of supported projects (Meyer-Krahmer, 1989). Clientele analysis attempts to determine to what extent the business target groups were supported. The study of acceptance seeks to discover to what extent the target groups utilise the innovation centres and to uncover possible entrance and acceptance barriers. Of central importance to an analysis of the effectiveness of innovation centres are the intended effects. These must be compared to positive and negative unintended effects. Comparing these effects with the targets set by the innovation centres, allows one to develop an opinion with regard to goal achievement. Learning curves represent changing attitudes and behaviours partially belonging to the intended effects of innovation centres. An evaluation of the underlying assumptions of innovation centres can clarify whether the facilities actually influence existing bottleneck situations by those receiving support or whether the requirements for funding even exist. A study covering implementation and administration issues can handle, for example, the interactive relationships between policy makers, innovation centre management and target groups. For commercialisation and diffusion, the central question is: to what extent are funded measures commercially successful and can they be found in the market?

Among the most central methods for analysing the effectiveness of a project, are before-and-after comparisons, using control-groups and the use of a case-study approach (Meyer-Krahmer, 1989). Before-and-after comparisons allow for the analysis of firms or regional characteristics prior to and following the use of the policy instrument (e.g. an innovation centre). The effects of various instruments should result in differences in regional and business development, which is what we aim to determine. Naturally, difficulties arise due to external effects (e.g. the economic context), which also impact the development situation.

The control-group concept compares characteristics from a group of firms or regions taking advantage of innovation centres and a control group not utilising innovation centre facilities (cross-sectional analysis). In accord with the strict standards of actual experiments, both groups should be selected randomly. In quasi-experimental cases a non-equivalent control group is selected, which is seen as comparable to the experimental group based on set criteria. The actual difficulty lies in the selection of the control group.

In the framework of the case-study approach, it is possible to analyse numerous aspects for few or several examples, whereby the problem of representative results must be addressed. Furthermore, econometric models exist, which so far have been plagued by methodological problems. For example, the use of quantitative methods requires that a computable model for innovation centres be developed with the help of a series of simplifying assumptions. Qualitative aspects tend to be ignored, despite their special value (e.g. the image effect) in the conception of innovation centres.

Study Design

The goals of this study are:

- to conduct a new survey of Western German innovation centres in the framework of a cross-sectional analysis based on available data from 1986 (see Sternberg, 1988), which makes a longitudinal sectional analysis possible. The main emphasis is in the comparison of two points in time (1986 and 1993-94) and in the interpretation of the changes over this period (see Behrendt, 1996);

- to derive initial estimations of IC effectiveness from a cross-sectional analysis of young innovation centres in Eastern Germany, in order to develop a basis for later longitudinal sectional analyses (see Tamásy, 1996);
- to develop a cross-sectional analysis of the successful firms, that have left innovation centres, and rate their time in the centres retrospectively (longitudinal sectional analysis) (see Seeger, 1997).

The ex post, empirical basis of this study includes four partial surveys of 1,021 firms and 108 innovation centres[1] allowing for representative statements about the German innovation centre landscape. Significant distortions with respect to regional aspects, business size and technology-specific factors do not exist (see Behrendt, 1996, Sternberg, 1988; Seeger, 1997; Tamásy, 1996). The longitudinal analysis also allows for a structural comparison of two time periods (1986 and 1993-94), and for drawing conclusions about the causes of changes. The central method for analysing effectiveness is the before-and-after comparison. While an explicit control-group analysis is not undertaken here, an indirect quasi-experimental method is employed. For example, the formation of various regional types leads explicitly to formulation of control groups (see Hembach, 1980). Similarly comparing firms no longer in innovation centres, with those still in the IC's offers some form of quasi-experiment approach. The approach to defining what constitutes a 'successful innovation centre' involved a pragmatic form of measurement. The success and the effect of innovation centres was measured in relation to the goals that these facilities had set for themselves. The three goals, 'support for business start-ups', 'creation of qualified jobs' and 'improvement of knowledge and technology transfers' are listed based on their order of importance.

Empirical Results

Demand Constraints

Our results show that around one-quarter of the innovation centres maintained under-utilised capacity in the 1990s, especially in rural areas. In Western Germany this development is partially determined by the price level of rental properties. Thirty five percent of facilities in 1993-94 and twenty six percent in 1986 in Western Germany reached local rental levels or were even somewhat higher. In Eastern Germany many innovation centres are still in the start-up and admission phase and the level of utilisation with respect to rental space was eighty four percent on average. In response to this lack of demand, many of the facilities in the East and West have reacted by downgrading the admission criteria. They have reduced the technological requirements posed on firms, have extended the period of stay or have abstained from using any selection criteria (as a last effort). At several locations the innovation centres even closed.

However, in spite of existing oversupply, around one-third of the innovation centres have engaged in physical expansion. Common wisdom claims that an efficient operating size is between 3,000 and 5,000 m^2. In practice, the average IC in Western Germany is 4,712 m^2 (Median: 2,600 m^2) and 2,682 m^2 (1,725 m^2) in Eastern Germany.

Due to the immobility of young, innovative firms, the option of deflecting demand from other regions, does not really exist. Three-quarters of the founders surveyed had already worked in the same city or region before becoming self-employed. The source or origin facilities (so-called incubators) are primarily other firms, in which 56% of the new entrepreneurs were employed prior to their founding, and also research facilities (universities, colleges of applied sciences) with technical and natural science orientations. In the absence of these incubating organizations, the likelihood of the development of spin-offs or spin-outs is low. The reason for the lack of spatial mobility of new firm founders is the existence of local information and contact networks in the private sector and future business field, which lower the risk of business start-ups. Even after leaving innovation centres around 65% of the firms locate in the same city. An additional 23% of the firms locate within 30 km of the city of the IC. The average distance of a new business location when leaving the IC was 35.7 km.

Marketing the Facility: Discrepancies between Claim and Content

The downgrading of admittance criteria due to demand constraints has led to a situation in various innovation centres, in which claim and content do not concur. In terms of the innovation level of tenant firms, success in populating the IC's with innovative enterprises, has only met with limited success (Table 7.2). At least one-third of the firms in innovation centres are active in low-value service sectors or commerce, with little innovative capacity. The firms in Eastern German innovation centres are even less technology oriented. The establishment of medium-sized firms, which do not necessarily have to be innovative in the East German IC's has much to do with the dilution of the IC concept.

Table 7.2 Technological orientation of firms in innovation centres/former tenants

Indicator	West		East		Former Tenants	
	No.	%	No.	%	No	%
R&D Expenditures in % of Turnover (n = 320/233/164)						
- no R&D expenditures	41	12.8	83	35.6	81	49.5
- less than 3.5%	8	2.5	20	8.6	6	3.5
- 3.5% - 8.5%	28	8.7	17	7.3	24	14.4
- more than 8.5%	243	75.9	113	48.5	53	32.6
Main Activities of Firm (n = 406/272/164)						
- Research	67	16.5	60	18.4	16	10.2
- Development	218	53.7	124	45.6	80	50.6
- Production	106	26.1	56	20.6	59	37.3
- Trade	79	19.5	38	14.0	41	25.9
- Services	295	72.7	185	68.0	97	81.4
Patents (n = 404/272/164)						
- no patents	310	76.7	214	78.7	121	73.8
- 1 - 2	62	15.3	38	14.0	18	11.0
- 3 - 5	27	6.7	17	6.3	18	11.0
- more than 5	5	1.2	3	1.1	7	4.3

In terms of firm age, 19% of the firms were over 2 years old, when they entered the innovation centres and, thus, technically were not new firms. They were admitted by the operating authorities in an attempt to avoid empty IC's and entered the IC's on the basis of temporary rental contracts.

Supply-side Orientation of Public Policy

Many municipalities have established innovation centres hoping that the rental space and consulting services will automatically generate demand for the IC's (supply-side policy). However, the facilities are often provided before an adequate estimate of required demand is undertaken. Often, the availability of public funding is the reason why an innovation centre is established.

For IC tenants, the centres offer the possibility of subsidized facilities and lowering costs and are, thus, passively effective. As noted earlier, rents in two thirds of Western German innovation centres are still lower than local rents. In Eastern Germany the supply factors that have made IC's attractive include legally-secured production space and adequate communications infrastructure.

The consulting services of the IC are not considered significant for the development of firms and are rarely seen as locational advantages (Figure 7.2). It would seem that either the firms exhibit little need for these services or that IC management does not offer appropriate manpower and facilities for facilitating business services. However, it could also be that new firms are simply unaware of their future needs in the area of business consulting and services as many of them have not yet penetrated the market.

A further problem relates to IC exposure and awareness of the facility. In many business locations potential and actual founders are not known by the innovation centres. Contacts with incubator facilities can improve interest in business start-ups, reduce information deficits, and lower inhibitions. In addition, the IC can create a demonstration effect by showing potential entrepreneurs, that success is possible. The initiating effect of IC's on innovation and motivation has been quite low, because technological innovation cannot be stimulated by the provision of physical facilities. Only 3% of the business founders reported that they would not have started their enterprises without the existence of innovation centres (see Figure 7.3). The majority of the founders took the IC support simply as an additional bonus.

Activities and Specialization

Today, service providers represent the largest group of tenants in innovation centres. Approximately two-thirds of all firms in innovation centres claim service provision as their major activity, followed by basic research, applied research, development, production and trade. In Western German IC's this service orientation has risen over time from 48% in 1986 to 73% in 1993-94. Concurrently R&D related activities have declined from 92% to 70%.

Very few IC's have any form of sectoral specialization or concentration in one area of innovative activity (e.g. research, marketing etc.). The few successful innovation centres (e.g. Heidelberg) with a clear specialisation are the exception to the rule. Currently, a wide range of areas of activity exist. Around two-thirds of the firms are involved in the following five major fields: 'information and communication technology including software', 'measuring and control technology', 'production and process engineering', 'consulting' and 'power and environmental engineering'.

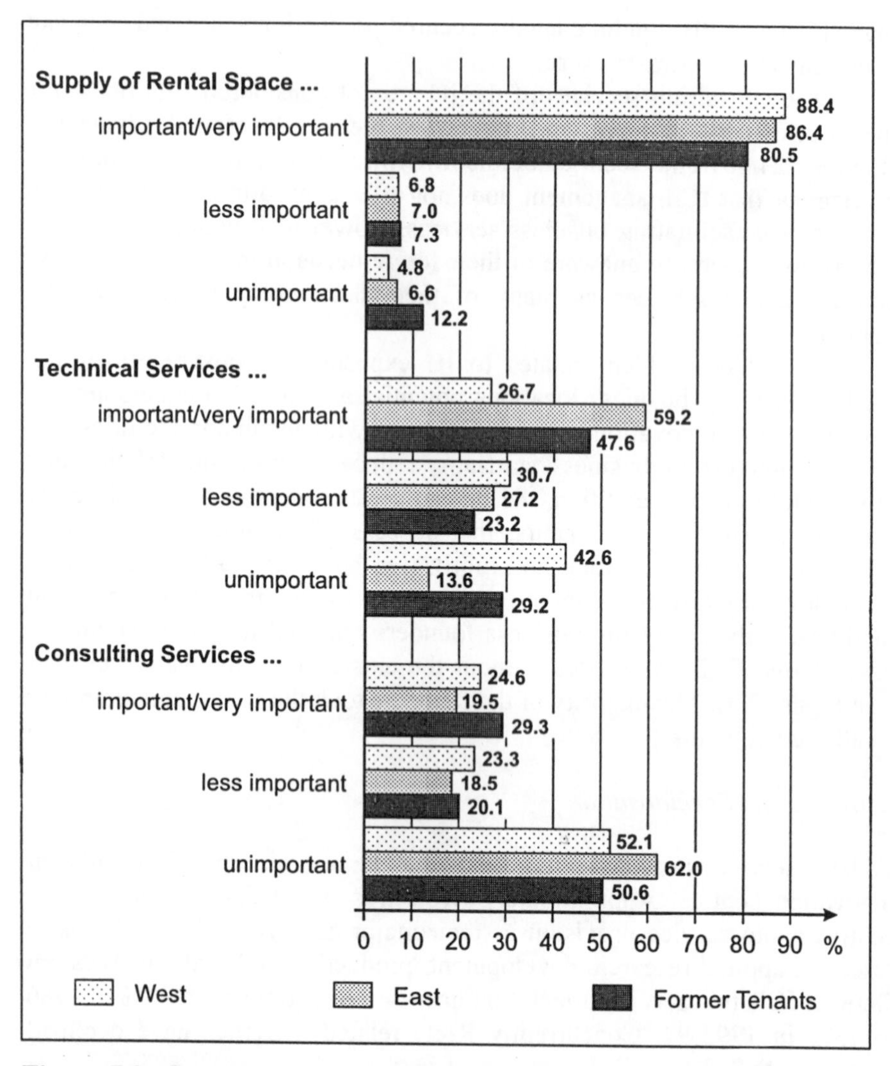

Figure 7.2 Significance of innovation centres for the development of firms

Operating Cost Considerations

The costs for building or renovating IC facilities totalled 1.3 billion DM (1993-94), of which 156.1 million DM went towards IC's in Eastern Germany. In addition to the investment costs, current operating for the Western German innovation centres, reached an average value of 1.0 million DM per facility and year. In the Eastern German innovation centres this figure reached 455,000 DM. Costs per square metre in Western German facilities were much higher (average: 3,054 DM/m^2) than in Eastern Germany (1,617 DM/m^2). This discrepancy is due to the much higher public subsidy in the case of the former, which encourages the building of new facilities, rather than the renovation of existing buildings. However, the level of investment costs and the degree of technological orientation of an innovation centre (measured by the share of innovative firms) do not indicate a statistically significant relationship. Lavishly built and equipped innovation centres lead to no more innovative activities and can be counterproductive, leading to negative public reaction.

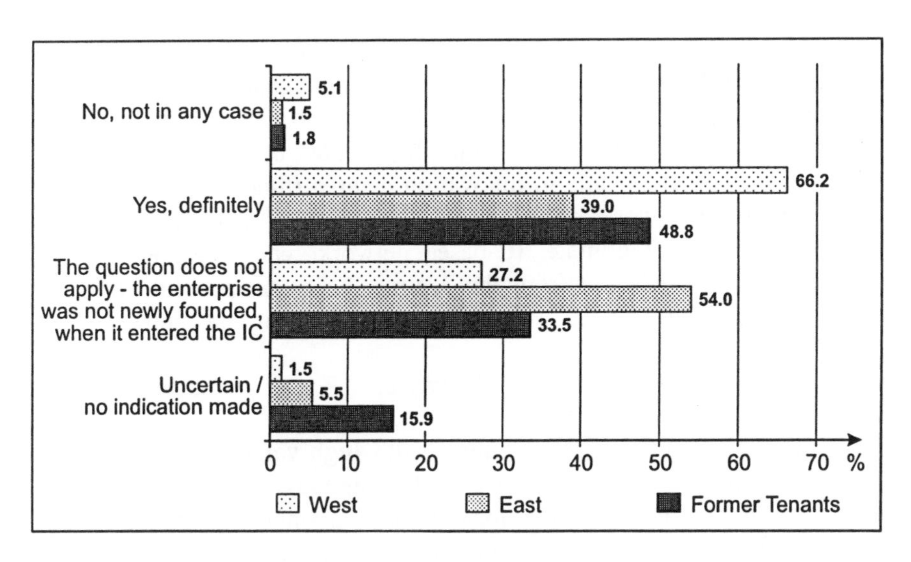

Figure 7.3 Would the enterprise have been started without the existence of the innovation centre?

Network Development

Innovation centres have a limited effect on the transfer of knowledge and technology within a region. Compared with other transfer facilities (e.g. innovation consulting offices from chambers of industry and commerce, transfer departments at universities) innovation centres are not competitive at most business locations due to a lack of consulting competency. IC management finds it hard to offer competent consultancy service in all fields (marketing, finance etc.) due to the variety of possible problems, the heterogeneous nature of the IC tenants and the personnel and time constraints. The IC often engages in consulting mediation acting as an intermediary between firms seeking assistance, on the one hand, and those offering support, on the other. Those innovation centres successful in hooking firms up to local and regional information and innovation networks play an important function. An important element of such a strategy is integrating the competencies that exist in the IC's of competent people and institutions into the strong operating or sponsoring organizations (municipalities, banks and chambers of industry and commerce) along with tenants' needs. In 70% of the facilities municipalities are amongst the sponsoring organizations and are involved in the introduction, construction and investment costs of the IC. In around one-third of the innovation centres, municipalities are the sole shareholders. The involvement of universities, however, has been low. Fifty-eight percent of the firms in innovation centres have established contacts to universities, most of these based on founders personal relationships and working experiences.

In terms of synergy affects within the IC, 56% of the firms report co-operation with other tenants in innovation centres (this includes informal contacts and joint work on projects). The frequency of co-operation depends on the work climate. IC-based networks can continue into the future even after departure from the centres. Forty six percent of firms, reported maintaining contacts with IC firms after leaving the centres.

Combination with High-value Rental Space as a Factor of Success

The combination of innovation centres with high-value rental space for 'graduating' firms has proven to be advantageous. It offers firms the possibility of staying with the location and maintaining contacts and networks. Twenty-nine Western German and sixteen Eastern German locations provide this kind of combination (often an IC with a technology or business park, e.g. Dortmund, Cologne, Bremen, Dresden and Frankfurt/Oder). In terms of total IC employment (close to 23,000) the effect of securing this employment locally through these means will hardly have any perceptible change on the German labour market. However, it does have an important demonstration effect locally and also ensures that part of the high share of university graduates (56% amongst IC employees), remain in the region.

Comparison of German Innovation Centres with Other Similar Initiatives

Definition, Concepts and Features of Science Parks

A comparison of German innovation centres with related facilities abroad such as science or technology parks indicates that most science parks are larger on average than German innovation centres measured by net operating or rental space (see Luger and Goldstein, 1991; Sternberg *et al.*, 1997). In addition, very few innovation centres in Germany enjoy a direct formal or organisational connection to universities and very few of them are private sector, profit-oriented initiatives, although private sector involvement is on the rise. Goals, strategies and motives of German IC's correspond strongly to those of GB and US science parks with 'support for technology and technology transfer', the 'founding and development of new businesses' and 'job creation' all featuring prominently.

In the US, 'incubator centres' or 'businesses incubators' exist often as separate entities to science parks (OECD, 1999). Most innovation centres in Germany fulfil this 'incubation' function. However, as they are often funded for political motives quick and visible success is of paramount importance. For this reason, they often fail to follow an incremental growth path. In many cases, as soon as an innovation centre is established, a complete list of services is already prepared.

Comparing the experience of US science parks with German IC's, we may note that the US findings that more mature parks in large agglomerations, are likely to be more successful (Luger and Goldstein, 1991) are very relevant for the German case. US findings suggest that science parks in smaller regions with less than 100,000 residents can develop into a source of agglomeration effects, which are not available otherwise is also a finding pertinent to the German experience with special relevance to Eastern Germany.

Findings from British studies (e.g. Monck *et al.*, 1988; Massey *et al.*, 1992; Westhead and Storey, 1994) suggest that the independent firms in British science parks grew (in employment terms) almost two times faster than independent firms outside such facilities. As in the German IC case, this employment growth is concentrated in a few firms. The five, fastest-growing firms on British science parks were responsible for 69% of the new jobs between 1986 and 1992; outside of science parks the top-five businesses generated 93% of the new jobs. In German innovation centres, 41% of the newly-created jobs were related to the top-five firms. The difference between the British and German results arises primarily due to the more limited growth prospects in German IC's compared to British science parks resulting in high-growth firms leaving the innovation centres relatively early.

Innovation levels of the firms on and off UK science parks are not significantly different. A similar situation exists with respect to German IC's. In addition, type and quality of connections to universities is not found to defer between on and off park firms in the US and between IC tenant firms and similar firms not located in innovation centres in Germany.

Conclusions

The results show clearly that it is high time to reconsider the underlying concept of innovation centres in Germany. Is the funding philosophy, which is tied to technology-based founders, too narrowly defined? Could greater job creation effects have been realized with an alternative use of public funds? Our results suggest that the potential of new facilities should be thoroughly investigated ex-ante. Cost aspects should be more highly considered, in an effort to avoid an uncritical distribution of public funding by municipalities. At the same time, innovation centres have to be demand orientated, in order to satisfy the real needs of young firms. New strategic elements like the combination of innovation centres with high-value business space and the establishment or use of networks, which have proven to be successful, should be firmly integrated into the IC concept. In the event predetermined goals cannot be reached, a strategic correction is necessary. Thus IC's with insufficient demand can be re-mandated as industrial parks. If correctly communicated, this can even be accepted positively by policy-makers.

In spite of contrasting conceptions, the results for the German IC's are not substantially different to those for Great Britian and the USA. In all three places the results point to the real-estate characteristics of these facilities. As the British studies show, firms on and off science parks do not differ notably in their levels of growth. The most significant difference between the businesses in innovation centres and those outside can be found in the higher satisfaction of the businesses within the facilities. As in the British and American science parks context, active support (e.g. information, consulting, etc.) in German innovation centres has shown little success so far. The success of all facilities is dependent upon the time of establishment and the location. The goal of creating jobs appears to be greatly determined by chance at least in a quantitative sense.

The level of similarity in the results hints at the fact that the goals and basic assumptions of innovation centres and science parks can only be successful at particular locations. There is no definitive answer to the hypothetical question whether innovation centres could have been more successful at other locations. It is rather obvious that the supply-oriented concept of science parks and innovation centres fails to promote higher growth in business start-ups. The target groups at most locations are too narrowly defined to achieve full-utilisation of the space available. In addition, the problems and needs of small firms are often poorly satisfied by the supply provided. Talented management can sometimes compensate for such basic deficits at some locations. However, it is perhaps time that some of the assumptions of this model be reconsidered and strategic goals re-formulated.

Note

1 This analysis involves surveys of 108 innovation centres (of which 72 are in Western Germany and 36 are in Eastern Germany) and of 1,021 firms (of which 585 are in Western German innovation centres, 272 are in Eastern German innovation centres and 164 have moved out of innovation centres). Thus, 92% of the innovation centres existing in 1993-94, 31% of the firms in innovation centres and 43% of the firms which have successfully departed innovation centres are considered.

References

Baranowski, G. and Raetz, G. (1998), *Innovationszentren in Deutschland 1998/99, Mit Firmenbeschreibungen*, Weidler Buchverlag, Berlin.

Behrendt, H. (1996), 'Wirkungsanalyse von Technologie- und Gründerzentren in Westdeutschland', *Wirtschaftswissenschaftliche Beitrage,* vol. 123, Physica, Heidelberg.

Hembach, K. (1980), *Der Stellenwert von Wirkungsanalysen für die Regionalpolitik. Eine Systematisierung der Problematik am Beispiel der regionalen Wirtschaftspolitik*, Peter Lang, Frankfurt/Main.

Luger, M. I. and Goldstein, H.A. (1991), *Technology in the Garden, Research Parks & Regional Economic Development*, University of North Carolina Press, Chapel Hill, London.

Massey, D., Quintas, P. and Wield, D. (1992), *High-Tech Fantasies. Science Parks in Society, Science and Space*, Routledge, London.

Meyer-Krahmer, F. (1989), *Der Einfluss staatlicher Technologiepolitik auf industrielle Innovationen*, Nomos, Baden-Baden.

Monck, C.S.P., Porter, B., Quintas, P., Storey, D.J. and Wynarczyk, P. (1988), *Science Parks and the Growth of High-Technology Firms*, Croom Helm, London.

Organisation for Economic Co-operation and Development (OECD) (1999), *Business Incubation, International Case Studies*, OECD, Paris.

Seeger, H. (1997), 'Ex-Post-Bewertung der Technologie- und Gründerzentren durch die erfolgreich ausgezogenen Unternehmen und Analyse der einzel- und regionalwirtschaftlichen Effekte', *Hannoversche Geographische Arbeiten*, vol. 53, Lit, Münster, Hamburg.

Sternberg, R. (1988), *Technologie- und Gründerzentren als Instrument kommunaler Wirtschaftsfürderung, Dortmunder Vertrieb für Bau- und Planungsliteratur*, Dortmund.

Sternberg, R., Behrendt, H., Seeger, H. and Tamásy, C. (1997), *Bilanz eines Booms – Wirkungsanalyse von Technologie- und Gründerzentren in Deutschland. Dortmund* (2. ed.), Dortmunder Vertrieb für Bau- und Planungsliteratur.

Tamásy, C. (1996), 'Technologie- und Gründerzentren in Ostdeutschland – Eine regionalwirtschaftliche Analyse', *Wirtschaftsgeographie*, vol. 10, Lit, Münster.

Westhead, P., Storey, D.J. (1994), *An Assessment of Firms Located On and Off Science Parks in the United Kingdom*, HMSO, London.

8 Creating University-Industry Collaboration in Hong Kong

Jerry Patchell and Tony Eastham

Introduction

University-industry collaboration has become an important part of national and regional innovation systems. Its primary objectives are improving economic prosperity and national competitiveness by capitalising on social investments in university research and speeding the process of know-how and technology transfer. University-industry collaboration is ubiquitous among the OECD countries (Bower, 1992), and now NIEs are using universities to close the technological gap with the developed countries (ASAIHL, 1997). The national innovation system literature acknowledges that universities catalyse technological advances and increasingly interact with industry through consortia and collaborative institutions (Freeman, 1992). Regional governments also engage in converting the benefits of academic research into industrial activity (Feller, 1992; Bowie, 1992). Investment in university-industry collaboration has increased greatly over the last two decades, as have expectations of its outcome.

In the geographic literature, however, the role of universities in regional development is recognised but not explored. Case studies, for example, note the relationship between regional development and universities (Cooke, 1996; Hudson, 1994; Nijkamp and Mouwen, 1987) and the impact of universities on the development of Silicon Valley, Route 128 (Saxenian, 1994) and Cambridge (Peters, 1988) is undeniable. This literature, however, only mentions the university, or lists attributes offered to industry. Perhaps it is because of this circumstantial approach that some geographers see the impact of exceptional universities such as Stanford as skewing our view and remain sceptical of their impact on development (Malecki, 1997).

To clarify the associations attributed to university-industry relations, in both the university-industry and regional development literature, we undertake an analysis of one university's impact on university-industry interaction. We believe that to understand how a university can impact a region depends on how it designs its interaction with a region. University-industry interaction has to be consciously fostered in a manner that balances commercially related activities with traditional academic roles. Importantly the design must take into consideration not only university and corporate incentives, but also faculty incentives because they are generators of innovation. Thus, our analysis is based on a model of university industrial liaison organisations (ILOs) acting as multilateral organisations mediating the interests of autonomous parties within the community (other examples include: industry associations, co-operatives, chambers of commerce). Parties join these organisations to decrease governance costs, and for concrete benefits such as public or commonly held facilities. To illustrate the model we describe a university liaison organisation designed for Hong Kong. We also evaluate the effectiveness of this attempt to foster university-industry interaction by measuring the extent of interaction and the willingness of faculty to engage in interaction. The use of the Hong Kong University of Science and Technology (HKUST) as a case study clarifies the problems of fostering university-industry interaction because it is a technology and business oriented school and because it is a greenfield university set in a region with little experience in university-industry relations.

Designing University-Industry Liaison

University-industry interaction originates in collaboration between German chemical companies and universities late in the 19th century (Freeman, 1997) and has been common in American universities throughout the 20th century (Rosenberg and Nelson, 1994). Broad acceptance and proactive development of U-I collaboration, however, did not occur until the late 1970s. In the US, stagnation in federal research funding and Japan's challenge to American industry stimulated industry and universities to work together. Coincidentally, links between research and commercialisation in biotechnology and computer science created profit-making opportunities for faculty and universities (Bower, 1992). The Bayh-Dole Act of 1980, allowing universities ownership of intellectual property created with federal funding, engendered further commercialisation of research (Bowie, 1994). Regional interests also pushed for technology transfer to promote competitiveness (Irwin, 1992).

Industrial liaison organisations played an important role in this history. Pioneered by Wisconsin and MIT, ILOs have evolved over the last two decades as universities, industry and government sought efficient means for knowledge diffusion and technology transfer. In the US, federal and state governments initiated experiments in U-I collaboration, innovative examples being the NSF's University-Industry Cooperative Research Centres Program and Pennsylvania's Ben Franklin Partnerships. Germany and Japan responded by developing their own forms of cooperation (Bowie, 1994), as did UK, Canada, Australia and others.

The ILOs that have evolved to deal with U-I interaction, however, are more than additional university departments. In their role of promoting U-I interaction, ILOs are faced with a diversity of incentive generating and regulatory tasks. Established to stimulate U-I interaction, ILOs must give companies and professors incentives to work together. The payback to working with academics is not obvious to firms, while academic autonomy and institutional pressures against collaboration make it difficult to cajole professors into working with industry. Any success in matching the interests between companies and professors, however, must be balanced by safeguarding the separation of the private and public sectors. In this regulatory role, the ILO must ensure a firm or a professor does not divert the university from its essential knowledge dissemination role. The ILO receives powers to regulate interaction between professors and industry from the university, but it also acts for the university, which is interested in increased income and prestige. In attempting to overcome the conflicting interests of these different autonomous parties through both incentives and regulation, the ILO performs the governance activities of a multilateral organisation. It is an organisation in that it has designated staff and facilities and units, and in that it draws together and regulates the relationships of other agents.

The process of multilateral organisation formation is depicted in Figure 8.1. The university's pre-eminent role is creating and disseminating information through research and teaching. Especially in its research role, information should be freely disseminated as a public good. Many, however, see this role compromised by entrepreneurial technology transfer and industrial collaboration. Indeed, throughout the evolution of university-industry collaboration, the conflict of interest debate has raged and the primary governance role of the university has been to strike a balance between its traditional roles and the new role of technology transfer. Essentially, this conflict pits the university and faculty members' roles as creator and disseminator of information against industry's desire for proprietary control over information. Firms may insist on delaying publication, confidentiality in academic relations, or holding back development of products for their competitive strategy. There is also concern about inhibiting professors from venturing "forth from the ivory tower to contribute in a wide variety of expert, policy, advisory, or litigious settings" (Feller, 1990). A re-orientation from basic towards applied research and use of public funds for private gain are other potential problems of U-I interaction. Neither is a university an impartial party, but must consider technology transfer alongside needs for impartiality and to securing funding.

The university governs these interactions by allowing or disallowing employees to use university time and facilities for collaboration. Many universities allow professors one day a week for consulting or pay them on a 10-month basis or to use facilities and support staff for interaction. Yet, universities differ in balancing these conflicts of interest. Reticent universities limit faculty-industry interactions to grants and research contracts. Others proactively establish ILOs with significant funding and personnel, granting independent legal status to handle patents and other commercial relations. Meta-governance of the university directly influences ILO operation, but the ILO can influence meta-governance by citing the needs of technology transfer.

The objectives of this institution building, however, are to overcome conflicts of interest through both regulation and incentives. The conflicts of interest are felt directly by faculty. Career advancement depends on research, teaching and service to the community, and rarely on industrial collaboration. Yet professors are increasingly exhorted to interact for research funding and to stay abreast of industry developments. These goals could be incorporated as incentives in faculty evaluation, but seizing commercial opportunities are difficult to reconcile as they pull a professor farther away from scholarship and research disclosure. However, despite these conflicts, an initiative of some recent technology transfer programs is to foster academic entrepreneurship.

Obtaining personnel, training personnel, access to facilities, prestige and public relations are some reasons why business will use universities. Companies, however, are primarily interested in spreading the risk of research costs. Mounting research costs, rapid technological change and shortening of product life make co-operation with universities a cost-effective competitive strategy (Carboni, 1992). MNCs use universities to inform their own R&D, whereas smaller companies expect rapid commercialisation of products. Companies pursue information, however, to obtain proprietary control and exclusive advantage. Control can be exerted through patents, but much information is unpatentable, compelling companies to try to control collaboration-derived trade secrets and information divulged to university members.

The lower part of Figure 8.1 depicts the main units of ILOs and governance used to mediate the conflicts between university, faculty, and companies and to promote technology transfer. The primary interface is direct interaction between faculty members and businesses, and includes donations, contracts, directed research, collaboration between company and university personnel, and consulting. These types are characterised by various degrees to which a company can control the objectives of research. Contract research for example allows companies to sponsor projects and define the deliverables. In affiliate programs, companies pay an annual fee to be kept abreast of occurrences in a unit or department. Co-operative centres identify core areas of expertise and facilities. Universities have many of these centres and they are the fastest developing forms of collaboration. In 1990, Cohen *et al.* (1994) surveyed 1,056 centres on 200 US campuses and found that they account for almost 70% of industry support of academic science and engineering R&D. Centres draw companies into direct relationships with university employees and they may be operated as consortia. In consortia, firms pool money to support research projects, and access government and university funding. Members gain access to faculty and students, tours through the centre, regular publications and newsletters, notification of developments, faculty visits, scientist/engineer residency and informal linkages. Universities also support firm incubation and entrepreneurial activities (faculty, students and staff, and their business partners and firms using university technology). Support includes facilities, space on campus, venture capital, management, filing of patents and selling of licenses.

Governance is crucial to the success of these activities, and the primary method is contractual. Foremost, the faculty employment contract controls consulting, rights to intellectual property and profits from commercialisation. Contracts are also key to corporate collaboration as they identify a principal investigator, what facilities and staff will be made available and deliverables. The contract can also stipulate confidentiality of information and notification when filing patents or publishing results. Usually 'right to publish' within a limited amount of time is assumed unless stipulated. Universities may assume intellectual property rights, although first right of use, limitations on other companies' use, or decreased licensing fees and other conditions can also be written in. Liability can be a contentious issue; university and professors wish to absolve themselves of consequences when their technology is used, while companies believe researchers should stand by their results.

UNIVERSITY

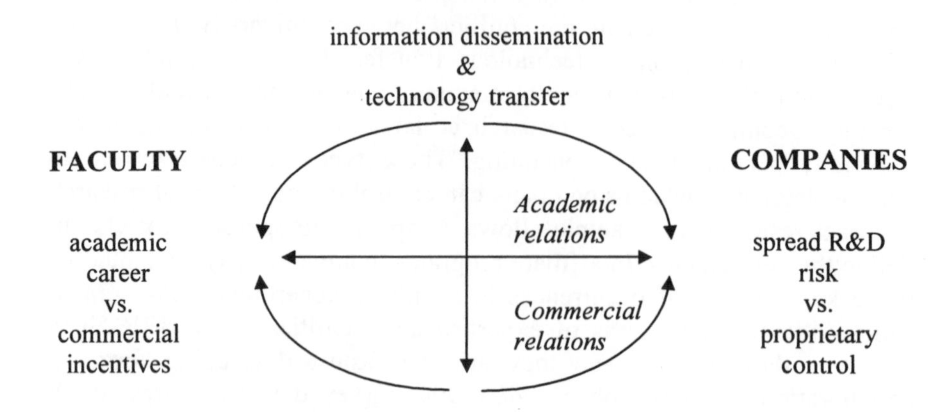

INDUSTRIAL LIAISON UNIT

Institution Building	Relational Governance
faculty-firm & centers/consortia & firm incubation	contractual & monetary & informal

Figure 8.1 Process of industrial liaison unit building

Consortia governance is more complex because corporate members sit on advisory boards and review research projects. Consortia mediate time periods for member rights, non-exclusive licenses to technologies, geographical and negotiated time limits on the right to use licenses. Directed research occurs with access depending on amount of funding, i.e. greater proprietary rights for larger sponsors. Protection of faculty, staff, and student property rights must be maintained.

All sides in these agreements are concerned about money. Corporations wish to minimise costs, and universities want to extract direct and indirect costs of research. Researchers want funding for the collaborative research, assistance for their primary research, and may also be interested in deriving financial benefit. University and liaison directors are responsible for defining these parameters, and (as noted in the discussion of conflict of interest) they vary widely among universities. For example, some universities invest in faculty start-up firms for university, faculty and regional economic interests.

Similarly, informal governance varies not only among universities but each department has a different attitude to the mediation of academic performance and industry interaction. Departmental pressure to interact (or not) with industry, impacts the success of formal liaison governance. Building trust and ensuring both sides of an agreement adhere to the spirit of collaboration is also important in firm relations. Corporations want confidentiality respected, while researchers do not want payments or future research compromised by pressure against publishing.

HKUST's Design for Collaboration

The Institutional Framework

Hong Kong is attempting to transform itself into a high-tech hub in the belief that it can no longer rely on the production of low value added goods and labour-intensive manufacturing. To achieve this goal it is calling upon its universities, and indeed is transforming its universities. Hong Kong's universities, however, face greater technology transfer difficulties than institutions in developed countries. They are charged with the task of transferring high technology capabilities in a business culture unfamiliar with not only university collaboration, but simply with taking fundamental or even applied research and transforming it into business opportunities. Yet, Hong Kong is adapting the US based model described by Figure 8.1 to its circumstances.

HKUST, from its inception, was intended to be a departure from practices of the existing universities and was charged with the responsibility to lead the transformation of Hong Kong's technological base. Not only was the University focused on teaching and research in science, technology and management but, as shown by the Ordinance containing its mission statement, the University was intended "to assist in the economic and social development of Hong Kong". This mission was interpreted proactively by the University's executives and governors, who established a separate R&D administrative infrastructure to promote technology transfer to the private sector. Although the R&D Branch is organisationally distinct from the academic branch, it has the dual role of facilitating academic research while promoting the application and commercialisation of advanced technical knowledge in the Hong Kong region. To that end, over the past decade, the R&D Branch has developed a multilateral organisation bridging the unique issues of university-industry-government relations in Hong Kong. These issues and their relationships are illustrated by Figure 8.2.

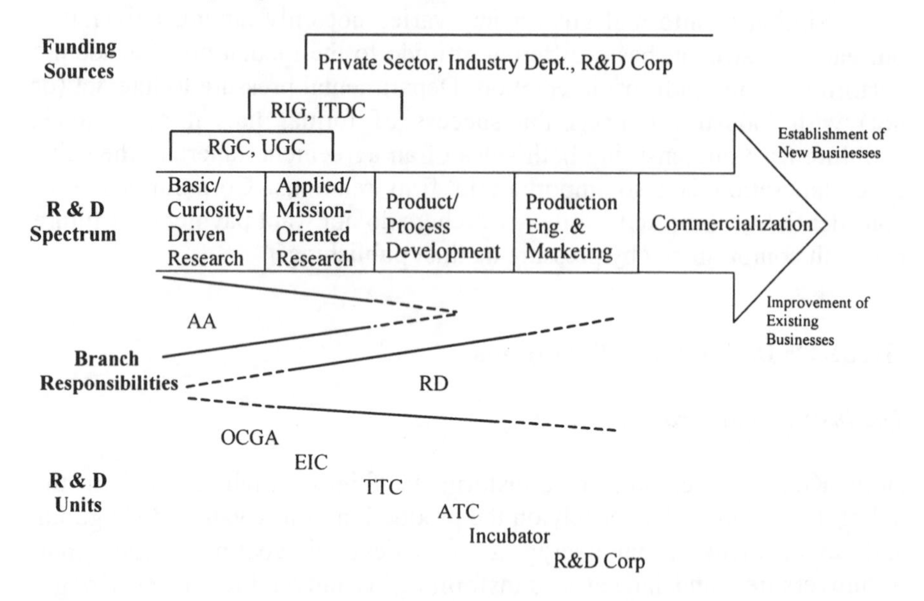

Figure 8.2 Industrial liaison organisation at HKUST

The arrow labelled R&D spectrum is a simplistic linear representation of what the R&D Branch intends to achieve for Hong Kong, i.e. convert basic research into outcomes capable of commercialisation and societal wealth generation. These outcomes can be useful for the upgrading of existing businesses or the creation of new businesses. The area labelled Branch Responsibilities schematises the changing roles that academics and business perform as an innovation moves through the developmental process and concomitantly the changing roles of the University's Academic Affairs (AA) and R&D Branches. The R&D Branch sees itself as picking up where AA and faculty members' expertise, interests and conventional roles diminish. To compensate for this diminishing activity, the R&D Branch created new operational units and secures expanded funding sources. The intention is to promote technology diffusion and transfer through interaction with existing business, provide technology to create new businesses, and to give faculty and students the opportunity to start-up their own businesses.

As schematised by Figure 8.2, various divisions within the R&D Branch assist faculty to take technology step-by-step to commercial fruition. Their roles are listed below.

- Office of Contract and Grant Administration (OCGA): identifies and promotes sources of funding; assists proposal preparation; administers funding; maintains information databases; reports on research activities; assists preparation of reports and publications.
- Engineering Industrial Consortium (EIC): technology partnerships with industries; professional development and training; collaborative research activities; technology diffusion.
- Technology Transfer Centre (TTC): establishes industrial contacts and co-operation; helps faculty develop proposals requiring industrial collaboration; assists technology development and transfer.
- Applied Technology Centre (ATC): bridges applied research in academic labs and industrial implementation by establishing technology development projects, demonstration programs, pilot plants, and pre-commercialisation activities.
- R&D Corporation: assesses, protects, markets and commercialises intellectual property as a separate business of the University; assists start-up companies; manages an incubation facility and venture capital fund.
- Research Institutes and Centres are not illustrated on Figure 8.2, but they identify research strengths to the business community. Twenty five Institutes and Centres include: Consumer Media Lab; Centre for Display Research; Electronic Packaging Laboratory; and Internet Business Consortium. Their roles are: promoting interdisciplinary collaboration; establishing critical mass of research capabilities in priority areas; attracting funding and managing projects; enhancing visibility and interactions with industry.

A full-time Director administers each unit within the R&D Branch. To ensure effectiveness and growth of the University's technology transfer capabilities, much of their effort is expended on managing the interface between the University and its funding sources. These sources, (top of Figure 8.2) like the R&D Branch units, describe a range of funding from government supported basic research to funds for commercialisation.

University Grants Committee: governing body providing general and recurrent funding; designates research funding to the Research Grants Council (RGC).

Research Grants Council: allocates government funds for university research (18% directly allocated to universities for seed funding and to support small-scale research projects; 80% competitive allocation, primarily by international peer review; balance to collaborative bids between universities for research facilities/equipment).

Research Infrastructure Grant (RIG): universities allocate approximately 2% of the UGC recurrent grant to develop new initiatives or enhance research infrastructure.

Private Sector, Industry Department, and R&D Corporation: direct promotion of industrial R&D through business funded contract research, consortia, consulting, licensing, etc; Industry Department matching grants; R&D Corporation supports faculty entrepreneurs by searching for venture capital or supplying it from its own limited resources.

This organisational framework incorporates many technology transfer mechanisms pioneered at other institutions, such as research contracts, industrial consortia, internships, incubators, and university supply of venture capital. More than most research universities, however, HKUST devotes managerial expertise and capital not only to converting basic research into applied, but also propelling it towards commercialisation. The rationale for this effort is the gap between research and the ability of local companies to accept and adapt research to their commercial needs. Particularly through the creation of the Applied Technology Centre, HKUST consciously develops closer linkages between research and industrial capabilities than is sought in universities in developed countries. Although all the units described were innovations to Hong Kong, they have stimulated the formation of liaison offices at other universities. HKU and CityU are using the R&D Corporation as a model. CityU has also established a consortium. However, the other universities have yet to devote full-time employees to these U-I interfacial issues.

Governance Framework

Reflecting the proactive stance taken in building liaison units, HKUST's governance of faculty-industry interaction is much more liberal than other universities in Hong Kong. This is especially true in its stance toward commercialisation and monetary rewards to faculty. It has adopted the one-day-a-week allowance for consulting work, and compensation is retained in full by faculty members. This contrasts with other Hong Kong universities, which levy a minimum of 15% on consulting irrespective of the use of university facilities. This levy is determined proportional to the perceived conflict with teaching and research duties. In those cases where university facilities are used, the levy on consulting is more than 35%. All other universities require prior permission to engage in consulting (in some universities Presidential permission is required), whereas HKUST requires this only for extensive consulting commitments. HKPU and HKBU do not allow faculty to form companies, and while it is allowed at CUHK, it is not encouraged, nor will the university become involved as a shareholder. HKU has no experience in the activity but will now entertain faculty entrepreneurs as long as the university can take a majority ownership. CityU also requires majority ownership and taxes income from ventures. These universities claim stakes and returns in companies as payment for allowing faculty involvement in a venture. In contrast, HKUST takes a small minority stake of 3% to demonstrate its support of the spin-off firm. At the same time, it provides a package of services including incubation facilities, to assist companies accepted into its Entrepreneurship Program. On the other hand, the University requires biannual disclosure of consultation activities to ensure compliance with its guidelines.

In terms of contractual governance, the University follows the standard practice of preserving the right to disseminate information, but can accept 90 days advance warning of publication or presentation and maintains the right to publish after six months completion of the research contract. It also protects its rights to intellectual property arising from research and from liability claims. The same conditions apply to consortia affiliates with direct technical projects. The consortium is governed by an advisory board, including industrialists, government officers, and academics.

An important aspect of HKUST's governance is that the R&D Corporation is incorporated as a separate legal entity. It takes over management of commercialisation of projects following initiation at the ATC or the TTC, and manages the Entrepreneurship Program and the Venture Capital Fund. Despite this strong lean to commercialisation, the University must concern itself with the protection of information dissemination and the private use of public resources. It therefore exercises disciplined control over commercial activities as set out in the University's principles governing commercial pursuits.

Formal governance would be ineffective if not accepted within the culture of academic departments. The liaison offices must spend considerable effort promoting industrial collaboration, especially in a research university in which most faculty are young and concerned about building their publication records. It does so through seminars, announcement of collaborative activities, and through departmental heads. So far, they have managed to involve about 20% of faculty in TTC activities. On the other side of the spectrum, faculty need to be occasionally reminded to stay within the consultation time limit guidelines.

Measuring the Effectiveness of Industrial Liaison

In this section we present evidence that HKUST has had success in applying its version of a liaison organisation within Hong Kong's challenging industrial milieu. The impact of a university on industry can be measured both objectively through the quantitative measures of patents, royalties, licenses and other income streams or subjectively, through the use of interviews to, for example, question corporate leaders about the impact of university R&D on their technological advancement (Rosenberg and Nelson, 1994) or determine specific outcomes of interactions between universities and industry (Cohen *et al.*, 1994). This section first employs some measures to determine HKUST's interaction with industry. In addition we measure how the institutional context enabling this interaction to occur, examining what incentives and disincentives have guided faculty behaviour. The objective measures of interaction outcomes are taken from university records and the subjective records from a survey of faculty-industry interactions.

Interaction Outcomes

The results of HKUST's interaction with business are impressive for a young university. Table 8.1 quantifies some of the results. With 20 patents granted and an additional 52 filed, HKUST is by far the leading patent producer in Hong Kong – private or public. As with any public or private institution, patents are no guarantee of successful commercialisation - indeed there are few successes and they may be a financial burden. HKUST realises this, but commits funds to the patent effort to raise the awareness of innovation in Hong Kong. However, in the future it will take a more business-like approach to potential costs and benefits. At 40, the substantial number of technology licenses granted provides a direct indication of the market success of the University's innovations.

HKUST has 7 formal (type I) spin-offs. One of these is Hong Kong's most successful university initiated business – Hong Kong's first internet service provider which, when sold, returned the University over ten-fold on its investment. The first incubator facility (developed in partnership with the Hong Kong Industrial Technology Centre) was opened in 1998 and assisted the growth of four firms. The following year (1999), HKUST introduced its Entrepreneurship Program and opened its second incubator facility; the Entrepreneurship Centre. By June 2000, this accommodated a further 17 start-up companies.

Table 8.1 HKUST's intellectual property protection, technology transfer and diffusion (to June 2000)

No. of patents granted	19 (USA) 1 (UK)
No. of patents filed, in addition to those granted	40 (USA) 5 (European) 7 (others)
No. of technology licenses granted	40
No. of technology companies created	(Type I*) 7 (Type II**) 17
Total contract research with industry (1995-2000)	HK $100 million
No. of Masters students graduated	2093
No. of PhD students graduated	160

* a spin-off company that has acquired HKUST technology and is now operationally independent of the University

** a start-up company established by HKUST staff or students (past or present) and which has or is being incubated at HKUST

Note: HKUST opened to students in October 1991

Contract research is another direct indication of interaction with business. The total for the 5 year period 1995-2000 stood at HK$ 100 million (US$12.8 million). Of that total, HK$ 30 million was contracted in 1999/2000, up from HK$ 4 million in 1994/95. Contract numbers have also grown from 40 in 1994/95 to 106 in 1998/99. There has been a progressive shift from testing jobs to more research oriented tasks with greater monetary and intellectual value to the University. The University is also fulfilling its more basic role of education and the supply of highly qualified scientists and engineers. In its eight years of existence, HKUST has graduated 2093 Masters and 160 PhD students. It had grown its graduate student population to 1,731 by the end of 1999, of which 283 are enrolled in Science, 745 in Engineering, 449 in Business & Management, 135 in Humanities & Social Science, and 119 in Joint Degree Programs.

Institutional Context

In the survey, two rounds of questionnaires were distributed to faculty members, in June 1999 and September 1999, respectively. There were 62 respondents in June and another 61 responded in September, with the latter questionnaire sent exclusively to those faculty members who did not respond to our original survey. These 123 faculty represent 28 percent of the total faculty population of 431. For the virtue of simplicity, the questionnaire designed for the September round deviated slightly from the original one. Fewer restrictions were imposed allowing respondents to answer rather specific questions (e.g. number of occurrences, how many times, etc.) in somewhat more abstract ways (e.g. a lot, a few, etc.). The main objective of this change was to simplify the questionnaire and be respondent-friendly – whilst not affecting the essence of the questionnaire. The results of the survey can be summarised as:

- research papers are the most common U-I interaction outcome, but also many product and process inventions and improvements on existing products and processes;
- higher level of interaction with Hong Kong and multinational firms, less with mainland China;
- few academic incentives for collaboration;
- most severe academic disincentives are neglect of industrial interaction in career evaluation and the time conflict;
- interest in commercial gain limited, with preference for consulting;
- commercial disincentives generally weak, but strong for a few individuals; and
- general satisfaction with liaison organization activities.

The specific interaction outcomes investigated covered a spectrum from primarily innovation oriented to various forms of intermediate outcomes. We also gauged the impact on Masters and PhD education. All faculty surveyed were asked to estimate the number of outcomes on a scale of Nil (0) to a Lot (6). When the responses are averaged out, almost every category has a mean less than 1. The lone exception is for research papers, which averaged 1.26. This low average, however, simply means that no faculty can have a finger in every pie. Table 8.2 excludes faculty with no interaction and shows the increased level of activity of half the faculty responding to the survey. Substantial innovation is occurring, as many new products, processes and prototypes are being developed. Yet, relatively few of the intellectual property rights of these innovations are being captured by the University or by faculty as there are only a modest number of invention disclosures, patent applications or patents issued. While these may be reassuring results for those believing universities should freely diffuse information, they are less encouraging to those who believe universities and faculty should capture income from innovation activity. The 39 faculty

Table 8.2 Outcomes excluding those respondents who reported no collaboration or any specific outcome

INNOVATION OUTCOMES	Observations	Mean	Standard Deviation
Research papers	39	2.62	1.60
New products invented	11	3.00	1.18
New processes invented	16	2.50	1.37
Prototypes	16	2.56	1.50
Invention disclosures	6	2.00	1.55
Patent applications	6	1.33	0.52
Patents issued	1	1.00	
Technology Licenses granted	4	1.75	0.96
Trade secrets	6	1.83	0.98
Copyrights	4	2.50	1.91
Policy advice (1)	15	3.13	1.92
Newspaper/magazine articles (1)	17	3.12	1.83
Television appearances (1)	8	2.75	2.05
INTERMEDIATE OUTCOMES	**Observations**	**Mean**	**Standard Deviation**
Improved existing products	14	2.21	1.58
Improved existing processes	24	2.38	1.50
Introduced new products	11	2.45	1.44
Introduced new processes	15	2.40	1.40
Improved corporate R&D efficiency	15	2.33	1.84
Introduced new R&D projects	23	2.57	1.56
EDUCATIONAL OUTCOMES	**Observations**	**Mean**	**Standard Deviation**
Master's degrees granted	16	2.56	1.36
PhDs granted	6	3.00	1.79

(1) Additional questions for the respondents in September 1999
Scale : Nil (0) to a Lot (6)

claiming research paper outcomes with a mean production of 2.62 and the high number of faculty still performing the important social role of arbitration through policy, newspaper, and television consultations also provide encouragement to advocates of the university's traditional role. The number of Masters and PhDs impacted by industry interaction is significant but not overwhelming. Our survey also revealed that 67 respondents had interactions with Hong Kong based companies, 35 with mainland (China) based companies, and 50 with multinational companies. These results indicate that the university is quite effective in interaction with local industry and multinationals, but interaction with mainland firms is not as yet as strong.

Yet, the level of interaction indicated by these results should be considered in light of what has been accomplished within existing institutional regulations and incentives and compared to what could be accomplished under others preferred by faculty. Thus to clarify the institutional context of the interaction outcomes, the next set of results measured the significance of existing incentives and disincentives. The significance of these was measured on a scale of Very Important (1) to Insignificant (7) for the academic and commercial incentives, and a scale of Very Severe (1) and Insignificant (7) for the disincentives.

The response of faculty to most of the academic incentives listed in Table 8.3 result in means that are close to neutral (i.e. 4). Exceptions falling away from the neutral position towards insignificant are: obtain equipment, access to industrial facilities, and bring industry into academic programs. These scores may reflect satisfaction with funding, facilities, and faculty strengths, or perhaps a weakness of similar factors in particularly local industries. Significantly, department or school expectations do not play an important role in providing incentives for industry interaction. This result reveals a dissonance between the goals of the R&D branch and departments, and perhaps from the standpoint of overall academic-industry collaboration governance, an appropriate check. Although not advancing far from neutral into positive territory, the need for R&D funds is the highest scoring academic incentive. In a positive perspective, the means may convey a satisfaction with prevailing incentives or at least an acceptance of their relevance. However, perhaps the more important aspect of this table is the large standard deviations. Clearly, various issues are given different significance by different faculty. The high scores in the other category also confirm a focus on single issues of importance. An interpretation of this variation is a need for an organisational variability to maintain a diversity of incentives. Academic disincentives and commercial incentives and disincentives were surveyed similar to academic incentives, and a summary of results follows.

Table 8.3 Academic incentives

ACADEMIC INCENTIVES	Observations	Mean	Standard Deviation
Dept./School expectations	94	4.26	2.20
R&D funds	100	3.37	2.24
Obtain equipment	97	5.14	2.07
Access to industrial facilities	92	5.33	1.83
Obtain information on industry needs	99	3.47	1.96
Opportunity to confer with industry	95	3.52	2.02
Expand / change research direction	99	3.67	1.93
Practical experience for students	99	3.65	2.00
Bring industry personnel into academic programs	96	4.80	1.91
Others	12	1.33	0.49

Note: the following is a list of 'others' as per the report of those 12 respondents: financial reward, technology unification, learn, profession, need to stay connected, personal interest, provide path to join the industry, promote HK high-tech industry, good for HK, relevant to consultancy fees, service to the society, social impact

Scale: Very Important (1) to Insignificant (7)

The neglect of external collaboration in career evaluation and the time conflict between external and academic roles are considered the most severe disincentives to industry interaction, yet not overwhelmingly so. Limited time horizon for research gained a severity score slightly beyond neutral. However, most issues fall toward insignificance, some substantially so. Importantly, although corporate restrictions on publication and restrictions on free information exchange are not insignificant, neither do they appear onerous. The story, though, for academic disincentives, is of generally large standard deviations indicating that not all researchers in the university feel that academic disincentives are negligible. These results may accrue from experiences with the demands of particular companies or of different departments. The severe scores for time conflict and career evaluation in particular may reflect a negative stance or a lack of recognition of industry interaction in some departments.

Commercial incentives may range from a very limited perspective of personal gain, through continued research funding, to benefits arising from patents or directorships. Research funding was most important, closely followed by income from consulting. Few had ambitions for company formation, obtaining patents or directorships. Thus while it seems most of the academics are interested in making extra money, few maintain great ambitions. Institutional policy should be informed with the knowledge that academics are most likely to be drawn into industry interaction by lower risk and modest monetary incentives and that few are likely to be compelled by entrepreneurial opportunities. These results accrue from a situation in which most academics are fairly happy with their moneymaking prospects. The commercial disincentives – such as too high a university share of returns, inability to profit from contract research, or mistrust of corporations – all score substantially toward the insignificant end of the scale. Again the standard deviation value is large signifying that some faculty may, for example, feel that they can not profit sufficiently from contract research or the university takes too great a revenue share from a venture. This variation in attitude toward commercial disincentives makes it easier to account for the seeming conflict between the insignificance scores, given the commercial disincentives of departmental prejudices and time restrictions and the higher severity weightings given to the academic disincentives of time conflict and collaboration neglected in career evaluation.

Finally, we examined whether faculty consider the R&D Branch's policies, organisational infrastructure and personnel effectively supporting industry interaction. We found that the university does remarkably well in satisfying its faculty stakeholders. In all categories, including the number and quality of personnel, participation of faculty in policy making and red tape, faculty find few problems. Again there is a large standard deviation and some people have obviously had a difficult time with some aspects of the institutional regime.

Conclusion

This chapter has argued that the impact of a university on regional economic development results from a complex interplay. Its expected economic role must be balanced with its academic roles, and fitted into its industrial context. The evidence suggests that HKUST has been successful in its endeavours, especially in consideration of its youth and the limited technological abilities of local industry. However, it is not simply the University, nor the faculty who achieved this success. HKUST has created liaison institutions and policies designed to provide faculty with both the means and incentives to interact with industry. Its patent applications, patents and licensing, or simply instances of interaction with industry are objectives of its performance. It is also necessary to compare the objectives of the liaison organisation against the effectiveness perceived by its faculty. The real benchmark to be drawn from this study is that of a research university that has created the basis for its faculty to interact with local industry despite the fact that faculty are academically rewarded on the basis of research and are of the calibre to draw substantial numbers of multinationals into collaboration.

Beyond the enumeration of patents etc., the analysis of incentives and disincentives devised for this study enables university and liaison organisations to be directly evaluated in regard to how they encourage or discourage interaction with industry. This evaluation is critical because it makes little sense to talk about the impact of a university on local industry if one does not consider the ability and willingness of faculty to engage with industry. From the perspective of HKUST's role in promoting economic development, there still seems to be some room for the integration of rewards for industry interaction into academic evaluation. And despite the fact that faculty are generally satisfied with commercial incentives and dismiss disincentives as insignificant, neither are they overwhelmed by the entrepreneurial opportunities offered them (this, however is a recent thrust). The large standard deviation warns, however, that there is a wide variation in what individuals consider important incentives and directs liaison design to offering a variety of incentives. From the perspective of those concerned with the integrity of the academic mission, the remaining discordance between commercial and academic incentives may be considered as an appropriate check against the commercialisation of the university. Perhaps this last point may indicate the greatest value of this type of evaluation. Although it is unlikely that the tensions between academic and commercial roles of a university can or should be resolved, there is great value in monitoring how academics perceive their incentives, and thus know when they are not adequately connected with economic objectives and when they are overly so.

Acknowledgement

The authors would like to acknowledge the assistance of Mr. Cecil Lui in undertaking the survey.

References

ASAIHL Seminar on University-Industry Partnership in Economic Development, Singapore 1997.

Bower, D.J. (1992), *Company and Campus Partnership*, Routledge, London.

Bowie, N.E. (1994), *University-Business Partnerships and Assessment,* Rowman & Littlefield Publishers, Boston.

Carboni, R.A. (1992), *Planning and Managing Industry-University Research Collaborations*, Quorum, Westport.

Cohen, W., Florida R. and Goe, W.R. (1994), *University-Industry Research Centers in the United States*, Carnegie Mellon University, Pittsburgh.

Cooke, P. (1996), 'Reinventing the Region: Firms, Clusters, and Networks in Economic Development', in P.W. Daniels and W.F. Lever (eds), *The Global Economy in Transition*, Longman, Harlow.

Feller, I. (1990), 'Universities as Engines of R&D-based Economic Growth: They Think They Can', *Research Policy*, vol. 19, pp. 335-348.

Feller, I. (1992), 'American State Governments as Models for National Science Policy', *Journal of Policy Analysis and Management*, vol. 11, pp. 288-309.

Hudson, R. (1994), 'Institutional Change, Cultural Transformation, and Economic Regeneration: Myths and Realities from Europe's Old Industrial Areas', in A. Amin and N. Thrift (eds), *Globalization, Institutions, and Regional Development in Europe,* Oxford University Press, Oxford.

Malecki, E.J. (1997), *Technology and Economic Development*, Longman Scientific & Technical, New York.

Nijkamp, P. and Mouwen, A. (1987), 'Knowledge Centers, Information Diffusion and Regional Development', in J. Brotchie, P. Hall and P. Newton (eds), *The Spatial Impact of Technological Change*, Croom Helm, London.

Peters, K. (1988), 'Universities and Local Economic Development', in H.J. Ewers and J. Allesch (eds), *Innovation and Regional Development*, Walter de Gruyter, Berlin.

Rosenberg, N. and Nelson, R. (1994), 'American Universities and Technical Advance in Industry', *Research Policy*, vol. 23, pp. 323-348.

Saxenian, A. (1994), *Regional Advantage: Culture and Competition in Silicon Valley and Route 128*, Harvard University Press, Cambridge, MA.

9 Promoting Local Growth in the Oxfordshire High-tech Economy: Local Institutional Settings

Helen Lawton Smith

Introduction

When clusters of firms with particular characteristics develop, a demand is created for a raft of professional services. In the case of clusters of high-tech firms, this demand includes venture capital, patenting advice, and prestigious premises. Demand is also created for scientific skills and, for some, intellectual capital from the local universities and research establishments. While most of these are provided by the private sector – even universities are commercial organisations participating in the market for technology - to varying degrees the public sector and voluntary organisations become part of the process by which the supply of innovation support is increased to meet the demand.

The high-tech sectors in Oxfordshire have become the leading source of employment growth in the local economy. They have tended to continue the manufacturing tradition for example in instrumentation and motorsport, but are developing in 'soft-science' markets including leading-edge software and biotechnology. Concomitant with this growth has been the development of a unique innovation support system. Unlike other centres of high-tech industry, which tend to have dispersed systems, the county has a clear focal point. This is the Oxford Trust, a charitable trust, established in 1985. Since its formation, the Trust's activities have expanded and incorporated a range of business organisations, local and central government and academics into its orbit. Thus the particular form of institutionalisation of innovation support takes the form of a coalition of different organisations, which both benefit from and contribute to the process by which the interests of local high-tech firms are promoted.

The success of the county's high-tech firms and of the Trust in co-ordinating innovation support has brought the county to a pivotal position in the national innovation system and in strategic thinking of both central government and of the Regional Development Agency, the South-East England Development Agency (SEEDA). The former advocates the development of clusters of firms (Competitiveness White Paper, 1998) and of biotechnology in particular (Sainsbury Report, 1999) with Oxfordshire as one of the three leading UK clusters. In this chapter, however, it is argued that while there have been considerable achievements and some moves to co-ordinate the different sets of activities, there is a lack of a coherent and strategic vision of what direction local economic development in Oxfordshire, as a whole, should take. As long as this is the case, efforts will remain piecemeal and the benefits of growth not absorbed throughout either the county or the country.

Innovation Support Processes

The literature on innovation systems (Edquist, 1998) tends to overlook innovation support policies although they are generally recognised as being one of the components of systems of innovation at various spatial scales - national (Lundvall, 1992; Nelson, 1993), regional and local systems (Braczyk *et al.*, 1998; De la Mothe and Paquet, 1998). Discussion tends to focus on such institutional factors as the structure of the research base, R&D spending and military expenditure and their relationship with the trajectories of industrial innovation in key sectors. When innovation support mechanisms are discussed in connection with local development they tend to be national programmes such as the UK's SMART awards (Simmie and Sennett, 1999) or science parks (Doutriaux, 1998). An exception to this is Howells (1997) who reviews 'best practice' and 'bespoke' policy mechanisms and reports on policies in the UK's South-East and in mainland Europe designed to encourage continuous improvement though learning.

Another literature builds on Alfred Marshall's classic work on industrial districts. This includes the GREMI school, those adopting a Neo-Marshallian approach (Amin and Thrift, 1992; Markusen, 1996), and evolutionary theories such as Maillat and Lecoq (1992) and Garnsey (1998). The institutionalisation of business interests into networks of power and information is a common theme, even though the causality for the growth of 'institutional thickness' (Amin and Thrift, 1994) is attributed differently. For example, while the state may be ascribed a key role in the establishment of agencies and measures designed to intervene in (national) local economic development on behalf of business, evolutionary economists favour a path-dependent explanation. Garnsey (1998) for example argues that there are processes of co-evolution at work, whereby interdependent, co-evolving units can transform their local environment without deliberate co-ordination or planning. It is endogenous economic growth which creates the possibility for networks and partnerships to be effective in bringing about local transformations in the economy, rather than the causality being the other way round. However, explanations of the types of processes which lead to the formulation and effectiveness of local innovation support strategies tend not to form part of the explanations for the *vitality* of clusters of high-tech firms (see for example summaries by Sternberg, 1996; Simmie and Sennett, 1999).

There has, then, been little research on the effectiveness of local growth policies. Moore (1998) argues that assessment of policies must be made in relation to the theoretical basis for policy intervention. The conventional (neo-classical paradigm) justifies or leaves space for policy only if there is market failure (inflexibility of markets, externalities, asymmetric information, etc.) that needs to be ameliorated. Examples of innovation policies are those adopted in the UK by the previous Conservative administration to overcome market failure in technology markets, for example the drive to exploit university research and create 'efficiency' in government laboratories (Lawton Smith, 2000). If on the other hand, the policy debate is couched in terms of 'institutionalist' or 'evolutionary frameworks' in which collective learning is central to locally constructed innovation processes, the intention should be to improve the learning process and the preconditions under which 'collective learning' can be enhanced. In the UK, current national policy orientation appears to espouse the institutionalist perspective encouraging networking at the regional and local level. In Oxfordshire, as this chapter shows, there is a third kind of objective – altruism – which is often referred to the US as 'civic entrepreneurship'.

With the purpose of providing a framework for evaluation, Moore (1998) asked five questions:

- Why do we need policy intervention?
- What are the objectives of policy?
- What are the targets of policy?
- What types of policy initiative are appropriate?
- What mechanisms are most appropriate to achieve desired aims?

Although these are key issues, others also need to be addressed. For example, how are the targets of policy decided? Why are some initiatives are more successful than others? Can it be assumed that there is consensus about the justification for the choice of targets? Moore (1998) has pointed out that the scale of delivery is dependent on the spatial scale at which policy objectives are designed to be effective. This chapter turns to Oxfordshire to provide some answers to these questions.

The Oxfordshire Economy

Thirty years ago Oxfordshire was a sparsely populated agricultural county with a famous university and a car factory somewhere in middle England. Now it is a key industrial county in the South East of England, part of the Thames Valley, itself the richest and most dynamic economy in the UK outside London (Economic Development Strategy for Oxfordshire, 1998/9). In October 1999, the overall unemployment rate was 1.3%, with the unemployment rate highest in the City at 2% and lowest in West Oxfordshire at 0.8% (Heart of England TEC, 2000). While the manufacturing sector has remained static in terms of total employment numbers since 1993, growth in manufacturing and services has been in a few high-tech sectors such as biotechnology and motor sport. High-tech industry comprises some 543 firms employing 19,465, with an average of 36 employees. These sectors account for less than three per cent of firms but just over six per cent of employment (Garnsey and Lawton Smith, 1998). Most of high-tech activity is located in Oxford City and Abingdon, to the south of Oxford, with secondary concentrations in the four country towns of Didcot, Witney, Banbury and Bicester especially, which are the designated growth areas in the county.

So why might the development of innovation support mechanisms be a policy issue in Oxfordshire? There are three main social and economic reasons. First, there is now a significant divergence between a high wage science/education sector and a larger low wage service sector (Oxfordshire County Council, 1998). Second, the county is significantly below average for the South East in terms of gross value added per head over the last 3-4 years, in spite of the high-density of high-value, highly skilled technological businesses in the area (Heart of England TEC, 2000). Third, there is the issue of complacency. Measures are needed to support economic development over the longer term to reduce the effects of downturns in the economic cycle (quote from Chief Executive, Vale of White Horse District Council, 1997).

Local Policy Processes in Oxfordshire

The particular forms of delivery of innovation support in Oxfordshire are explained by historical attributions of local power and responsibility and choices made within a range of possible responses on the part of actors (Markusen, 1999). A key role is played by the Oxford Trust and the networks of key individuals focused on it. This counteracts the negligible contribution of local councils, except through the planning process. Until recently the Oxford Trust and the local authorities only interacted through contact with key officers such as the County Council Economic Policy Adviser, rather than with the elected representatives. Since the late 1990s, however, there have been a series of measures directed towards co-ordination of the different activities to create a more inclusive system designed to capture locally the multiplier effects from growth of the high-tech sectors. However, these are still underdeveloped and represent the potential for multi-agency competition (Roberts and Lloyd, 2000) rather than coherence because of differences in objectives. To put these comments in perspective, The Oxford Trust is introduced, the agendas of local authorities are described, and some of the efforts to increase co-ordination between institutions are outlined.

The Oxford Trust - the Prime Agent

The Oxford Trust was established by the founder of Oxford Instruments, Sir Martin Wood, and his wife Audrey. It is a charitable trust whose aim is 'encouraging the study and application of science and technology'. It achieves its objectives through the management of incubation centres, running seminar programmes, schools initiatives and managing a number of innovation support activities. In October 1987, it established Oxford Innovation as a commercial activity. Oxford Innovation operates in the field of technology transfer, undertaking consultancy and property management, including the Oxford Centre for Innovation, where the Trust is located, and manages Curioxity, a hands-on science exhibition centre.

The Trust is the focal point of innovation support activities in the county and of a network of people who bring their expertise to these activities. The network is central to the delivery of innovation support in the county. Its members include senior local businessmen, the County Economic Policy Adviser, property developers and the Director of Commercial Services at Oxford Brookes University. The Trust also develops links with organisations in other places with similar objectives in order to increase capacity and resources. This has happened with St John's

Innovation Centre in Cambridge and with Cranfield University within the framework of the 'technology arc concept'. The development of links between Oxford and Cambridge was identified as a strategy in the draft EEDA strategy document and then dropped. The Trust, St John's Innovation Centre and SEEDA are now taking this idea forward.

Local Authorities

Policy intervention is a complex process, which is in part determined by the political agendas of the ruling political parties and the strategies adopted by local government officers. These are not necessarily the same. The lack of engagement in the interest of high-tech industry by councillors in the county (Conservative controlled) and Oxford City (Labour controlled) has led to a statutory vacuum and a lack of resources for economic development. This has been filled by The Oxford Trust and its network. Both the County and City Economic Development units are small and under-resourced. The County has one of the smallest innovation support teams in the country, and indeed the county is the second lowest spender behind Cornwall on everything. Even under the new Labour government is has been rate capped. In order to overcome resource limitations, the county's economic development department's strategy has been to develop partnerships with regional agencies such as the Thames Valley Economic Partnership and other agents of the state - Government Office of the South East (GOSE), and the Heart of England Training and Enterprise Council (TEC).

The County Council has the most explicit interest in the high-tech economy, largely because of the Economic Policy Advisor who is the driving force for supporting the sector. The stated aim of the 1998/9 Economic Strategy is, 'to encourage the development of a high wage, high skills, high value added economy which enhances and protects the quality of life of the residents of Oxfordshire and enable them to fulfil their potential'. In the 1995/6 version, the words 'high-tech' were dropped from the list on the instructions of the county councillors because it was felt that this description was insufficiently inclusive. From October 1999 more resources have been directed to strengthening the county's capacity to deal with economic and social issues. The economic policy adviser was appointed to a new post of Economy & Environment Manager, with responsibilities for policy in the areas of economic development, regeneration, social exclusion, crime/community safety, health, Europe, regional policy and the environment.

The policy objectives of the Labour dominated Oxford City Council are those of welfare and unemployment. Until the end of the 1990s, the targets of policy had been responding to problems caused by the then decline of the car industry which has left a low-wage economy based on public sector jobs in health, tourism and leisure and local government. This emphasis is recorded in the City of Oxford Local Plan Review (1997), which highlights the necessity of 'identification of underlying problems in local economy, unemployment in black communities, disability groups, the city's high rate of unemployment'. It emphasises population and employment restraint. However, high-tech industry and the potential for technology transfer are recognised as important. Note 4.40 p. 89 says, "the development of science-based industries especially those connected with research and innovation in establishments like Oxford University, Oxford Teaching hospitals and Brookes University is important to the national and local economy". On networking, policy is that the City "will co-operate with organisations such as the Oxford Trust' for the provision of premises for small businesses" (4.47).

An attempt to incorporate the interests of high-tech sector into a broader economic strategy of Oxford City, which failed, was *The Economic Development Forum.* This was an inclusive body comprising representatives from most parts of the Oxford economy including the universities, the College of Further Education, the motor industry, the hospitals, the Oxford Chamber of Commerce, the County Council, the Oxford Trust and small and large high-tech businesses. It was formed because the City Council has a duty to consult about how local taxes should be spent. When responsibility for the business tax was removed from local councils, the existing committee was renamed and re-badged. However, it had no formal powers and as it was not linked to the economic development committee, did not feed into the economic development system. This has been replaced by the *Development Agency for the City of Oxford* formed in 1999. Its purpose is, "to develop a comprehensive strategy and action programmes to promote and sustain the economic, social and environmental well-being of the city through a more collaborative approach involving the public, private and voluntary and community sectors in Oxford". One of its strategic priorities is to improve education and training by promoting co-operation between the universities and local business. The City Council "now wishes to make partnership and collaborative working a central rather than an incidental approach to the economic and social development of the city" (p. 2).

The potential for co-ordination between the districts is limited by the sharp contrast in the objectives of economic growth strategies of districts to the north and south of Oxford City. For example, Cherwell District's interest in the high-tech sectors is as a means of diversifying the economy away from traditional sectors, primarily through inward investment of high-value added, new technology businesses in the District (Economic Development Strategy, 1999/2000, 8). Cherwell have reduced emphasis on inward investment and increased emphasis on endogenous growth over the last 5 years. Their strategy has been outstandingly successful - probably the best in the country – over the last 10 years measured by fall of unemployment (15% down to 1.0%). Oxford Innovation has played a major part in putting their strategy into action, particularly in managing their incubator on the former US airbase at Upper Heyford.

To the south of the county, the Vale of White Horse District Council has been active in supporting the development of property initiatives for the high-tech sector, for example the Harwell International Business Centre, owned by the UKAEA. In contrast, its neighbour South Oxfordshire District Council (SODC) has a local reputation of discouraging economic development per se. The aim of SODC is 'to provide conditions in which the economy of South Oxfordshire can develop and prosper without adversely affecting the attractive countryside, villages and towns of the District' (Economic Development Plan, 1999/2000, 2).

A recently created organisation, the Oxfordshire Economic Partnership (OEP), formed in 1998 has the role of uniting non-statutory and statutory bodies. Its objective is "to inspire and motivate private and public sector partners to align their plans and activities towards the achievement of the shared goals for the economic success of Oxfordshire". Although it incorporates local authorities and the local TEC, it was at the outset focused on a key individual, Sir Martin Wood, who initially chaired the group. For the first time the high-tech innovation support system is harnessed to broader social objectives.

Local Innovation Support Initiatives

The diversity of organisations which interact through this network are drawn together as an 'elite coalition' to represent the interests of the high-tech community. They intersect to manage the policy agenda and to promote the development of local resources (Jessop, 1994). Collectively these institutions create an innovation support system, which mobilises resources to secure the specific outcome of enhancing the capacity of high-tech firms to utilise available factors of production – land, labour, capital and knowledge. Examples of how these are made available in Oxfordshire are shown in Table 9.1, with their objectives and targets, and leading actors. The first three initiatives are collective endeavours and with one exception involve The Oxford Trust. The fourth is a sector-based initiative, which combines elements of the first three and is again based on the Trust. These are mostly directed towards small and medium-sized enterprises (SMEs). The concern of the fifth is technology transfer from Oxford University to industry.

(a) Land

Since the early 1970s universities, property developers, local authorities, and in some countries, regional and central governments have initiated the development of prestigious property developments. Some of these are science parks, others are up-market business parks. In Oxfordshire, the majority of these parks are integrated into a broader innovation support system.

The Science Park System The dispersal of high-tech firms throughout the county on some twenty science parks, business parks and incubators results from a combination of factors. Planning policy has been to restrict growth in Oxford City. An outcome of this was that it took a decade for a suitable site to be developed as a science park. After many years of failure by a number of Oxford University colleges to establish science parks, Magdalen College eventually succeeded in building the Oxford Science Park (OSP) in 1991, some twenty years after the founding of the Cambridge Science Park. While other proposals failed because of restrictions on the use of green-belt land, part of the Magdalen College site located to the South-East of the city already had permission for industrial use. Networks are very much part of the story of the existence of the OSP. Path-clearing for the Magdalen science park was facilitated by the Oxford Science Parks Committee, set up in the late 1980s and based at The Oxford Trust. The committee brought people with different interests together and, because it had no commercial interest, the Trust was able to mediate between them.

The high cost of accommodation on the Oxford Science Park has meant that many firms have chosen to locate on other science parks and business parks. From the mid-1980s the choice of competing sites with different kinds of locational advantages expanded rapidly with the development of new parks and incubators and the upgrading of existing sites. The most important example of the latter is Milton Park, a business park located near Didcot to the south of Oxford, which differs from OSP in that it encourages manufacturing activity. Its owners are determined to make Milton Park the 'Science Factory' of the region.

Table 9.1 Innovation support initiatives in Oxfordshire

INITIATIVE	OBJECTIVE/TARGET	LEADING ACTORS
Land		
Science parks and incubators (1991-)	To increase supply of dedicated premises for new and established high-tech firms	Milton Park, Oxford Science Park, The Oxford Trust, Oxford Innovation
Finance		
Oxford Investment Opportunity Network (OION) OION IPR package (1998)	To increase supply of venture capital, mainly new and small firms guidance on IP issues	The Oxford Trust, business angels, plus local law firm
Labour		
Motorsport degree Physics-based degree programme Professional Development	To increase supply of trained motorsport engineers to local motorsport firms tailored for needs of physics based high-tech businesses	Oxford Brookes, Heart of England TEC, The Oxford Trust, Oxfordshire Economic Policy Adviser Oxford Brookes, Heart of England TEC Oxford University Department of Continuing Education
OUTINGs	To familiarise graduate students which small high-tech firms	Oxford University Physics Department
Sector		
Oxfordshire BiotechNet (1997)	To double the rate of growth of firms in the biotechnology sector by making it easier for scientists to exploit their ideas.	The Oxford Trust, Business LINK
Knowledge		
ISIS Innovation Oxford Innovation Society (1990)	To promote the commercialisation of research ideas generated by Oxford academics and owned by the University. Exists so that industrial companies can benefit from ISIS Innovation's activities by having a 'window' on Oxford Science.	Oxford University
Regional Liaison Officer (1999)	To foster university links with the local region	Oxford University

One of Oxford Innovation's major activities is managing incubators. It now manages six incubator centres, including Oxfordshire's first bioscience specific business centre, which was opened officially on June 28

1999 on the site of the Yamanouchi Research Institute. Two opened in 2000, one of which is on the Harwell Business Park. This is significant in local terms because it represents the integration of the Atomic Energy Authority into local networks from which traditionally they have been (voluntarily) excluded (see Lawton Smith, 1990). The other is owned by Oxford University.

Inward Investment The Cherwell M40 Investment Partnership was established in 1991 to encourage inward investment to Cherwell District. From June 1999 there has been a formal system of networking between the districts and the county council and the Heart of England TEC, with Cherwell at the heart of the network. Enquiries on inward investment received by SEEDA are passed to Cherwell who, acting on behalf of OEP, circulate the information to these other organisations. The main focus of inward investment activity has shifted back to the county from the Reading/Slough area. Some of the districts, Cherwell for example and the County were members of the Thames Valley Economic Partnership, which was set up to encourage inward investment, then re-investment in the Thames Valley. They have withdrawn from that organisation on the grounds that Oxfordshire's interests were insufficiently represented, being dominated by those of Berkshire.

(b) Finance

One of the characteristics of the UK system of finance is the limited amount of both small scale venture capital (less than £0.5 million) to fund early stage and start-up finance and large scale investment finance. Moreover, the informal venture capital market (business angels) is both inefficient and underdeveloped, both problems stemming from an information problem (Harrison and Mason, 1996). A local initiative OION was established to increase the supply of local venture capital through a formal mechanism to run along side investments made by local business angels.

OION is a company limited by guarantee. Its role is to act as a marriage broker bringing potential investors and firms seeking investment together. By 1998 the best guess was that there was a pool of £4 million available from the 50/55 investors in the Network. Investors range in size from individuals to national banks, venture capital companies and local corporate investors. Some 30 companies per year go through the system. The Network's steering group includes representatives from Business Link, banks and venture capital companies. The Oxford Trust is involved in OION and this support lends weight to the venture.

(c) Labour

The development of a science oriented, flexible labour market is a prime condition of a successful knowledge-based economy (Carnoy *et al.*, 1997). Of major importance is how local institutions working together can effect a change in the quality and composition of the local labour market. In Oxfordshire, the upgrading and training of local skills to meet the needs of the high-tech sectors has been addressed by both statutory and non-statutory bodies.

The Motorsport Initiative and the Motorsport degree programme at Oxford Brookes University resulted from collective and separate actions on the part of The Oxford Trust, Oxford Innovation, Oxfordshire County Council, Cherwell District Council, and the TEC. The Initiative was designed to assess the strengths of North Oxfordshire in relation to an innovation centre at Upper Heyford. This was an EU funded study, led by the Chief Executive of the Oxford Trust, Paul Bradstock. At the same time Oxfordshire's Economic Policy Adviser had identified that there was a huge concentration of performance car research testing and manufacturing in Oxfordshire, which led to the creation of the Performance Car Forum. When people from the industry first met, 'they all saw each other as the enemy'. This was because there was a low level of relevant skills in the economy and firms were constantly poaching from each other. Oxford Brookes has gained from the effect of increasing demand for places on all engineering courses, the raised profile of the university as a result of the publicity given to this initiative, and from closer ties with local industry.

Other programmes led by the TEC are a physics based high-tech industries degree course at Oxford Brookes University, courses run by Oxford University Continuing Education Department for professional development for R&D scientists and engineers' management skills, and the local skills task force. The physics department of Oxford University and the Heart of England TEC run the OUTINGs. The scheme allows a limited number of graduate students to spend a week in a local high-technology company.

(d) Sector

So far the Oxfordshire BiotechNet is the only example of sector-specific innovation support programme, although others are planned. This is a government sponsored initiative implemented at the local level. Local sponsors include 3i plc, Barclays Bank, Oxford Instruments and the Yamanouchi Research Institute. In February 1998 an award of £400,000 was granted to the *Oxfordshire BiotechNet* by the Department of Trade & Industry from the Biotechnology Mentoring and Incubator Challenge. The BiotechNet offers a mentoring network, providing business and technical advice for start-up bioscience companies, from the foundation of a business idea through to business expansion. The incubator centre - the Oxford BioBusiness Centre – offers technical facilities such as laboratories and on-site support services.

The spill-over effect and circularity of influence of networks is demonstrated first by the recent trend of investors to network amongst themselves and develop new personal networks; and by many of the same people being involved in the *Oxfordshire BiotechNet* and OION. In spite of Oxford University being a centre of excellence in bioscience, until recently, it was not represented in the Oxfordshire BiotechNet.

(e) Technology

The main channels for the commercialisation of research undertaken at Oxford University have changed in recent years. Until 1997, technology transfer resulted from choices being made by individual academics and by the activities of the Research Support Office. After 1997, Dr Tim Cook was appointed to head ISIS Innovation, the university's commercialisation company. ISIS Innovation was formed in 1988. One of its remits is to seek licensees interested in securing a proprietary position over a wide range of inventions developed by Oxford's scientific staff and patented by ISIS arising out of research. ISIS also exploits the intellectual property of the University by setting up individual companies using venture capital or development capital funds. As part of that remit it assists in finding investment for start-up companies. These are funded by local companies, banks and overseas venture capitalists. ISIS runs the Innovation Society which holds three major meetings each year and specialist seminars. Members receive personal notification of all patents in their areas of interest registered by ISIS Innovation on behalf of the University as well as a regular newsletter and portfolio of the patents currently available for licensing.

Until Dr Cook was appointed in 1997, ISIS had a staff of three and had made a very small contribution to the formation of spin-off companies. Under Dr Cook it has expanded and has adopted a high-profile strategy for facilitating patenting and entrepreneurship, which includes interacting with other key organisations. It now employs 18 staff. ISIS has at its disposal a small pre-seedcorn fund for paying the costs of protecting intellectual property rights and for taking work to a stage where its potential can be assessed. It is currently managing about 100 patents and patent applications and has signed over 25 licence or option deals since April 1997. On average ISIS files one new patent application each week. It established 14 companies between February 1998 and July 2000 (ISIS Innovation web site).

The far more positive attitude of Oxford University towards the region and its propensity to engage in local networks is signified by the appointment of a Regional Liaison Officer in 1999 (on secondment for 2 years from Barclays Bank). In addition, the university commissioned a report on 'outlining its impact on the local economy' to be published by the External Relations Department and the 'Oxford Day in London' in November 1999 organised jointly by Oxford University and SEEDA. This was designed to bring to the attention of government to the innovation potential of the county and its institutions.

Conclusions

This chapter has shown that a range of actors playing different roles have combined to promote the development of local resources specifically for the mainly small firm high-tech sector in Oxfordshire. The key players, Sir Martin and Lady Wood and The Oxford Trust, recognised the need to support emerging sector, set in train a process of identifying what resources were needed and how to bring together the people who could help firms at their start-up phase. The Trust's activities have been developed in conjunction with local organisations with differing agendas ranging from profit as in the case of local business, fulfilling its educational remit while enhancing its reputation (Oxford Brookes University), and statutory responsibility (Heart of England TEC). The strengths of the network lie in its ability to determine the policy agenda by focusing on particular objectives, to bring together the complementary core competences of individual members and, these achievements promote the increase local resources dedicated to supporting the high-tech sector.

Most initiatives have been successful, with the development of dedicated property being one of the best measured by the number of sites and rate of occupancy. Moreover, the importance of these initiatives is greater than the provision of space because they provide a focus or medium for further networking activities. There is, however, a caveat to this story of success. The Oxford Trust even with its partners, has limited capacity to have a major impact. There is a need for policy intervention, and one which recognises both the need for co-ordination and support for specialised groups. Without key players with real power, such as the local authorities and Oxford University, making major commitments, initiatives will be small scale and the potential growth from the high-tech sectors benefiting both the county and UK as a whole limited.

The development of new institutions such as SEEDA and OEP potentially have the capacity to provide this policy coherence and reconfigure local networks. They have strategic roles being created for the explicit purpose of co-ordinating existing initiatives. Both are designed to be more inclusive by uniting a more comprehensive spectrum of interests. A major problem lies in the lack of clear vision of what Oxfordshire needs to be in the longer term - whether manufacturing or service orientated, whether it should encourage inward investment or the growth of more small firms. OEP represents the opportunity to provide the vision, but only if entrenched parochial positions of local authorities can be changed. The danger is that SEEDA and OEP will create extra layers of bureaucracy while the real work will be continued by the Oxford Trust and its partners.

References

Amin, A. and Thrift, N. (1992), 'Neo-Marshallian Nodes in Global Networks', *International Journal of Urban and Regional Research*, vol. 16, pp. 571-587.

Amin, A. and Thrift, N. (1994), 'Living in the Global', in A. Amin and N. Thrift (eds), *Globalization, Institutions and Regional Development in Europe,* Oxford University Press, Oxford, Chapter 1, pp. 1-22.

Braczyk, H.J., Cooke, P. and Heidenreich, M. (eds) (1998), *Regional Innovation Systems*, UCL Press, London.

Carnoy, M., Castells, M. and Benner, C. (1997), 'Labour Market and Employment Practices in the Age of Flexibility: A Case Study of Silicon Valley', *International Labour Review*, vol. 136, no. 1, pp. 27-48.

Cherwell District Council (1999), *Economic Strategy, 1999/2000*, Cherwell District Council, Banbury.

De la Mothe, J. and Paquet, G. (eds) (1998), *Local and Regional Innovation Systems of Innovation*, Kluwer Academic Publishers, Dordrecht.

Doutriaux, J. (1998), 'Canadian Science Parks, Universities and Regional Development', in J. De La Mothe and G. Paquet (eds), *Local and Regional Innovation Systems of Innovation*, Kluwer Academic Publishers, Chapter 15, pp. 303-326.

DTI (1998), *Our Competitive Future: Building the Knowledge Driven Economy,* Department of Trade and Industry, London.

Edquist, C. (ed.) (1997), *Systems of Innovation*, Pinter, London.

Garnsey, E. (1998), 'The Genesis of the High Technology Milieu: A Study in Complexity', *International Journal of Urban and Regional Research*, vol. 22, no. 3, pp. 361-377.

Garnsey, E. and Lawton Smith, H. (1998), 'Proximity and Complexity in the Emergence of High Technology Industry: The Oxbridge Comparison', *Geoforum*, vol. 29, no. 4, pp. 433-450.

Heart of England Training and Enterprise Council (2000), 'Labour Market Analysis Quarterly Update', Issue 3, January 2000, Abingdon.

Howells, J. (1997), *Research and Technological Development in South East Britain*, South East Development Strategy, Harlow.

Jessop, R. (1994), in A. Amin (ed.), *Post-Fordism: A Reader,* Blackwell, Oxford, Chapter 8, pp. 251-279.

Lawton Smith, H. (2000), *Technology Transfer and Industrial Change in Europe*, Macmillan Press, Basingstoke.

Lundvall, B.A. (ed.) (1992), *National Systems of Innovation: Towards a Theory of Innovation and Interactive Learning*, Pinter, London.

Maillat, D. and Lecoq, B. (1992), 'New Technologies and Transformation of Regional Structures in Europe: the Role of the Milieu', *Entrepreneurship and Regional Development*, vol. 4, pp. 1-20.

Markusen, A. (1996), 'Sticky Places in Slippery Space: A typology of Industrial Districts', *Economic Geography*, pp. 293-313.

Markusen, A. (1999), 'Fuzzy Concepts, Scanty Evidence, Policy Distance: The Case for Rigor and Policy Relevance in Critical Regional Studies', *Regional Studies*, vol. 33, no. 9, pp. 869-884.

Moore, B. (1998), 'Collective Learning and Policy', in D. Keeble and C. Lawson, *Report on Presentations and Discussions*, at Networks, Collective Learning and RTD in Regionally-Clustered High-Technology SMEs Goteborg Network Meeting, April 17&18, 1998 ESRC Centre for Business Research, University of Cambridge.

Nelson, R. (ed.) (1993), *National Innovation Systems: A Comparative Analysis*, OUP, Oxford New York.

Oxford City Council (1997), *Oxford Local Plan Review*, Oxford.

Oxford City Council (1999), *A Development Agency for the City of Oxford: Proposal and Outline Plan,* Planning Policy and Economic Development Department, Oxford City Council, Oxford.

Oxfordshire County Council (1998*), Economic Development Strategy for Oxfordshire 1998/99*, Oxford.

Peck, J. (1995), 'Moving and Shaking: Business Elites, State Localism and Urban Privatism', *Progress in Human Geography*, vol. 19, pp. 16-46.

Roberts, P.W. and Lloyd, M.G. (2000), 'Regional Development Agencies in England: New Strategic Planning Issues?', *Regional Studies*, vol. 34, no. 1, pp. 75-79.

Sainsbury Report (1999), *Biotechnology Clusters; Report of a Team Led by Lord Sainsbury, Minister for Science*, August 1999, HM Treasury, London.

SEEDA (1999), *Building a World Class Region: Towards an Economic Strategy for the South East of England*, SEEDA, Guildford.

Simmie, J. and Sennett, J. (1999), 'Innovative Clusters: Global or Local Linkages?', *National Institute Economic Review*, October, pp. 70-81.

South Oxfordshire District Council (undated), *Economic Development Plan 1999/2000*.

Sternberg, R. (1996), 'Regional Growth Theories and High-tech Regions', *International Journal of Urban and Regional Planning*, vol. 20, no. 3, pp. 518-538.

10 Technological Change, Local Capabilities and the Restructuring of a Company Town

Eirik Vatne

Introduction

The purpose of this chapter is to begin to develop an understanding of the processes involved in the restructuring of Rjukan, a company town in Norway. In a recently published study on new firm formation, small firm growth and regional development in Norway, Isaksen and Spilling (1996) have shown that the ratio of new firm formation to working population tends to be lowest in single industry towns. Importantly, the study concludes that the rate of new firm formation seems to be a function of a town's existing industrial structure. In other words, an economic structure is reproduced mainly through the creation of new businesses in the lines of already existing industries, and where large firms dominate few new firms are set up. It is argued here that traditional approaches to understanding economic change based on static analyses of production functions and factor inputs, only weakly explain this pattern of change. It is suggested, instead, that a dynamic approach to understanding value creation is more revealing, drawing on structuration theory and theories of 'learning' and knowledge production.

Economic geography has long been concerned with how economic activities are connected to specific locations, and how firms use spatially uneven resource endowment in their struggle to earn a profit. In the search for explanation, the discipline has moved beyond normative, marginalist economics and structuralist approaches to embrace ideas from evolutionary economics, economic sociology, institutional economics and social theory that see social behaviour and economic dynamics as embedded in place-based social relations – i.e. that local economies are socially constructed. The analysis presented in this chapter seeks to build on the ideas of structuration theory and theories of learning and knowledge production to understand the restructuring of a company town in terms of the dynamics of value creation through learning, innovation, networking and competencies.

Socio-Economic Perspectives on Local Economic Growth

New socio-economic perspectives on local economic growth, have brought new challenges and new directions to economic geography. They have sparked renewed interest in the role of 'place' in human geography, and they have created a 'new' regional geography that situates 'places' within their wider social contexts. They have revitalised interest in local societies as a starting point for studying endogenous growth and change, and they have intensified interest in the role of culture in shaping economic life (Allen & Massey, 1988; Cooke, 1990; Bagguley *et al.*, 1990; Storper, 1997; Gertler, 1997). Economic action and economic life is now seen as deeply rooted in social systems. It is difficult to understand economic processes without including social actions and social institutions at the heart of a working economy (Nelson & Winter, 1984; Hodgson, 1988; Block, 1990; Granovetter & Swedberg, 1992). And, to understand economic behaviour, analyses must be sensitive to historical context and open to factors outside the realm of pure economics. In the modern, globalised world it is also recognised that there is continuous, reflexive interaction between regional, national and even global institutions or cultures.

For the analysis developed in this chapter, two areas of theory are drawn on to begin to understand the restructuring of a company town:

- structuration theory (Giddens, 1979, 1981, 1984; Gregory, 1994), which offers a perspective on the role of structure and agency in the constitution of social life which emphasises the importance of history and space; and
- theories of knowledge production, 'learning' and local innovation systems (Lundvall, 1992; Maskell *et al.*, 1998; Braczyck *et al.*, 1998), which emphasise the recursive role of place-based technology and knowledge in shaping local growth.

Each offers a different but complementary interpretation of the dynamics of local economic change and restructuring from a socio-economic standpoint.

Structuration Theory and the Development of Places

Giddens' (1979, 1981, 1984) theory of structuration offers a perspective on economic change that emphasises the two-way relationship between individual agency and structure, which also includes the importance of history and space, in the constitution of social life. The starting point of structuration theory is that actions create structures, and structures restrict actions (Giddens, 1984). In other words, structural elements, as established institutions, relations of power, or systems of norms, will restrict human agency and 'guide' action in certain directions, but not in a deterministic or 'law-like' way.

Economic life is deeply integrated into social life, and social routines, conventions and cultures are produced by a history of *past* actions. At the same time, innovations, change and *new* ventures are at the core of capitalist development. External shocks can alter a firm's competitive strength, and radical discoveries of new materials or new organisational principles can change the way business is done. But, adaptation to such events and the ability to develop an endogenous innovative power is still very deeply rooted in the history and the present environment of the firm. Even for innovative activities, context is of paramount importance, and that context has spatial and temporal dimensions.

Individuals' actions are connected to the constitution of societies through what Giddens calls social integration and system integration. Social integration is formed by routinised interaction between agents who are present both in time and space. Social systems are formed in a society by social rules and control over important resources. And, those rules and resources, in turn, shape social practises. Through communication, power and sanctions, interaction systems are guided in specific directions. This implies that actions in everyday life are connected to the long-standing development of social institutions. Power relations in such systems will often result in a reproduction of the existing social institutions. To break away from established norm systems and change important institutions, conflicts between individuals/groups of individuals are a necessity, or contradictions between what Giddens calls structural principles must appear.

In Giddens' terms, a basic feature of a social system is the clustering of institutions in time and space to achieve co-ordination. Such concentrations follow from what Giddens calls structural principles. It is, nevertheless, problematic to identify the borders of such a society, and also the closedness or openness of such a society in time and space. Indeed, structuration theory has been criticised as 'grand theory' that is difficult to apply in concrete situations.

In human geography structuration theory has attracted a lot of interest because it is one of few attempts in social theory to include the spatiality of social life. The theory tries to show how the limitations of individual 'presence' are transcended by the stretching of social relations across time and space. For Pred (1985), structuration processes help in recognising the 'the constantly becoming of places'. And, because these processes act not only as barriers to individual or collective human action, but are also fundamentally involved in the production of those actions, the 'becoming' and 'maintenance' of places can be understood as historically contingent.

Gregory, however, is more sceptical of Giddens' understanding of spatiality because structuration theory says little on the issues of 'sense of place' and the role of symbolic landscapes in the reproduction of social life (Gregory, 1989). In Gregory's mind, Giddens underplays strategic intentionality and discursive consciousness by avoiding discussion of the influence of technical change and the diffusion of innovations, and also by failing to appreciate discursive knowledge or the unknown. As such, structuration theory is best equipped to conceptualise traditional, routinised and static societies, rather than open, complicated, fractured and non-routinised societies in the context of a globalised world (Gregory, 1989). Also, while Giddens sees the stretching of social relations across time and space (time-space distanciation) as essentially progressive and gradual, Gregory sees far greater volatility in contemporary capitalism.

> The landscape[s] of contemporary capitalism ... are riven by a deep-seated tension between polarization in place and dispersal over space. On the one side, constellations of productive activity are pulled into 'a structured coherence' at local and regional scales, while on the other side these same territorial complexes are dissolved away through the restructuring and resynthesis of labour processes. The balance between them - the geography of capital accumulation – is drawn through time-space distanciation as a *discontinuous* process of the production of space. (Gregory, 1989, p. 207)

Space seems to be a barrier for the circulation of capital, but this barrier can be transcended through the production of fixed and immobile spatial installations. Here lies a contradiction: in order to overcome space, spatial organisation and immobile configurations are necessary. In Harvey's (1982) mind, this fact explains why regional configurations are chronically unstable. In this manner, "time-space distanciation is closely connected to spasmodic sequences of valorization and devolarization and must be embedded in a theorization of locational structures of production and reproduction" (Gregory, 1989, p.208).

Technology and Knowledge Production

The limitations of structuration theory are to a considerable extent addressed in the emerging theorisation of technology and knowledge production (Maskell *et al.*, 1998). Renewed interest in knowledge and learning and the institutionalisation of knowledge represents a shift in thinking on economic change away from linear development paths towards a more chaotic and change-orientated perspective, where innovation is crucial for economic of competitiveness. The debate on local growth is focused on the technological aspects of knowledge, with technology seen as the most dynamic element in economic activity. Technological knowledge is regarded as mainly experience-based and tacit and, therefore, specifically connected to the environment within which it is developed. As such, the transfer of technological knowledge is difficult, and the cost of its adoption high. It is exclusive, firm-specific and, in some cases, regional-specific. It is very much a product of non-institutional endeavours mainly based on learning processes which in most senses are 'localised', and are specifically linked to the history and experience of individual entrepreneurs (Antonelli, 1995).

This localised technological knowledge is highly idiosyncratic. It is developed through everyday practises in factories and offices, through interactions between individuals' experience-based knowledge and purchased machinery and processes. It is developed through problem solving and the development of tools and equipment, and through interaction with customers, other producers, and suppliers of capital, materials and equipment. Such knowledge is mainly a result of learning-by-doing, learning-by-using or learning-by-interacting. Even formalised R&D will in this case be characterised as experimental problem solving for the development of new products or processes, or a learning process capitalising on previously developed knowledge.

In contrast, generic technological knowledge consists of general principles usable under different circumstances and by many different users. It is codified and, because it is often embodied in material products such as machinery, it can be more easily transferred. Even so, the use of such machinery, and its integration with existing equipment and procedures, is often difficult and costly, and again requires access to tacit, experience-based knowledge (Gertler, 1995). The development of knowledge is incremental and cumulative, though radical innovation may occur from time to time. Industrial praxis and established production technologies are, therefore, shaped and maintained by the actions of individuals. Those actions, in turn, are integrated into a specific context where agency is 'guided', and the development of firms and even localised places are structured, by 'path dependence' and 'embeddedness' (Dosi *et al.*, 1992).

From this 'learning' and innovation perspective, the growth and development of a local economy is based on human action, local social structures and institutions, former experiences and practise, daily learning experiences, and the adoption of generic knowledge. Some local environments and cultures may facilitate change, but others may act as a dead hand thwarting entrepreneurship and stifling change. It is important, therefore, to identify which place-specific actions and institutions, past experiences or extra-regional linkages are important in determining whether a place, its institutions and individuals, will be able to change in the face of globalising capitalism, or whether they will stagnate and decline through 'lock-in'.

The Restructuring of Rjukan: A Norwegian Company Town

The theoretical propositions on structuration and technology and learning developed in the previous section can be used to develop a socially sensitive interpretation of local restructuring that recognises issues of time, space and place, and global-local tensions. The place chosen to test these ideas is Rjukan, one of the oldest company towns in Norway, established by a few Norwegian entrepreneurs who had good access to foreign capital. The material base for economic development of the town was a large waterfall that was easy to convert to produce hydro-electric energy. The social foundation of development was an invention that used electrical energy to extract nitrogen from the air which could be further processed into nitrogen fertiliser. At that time, at the turn of the nineteeth century, commercial agriculture was in a first development phase and the market for artificial fertilisers was tremendously strong. At the same time, new technologies were being pioneered in the chemical industry. Many competing technologies were fighting to win a share of the huge profits to be won. So, the location of a nitrogen fertiliser factory and other associated works in a remote Norwegian valley can explained, in part contingently, through a Norwegian scientist's discoveries, and in part rationally, in terms of location economics and territorially fixed energy resources which it was not possible to transfer at that time.

From 1907 to 1967, the factories gave work to between 1,200 and 1,600 people, and the town's population peaked at 12,000 people. Through the First World War and the turbulent 1920s and 1930s, the plants and the place experienced crises, redundancies, strikes and high unemployment. Nevertheless, in spite of technological change, economic crisis, social unrest and political struggle, the town has provided over the years a fairly stable life for many families and individuals.

This uneasy stability was shattered in 1962. A new radical innovation in process technology revolutionised the fertiliser industry. Factories based on coal or hydro-power lost out to a new process based on oil (later gas) that needed only one tenth of the site, one tenth of the electrical energy, one third of the labour force and one half of the capital required by the existing technology. The new technology also offered strong economies of scale in a way that the old technology had not, and its prime input, oil, necessitated a port location, at least in Europe. It is unsurprising, therefore, that most investment in new capacity went to new sites. The old technology and the places associated with it faded away, some swiftly others more slowly (Vatne, 1981). At Rjukan, the company's decision was to restructure its activities and to gradually shed capacity and labour from 1967 to the late seventies when a planned close down was to be completed.

Since 1962, the dominant company, individuals, the local and national government and a range of public agencies have worked to restructure the town and the remaining activities and other businesses. As the figures in Table 10.1 indicate, these efforts have not been very successful. Almost two thirds of the jobs in manufacturing have disappeared and the population has shrunk almost continuously.

The most striking features of Table 10.1 are the extent of the external control of existing manufacturing firms in this locality and the almost total absence of successful local job generation initiatives. Notwithstanding the fact that the local economy has experienced nothing short of a small revolution, the overall picture of employment relativities is more or less the same as it always has been since the place was founded. It is here that structuration theory and theories of 'learning' and knowledge production throw light on processes that appear to defy change.

Two structural elements, international industrial capital and new, generic chemicals technology, shaped the development of Rjukan. In this concrete situation, they created a large, capital-intensive, process technology organisation in a peripheral, sparsely populated and traditional agricultural society. It brought with it an influx of labour and laid the base of a social system that ordered and organised everyday practices, and shaped the norms system and power relationships within the local society.

Table 10.1 Manufacturing employment in Rjukan, 1960 to 1995

Year	1960	1970	1975	1980	1985	1990	1995
Ownership							
Dominant firm	1600	800	570	626	565	292	300
Owned by the dominant firm						44	
Small plants controlled from elsewhere in Norway		293	461	480	247	182	148
Small plants under foreign control			32	52	23	80	79
Locally owned small firms	26	16	30	26	35	59	63
Total	**1626**	**1109**	**1093**	**1184**	**870**	**613**	**634**
Tinn Municipality	**9614**	**8478**	**7788**	**7482**	**7270**	**6899**	**6775**

Note: Growth in private service jobs has not been large in the town, as it has no hinterland and central service functions. Public activities in school and health care have expanded as elsewhere in the country. Manufacturing and later back office producer services (telemarketing, travel agency), therefore, are still fundamental to the local economic base. Because of the gradual closedown of the dominant company, its administrative functions have survived longest. What is left of the plant, which is included here, is now mostly service functions

In most place-based societies only a few institutional projects will dominate the time resources of individual agents. Paid work and economic institutions are the most common and influential elements in this structuration process. Dominant economic projects are, in this way, a driving force in the production of economic and technological experiences and knowledge in a society. They strongly influence the local, and they influence the social stratification of society as a whole. They knit together what is 'present' and what is 'absent' in time-space. Economic projects are, therefore, institutions of special importance in the systemic integration of a social system.

Since social integration primarily takes place in situations where individuals meet face-to-face, there exists a sort of territorial barrier for such interaction, making local societies important arenas for social integration. And, in a limited area, there is a limit on how many institutional projects can run at the same time, be they economic, cultural or political in character.

In a causal explanation we will argue that social life starts with human agency. Through experiment and social struggle, routines are established and practices evolve over time which develop into institutions and important social objects. In this way, a society will be shaped from the bottom up in correspondence with collective interests or in other cases through the uneven power relations present in society.

The old agricultural society in the valley where Rjukan is located was a traditionally based social system, with strong links between an individual's occupation, fairly equal control over agricultural land, and the rules and sanctions governing the society. The result was the repetitive reproduction of old structures through well-established property rights and old traditions, and through a common identity and belonging to a place-based society. As a result, a stable and static society was produced with strong ties to history and internal relations, but weak ties to present and external social systems.

Industrialisation brought to the valley a new set of principles for social and system integration that were completely at odds with the existing social order that had been built over generations. The commercial industrial organisation that came to occupy the hillside and bottom of the valley was the creation of knowledgeable, external agents, helped by farmers who sold their land and property rights over water resources. Also, the people whose everyday lives became connected to this institutional project came from outside. In other words the new and dominant project in the valley did not grow organically from local action and struggle. It was established as an externally generated action, physically 'present' in the local society, but where the men that drove it were 'absent'.

That, in the sequence of development the institutional project came first, had a decisive influence for the structuring of Rjukan as a localised society. The migrating workers who sought jobs in the new electro-chemical factories had in common that they were poor and wanted to sell their labour power. The company, Norsk Hydro, owned the land and the work place, the apartments and the shops, the means of transportation and the funeral home. They controlled the power elite of the place - the engineers and supervisors – many of whom were only visiting the valley as part of their careers in a multi-plant, and later multinational, company. They were socialised into Norsk Hydro's social system as 'company men' and had few ties to the place-specific social system that developed in Rjukan. The workers, in contrast, stayed in the local community and soon developed a common identity; first as underpaid labour, next in negation to the conservative agricultural society, and later as settlers in a unique landscape. Through this common identity a new local culture emerged based on collective action that was quite different to that of the existing agricultural society. These two social systems were spatially co-present but socially disintegrated.

Through common identity and collective action, the power of the dominant project could be resisted. Through their own, routinised actions, individuals could collectively affect the structuration of the society. The agents behind the dominant project controlled most of the allocative resources in the community, but the common people controlled the authoritative resources in the shape of the local political system. The industrial society was rooted in the town of Rjukan, the agricultural society in the rest of the local municipality. In local government, the labouring class had the majority, and as a consequence the rest of the community was often in strong opposition to 'the ruling class'.

The production of these institutions was closely connected to the conscious actions of a political elite among the blue collar labourers in Rjukan, who also had external ideological and organisational links to a larger, radical, social democratic movement at that time. Locally, the same person could be the leader of the union, the leader of the local social democratic party and the mayor of the community. After a while, however, the reproduction of the social system seemed to be produced through routinized and often unconscious actions of the local majority.

An important impact of the dominance of the two institutions of 'big capital' and 'social democrats' was to block the emergence in Rjukan of alternative institutional projects. Small scale capitalist projects had difficulties for several reasons. The dominant ideology supported co-operative shops. Private shopkeepers were perceived as profit makers. Norsk Hydro had almost a total responsibility for the infrastructure of the town and most of the stock of dwellings, and serviced it with their own employees. This closed a potential market for private craftsmen and service firms. As 'private' service functions were integrated into the large firm, outsourcing and privatisation of these functions were out of the question. The dominant labour culture also sanctioned individual behaviour that differed from the dominant collective values. This happened inside the gates of the factory. It controlled labourers flirting with the values and norms of both the agricultural society and small-scale business.

The repeated actions of everyday life were the foundation of local conventions; routines, practices and rules. At Rjukan, the dominant economic project drilled the adult population and habituated it to a specific form of everyday action. In the plant, the different operations and the division of labour were strongly structured by the production process and it's demand for close supervision and regulation. The danger of malfunctioning and explosion, with consequences for the lives of individuals, the installation and downstream activities, meant that work was highly regulated according to strict routines and procedures. Factory practices involved adhering closely to written manuals or rules learned under strong supervision. These instructions specified the actions to be taken if any problem occurred in the flow of production. In such a hierarchical structure, action is first taken only after instruction from higher levels in the organisation, instructions that are codified and embedded in manuals and routines or by regular orders.

The end result of this socialisation process was that the local society was inhabited by a large group of clever production workers who knew their routines and could act smoothly in accordance with the logic of the highly idiosyncratic technology operated in this society. Virtually no one this society had any experience of other economic activities: the development of products; sale and marketing; contact with the needs of the customers; or understanding accounting and economic control. A homogenous, deep and narrow knowledge capital was created by Norsk Hydro through daily factory activities that was reproduced day after day, generation after generation. The cultural context of the place also acted as a filter governing the language people learned, their understanding of life's problems, and their very personalities. This context created an implicit, unarticulated ideology that governed people's attention and their ability to 'read' actions in the local or extra-local environment. Also the material context of Rjukan, with the factory as a centre in a dramatic landscape, isolated from other social communities, came to affect individuals and advanced a common identity.

Process work is shift work. Work in the process industry has traditionally been men's work. For the female population there were only a few openings in the labour market in Rjukan. A few female jobs existed in the offices and in the cleaning and catering department of the dominant employer. It was primarily in the public sector and in retailing that women found a limited supply of jobs. This limitation in the labour market combined with the symbiotic relation between the factory, the local public sector and the retailing sector, lead to the men's world and daily activities also structuring women's ideology and social understanding.

To sum up, this small localised society and it's structuration process has been one-sided both in its social and system integration. Norsk Hydro, the local union and local Labour Party, were the dominant players in the forming of the social system. The same agents controlled links to individuals and institutions outside the local community, and to the historically dependent contradictions and forms of understanding of the social order. In other words, Rjukan was a homogenous society in spite of strong conflicts between labour and capital. The society remained coherent and stable through the reproduction of local power relations, language, practices, conventions and knowledge, divorced from any knowledge of alternatives. It was to take a profound external shock, involving new technological and institutional processes, to change this system.

However, although a society such as Rjukan's might at first sight appear conformist and static, a deeper view reveals a lot of innovative activity and learning at a micro-scale inside the main company. Production may have been based on generic chemical process knowledge, but the transfer of this process from a small-scale prototype plant to a large-scale commercial unit was not straightforward. Indeed, in the company's first twenty years, a radical innovation had to be further developed and fine tuned on site to attain satisfactory product quality and sound economic performance. To achieve this goal required an enormous input of manpower and knowledge, learning by trial and error, the refunding of the business and a technology transfer agreement with a competing German company. After that, the future brought many further technological challenges, including a radical change in nitrogen production, adjustment and rebuilding new and older equipment, and many incremental improvements in the efficient use of energy or catalysts, working routines and the product range and quality.

There has, in fact, been a continuous knowledge production inside the Norsk Hydro plant and new knowledge brought in from local production engineers, from other parts of the larger corporation, from consults and competitors. People on the shop floor helped to fine tune the technology and reorganise the work process. But, the new knowledge they created grew out of existing knowledge and the path that knowledge development could take was already defined by the technology brought in with the construction of the first plant in 1907. The whole town and several generations of workers have, as a result, been carriers of a specific type of knowledge concerned principally with the supervision and maintenance of a large and complex technical installation. The place, the factories and their actors developed in a sort of symbiotic relationship in spite of clashes of interest and many conflicts. The result was a very narrow technological and knowledge development path – a lock-in, vulnerable to the new technology that came along in the 1960s.

Restructuring Demands, New Competencies and Entrepreneurship

Increasingly, it is acknowledged that technological change and innovations are endogenous processes that develop along lines guided by experience-based knowledge, routinised action, and the merging of 'new' imported knowledge with already established knowledge. The development of innovative capabilities and innovative milieux demands access to diverse information and experience. In environments characterised by narrow and one-sided competencies, few competing projects and weak relations to other social systems, the ability to be entrepreneurial is severely restricted. At best, innovation in such environments often imitates existing projects.

In the Rjukan case, even the ability to imitate was strongly restricted. First, this was because job training inside the dominant project only gave participants insight into and knowledge of a few, discrete technological functions, and provided no understanding of the whole system of different technologies. Second, the social context gave no access to commercial training and knowledge on how to run a firm. Third, the capital intensity of chemical production acted as a barrier to the entry of competing small-scale operations. Fourth, the local ideology was hostile to private entrepreneurship and individuals being responsible for their own employment. Finally, the opportunity to accumulate capital for later investment was problematic for individuals dependent on waged labour.

Successful industrial districts in Italy and Germany for example are characterised by narrow and specialised knowledge bases developed over long periods of time. They also have strong interrelationships between the public and private sectors. But, in contrast to the Norwegian case, their social knowledge capital is more widely spread, with the division of labour making it possible to subdivide functions of production and, where barriers to entry are low, to open up markets to entrepreneurship and small business development. In such an environment competition, even locally, is strong and acts as a stimulus for the development of firms. The opportunity to learn economic competencies 'in action' is good, creating the local capacity for continuous technological up-grading and organisational change. In successful industrial districts, just as in one-sided industrial communities, 'learning' is regulated by the ideology, conventions and sanctions of the local society. But, while in a one-sided industrial community that regulatory consensus is antagonistic, in an industrial district it is organic.

When the big shock came to Rjukan in 1962, there was no flexibility in the social and material structures that had been developed over the preceding sixty years. No skills, economic and entrepreneurial resources were available to build a new existence in new economic directions. The ideology of the local society was incapable of individualising the responsibility for the maintaining the place. On the contrary, the ruling ideology would collectivise and externalise the action needed to bring in alternative jobs for the inhabitants. The collective of workers demanded that the large company should create new jobs as the existing jobs disappeared. The local community also expressed through the local government a moral right for local use of the energy resources produced in the valley, even though the legal owners were Norsk Hydro (a semi-public company) and Statkraft (a public utility company). Through the control of authoritative resources locally and strong links to the national political system, local action succeeded in fixing geographically the energy inputs for manufacturing. Norsk Hydro was forced to use a lot of the locally produced energy at site or sell it to other manufacturers for local use. In fact, at that time, a public agency (Statkraft) was given a monopoly to build and operate the interregional/national electricity grid in Norway. Norsk Hydro also had to guarantee 600 jobs in the community and to fund part of the activities needed to bring in new firms and to create new jobs. The common feeling among ordinary people in the valley was that the only alternatives were to prolong the existing structures or to establish new projects in accordance with the technological and institutional framework developed through the generations. This was the only way the knowledge capital and the skills of the male population could be used.

As Table 10.1 shows, the results of the restructuring process were mixed. Because of their age, many Norsk Hydro employees were given early retirement. Other workers, especially the engineers and the highly skilled, were able to move to new jobs and some accepted jobs with Norsk Hydo in other locations. For the rest, work was provided through the politically prolonged life of the factory and in new local but externally controlled firms. And, those alternative local jobs have partly been created as a result of incentives offered by the public sector and Norsk Hydro.

The most striking feature of the restructuring that has occurred is the almost complete absence of local entrepreneurship. Only five, small, locally controlled, manufacturing firms have been set up in thirty years of restructuring, and two of these were relocated firms that were taken over by local employees after they went bankrupt. The first genuine local entrepreneur emerged in 1985, twenty-two years after the restructuring started. His experience and knowledge came from working in some of the externally owned production plants in the area. The other locally based entrepreneurs set up businesses in 1988 and 1992. They all belonged to a generation that had grown up after the down-sizing of Norsk Hydro, and had no experience from working in that project. They had not been socialised into a system of instrumental factory work, union membership and social democratic ideology. But still 90% of manufacturing jobs in Rjukan are externally controlled, and there is still a long way to go before the town ceases to be a vulnerable branch-plant location.

Looking back at the branch plants that have been established in the valley in the past 35 years, the turbulence has been high. Many have come and many have moved further on to cheaper locations in places like Portugal or Malaysia. A majority of these new jobs created have not been for former Norsk Hydro employees, but for the under-employed female population in the textile or footwear industries. A few, more stable, plants have been established in energy intensive production, and these have provided openings for former Norsk Hydro employees. New plants have also moved in working in the mechanical and plastic industries. The jobs created in these plants have been mostly for male workers, but principally for the young, straight from school. Former Norsk Hydro employees have been described as unstable, not willing to learn a new trade, and unable to perform the functions and work processes of a small factory. They preferred the well-known routines, the strongly unionised collective and the distance between owners/managers and the shop floor in large plants – the ordered world they understood well. They returned to that world as soon as opportunities arose.

Conclusion

This study of the restructuring of a company town shows quite clearly that knowledge built up over generations was of no commercial value when the dominant project closes down and a different knowledge bas is asked for. It was the women of the society, with no established working habits, who first capitalised on the influx of new opportunities. Next, the young were recruited to the new plants. Both groups represented unskilled labour with no previous experience in manufacturing production. The migrating firms did not want skilled workers, only a cheap and stable labour force for standardised, repetitive manual work. Most new firms did not need the highly skilled and well paid labourer that had formerly worked for Norsk Hydro. The skills they had were very specialised, and the industry they had supported had created a specific social structure that made it difficult for the workers to retrain and unlearn practices, habits and attitudes to work. When the rules of the game changed, the social web, the former structuration of the society and the continuing ideology trapped individuals in an understanding of social relations that was no longer of any value.

The most serious barriers to change in the town were the failure to accept the need for private entrepreneurship and innovation as a part of capitalism, and the failure to understand how business is done in a competitive world. No entrepreneurial role models existed in the society, and very few had any experience of small business operations, or knowledge of markets, customers and meeting demand. A new generation was, and is still needed to produce agents capable of developing new practices and new knowledge in alternative settings. Rjukan is now slowly transforming. The aspirations of the younger generation are different and the strongly conformist social system of former times has dissolved (Henriks, 1992). Young people of the new generation aspire to education and jobs not available locally and as a consequence migrate to other and more prosperous places in the country. The present dominant economic and social structure still seems to reproduce the old industrial structure in conflict with the aspirations of the new generation.

The case study reported in this chapter shows that the systems of norms, established conventions and dominant institutions in a place, hinder dynamic restructuring and the adoption of new ways of work and interaction. Local social systems with no capacity to innovate and generate new and alternative lines of knowledge are much more vulnerable to change than a dominant project, such as Norsk Hydro, that creates and moulds them. A firm has the opportunity to move on to another site and still be in the same line of business, following the same technological path. Norsk Hydro did this when it built two new nitrogen and fertiliser plants on the coast, took over major national fertiliser companies in many other countries, and restructured its entire business on a European scale. It is now the largest producer of nitrogen fertiliser in the world and is in good health. Rjukan, as one of the firm's first plant locations, is still alive, but not in good health. The place is still striving to overcome the traits embedded in its society after 80 years of conformity. Time and space are deeply implanted in the social life of this community, as in most societies.

For economic geography, this case study demonstrates that the acquisition of new institutions and economic incentives is only a small part of the work involved in restructuring a place-related society. Of equal importance is the social system of the place and how its reaction to restructuring process. More attention needs to be paid to these aspects of local economic performance: not just the 'sunny side' of innovation and successful entrepreneurship, but also the 'dark side' of factors hindering innovative social action, learning and access to knowledge in place-specific societies.

References

Allen, J. and Massey, D. (eds) (1988), *The Economy in Question*, Sage, London.

Antonelli, C. (1995), *The Economics of Localized Technological Change and Industrial Dynamics*, Kluwer Academic Press, Dordrecht.

Bagguley, P. *et al.* (1990), *Restructuring: Place Class and Gender,* Sage, London.

Block, F. (1990), *Postindusrial Possibilities. A Critique of Economic Discourse,* University of California Press, Berkeley.

Braczyk, H-J., Cooke, P. and Heidenreich, M. (eds) (1998), *Regional Innovation Systems*, UCL Press, London and Bristol PA.

Cooke. P. (ed.) (1990), *Localities,* Hutchinson, London.

Dosi, G., Pavitt, K. and Soete, L. (1990), *The Economics of Technical Change and International Trade*, Harvester Wheatsheaf, London.

Gertler, M.S. (1995), *Manufacturing Culture: Regional and National Systems of Regulation*, Unpublished seminar paper presented at the University of British Columbia, August 23-25.

Gertler, M.S. (1997), 'The Invention of Regional Culture', in R. Lee and J. Wills (eds), *Geographies of Economies*, Arnold, London.

Giddens, A. (1979), *Central Problems in Social Theory. Action, Structure and Contradiction in Social Analysis*, MacMillan, London.

Giddens, A. (1981), *A Contemporary Critique of Historical Materialism. Vol.1: Power, Property, and the State*, MacMillan, London.

Giddens, A. (1984), *The Constitution of Society*, Polity Press, Cambridge.

Granovetter, M. and Swedberg, R. (1992), *The Sociology of Economic Life*, Westview Press, Boulder.

Gregory, D. (1989), 'Presences and Absences: Time-Space Relations and Structuration Theory', in A. Held and J.B. Thompson (eds), *Social Theory of Modern Societies: Anthony Giddens and His Critics*, Cambridge University Press, Cambridge.

Gregory, D. (1994), 'Social Theory and Human Geography', in D. Gregory, R. Martin and G. Smith (eds), *Human Geography. Society, Space and Social Science*, MacMillan, London.

Harvey, D. (1982), *The Limits to Capital*, Blackwell, Oxford.

Henriks, C. (1992), *Hvor blir det av Rjukan-ungdommen*? (Where Did Rjukan's Young Generation Go?), Hovedoppgave i geografi, Institutt for geografi, UiB.

Hodgon, G.M. (1988), *Economics and Institutions: A Manifesto for a Modern Institutional Economics*, Polity Press, Cambridge.

Isaksen, A. and Spilling, O. (1996), *Regional utvikling og smaa bedrifter* (Regional Development and Small Firms), Hoyskoleforlaget, Kristiansand.

Lundvall, B. (ed.) (1992), *National Systems of Innovation. Towards a Theory of Innovation and Interactive Learning*, Pinter Publishers, London.

Maskell, P., Eskilinen, H., Hannibalsson, I., Malmberg, A. and Vatne, E. (1998), *Competitiveness, Localized Learning and Regional Development. Specialization and Prosperity in Small Open Economies*, Routledge, London.

Nelson, R.R. and Winter, S.G. (1984), *An Evolutionary Theory of Economic Change*, The Belknap Press of Harvard University Press, Cambridge, MA.

Pred, A. (1985), 'The Social Becomes the Spatial, the Spatial Becomes the Social: Enclosures, Social Change and the Becoming of Places in Skaane', in D. Gregory and J. Urry (eds), *Social Relations and Spatial Structures*, MacMillan, Basingstoke.

Storper, M. (1997), The Regional World: Territorial Development in a Global Economy, Guildford, New York.

Vatne, E. (1981), 'Teknologisk nyskaping og regional endring - en historisk analyse' (Technological Innovation and Regional Change), in P. Friis and P. Maskell (eds), *Teknologi - og regional utdvikling - en nordisk antologi* (Technology and Regional Development: A Nordic Anthology), Kritisk samfundsgeografi - bind 1, Roskilde Universitetsforlag, Roskilde.

11 Ethnic Minorities' Strategies for Market Formation: The Israeli Arab Case

Michael Sofer, Itzhak Benenson and Izhak Schnell

Introduction

Inter-firm linkages and networks are often treated as indicators of business growth. Intensive linkages may generate economic multipliers along with other economic benefits to the local and the regional economy (Scott, 1991; Felsenstein, 1992; Staber and Schaefer, 1996). A key issue in network studies is the degree to which firms are embedded in various markets through their relationships with competitors, suppliers, regional and national business organisations, and public decision making forums (Best, 1990; Harrison, 1992; Markusen, 1994; Lakshmanan and Okumura, 1995). In this context, particular attention is devoted to barriers that ethnic entrepreneurs are forced to overcome in developing business linkages and networks (Aldrich and Waldinger, 1990; Barrett *et al.*, 1996). Ethnic entrepreneurs may find themselves trapped within ethnic enclaves or they have to overcome barriers that stem from their ethnic origin and/or location in the national, regional and urban periphery. Taking into consideration the key role of entrepreneurship for minorities' socio-economic mobility, their success in developing wide business linkages and networks is crucial for their integration into the economy as well as society.

Previous research on Arab entrepreneurship in Israel has shown that up until the 1980s Arab entrepreneurs were restricted mainly to intra-ethnic and local networks. Since then, they have exhibited increasingly high motivation and determination to exploit any slight chance to enter new markets (Schnell *et al.*, 1995). In investigating issues of purchasing and information networks, we have discovered that more than three-quarters of

Arab entrepreneurs continue to develop linkages with one or two types of destinations (Schnell *et al.*, 1999). In comparison, sale linkages seem to be more complex. Arab plants sell to a variety of markets ranging from limited local to more sophisticated metropolitan markets (Schnell *et al.*, 1999).

This chapter has three complementary goals: to analyse the way in which individuals' plants participate in the market; to discover major routes of expansion and contraction in markets; and to investigate the viability of the barriers of ethnicity and peripherality. We try to investigate the set of forces that shapes the structure of people's opportunities by comparing different entrepreneurs' choices. More particularly, we compare the order of entrepreneurs' expansion into markets using four independent variables: religious community, generation, education and industrial branch. We hypothesise that entrepreneurs of different religious origins, generation, education or industry within the Arab community in Israel, may have different sets of opportunities. The Israeli Arab case can be treated as an example of the impact of ethnicity and peripherality on the economic growth of marginal groups.

The first section of this chapter deals with the development of Arab industry in Israel and the structural setting of this development. The second offers a conceptual framework for understanding market expansion, followed by a discussion of the research methods employed in the study. The fourth section analyses the participation of individual plants in the different markets of the Israeli Arab space-economy. We conclude with a discussion of the selective impact of the two sets of structural barriers – ethnicity and peripherality – on the chances of market expansion for ethnic entrepreneurs.

Arab Industrialisation in Israel

The Arab population of the State of Israel (not including the Palestinian Arabs living in the Gaza Strip and West Bank) constitutes a minority of about 1.2 million or 20% of the total national population of some 6 million. They are concentrated in three major peripherally located regions. About 130 towns and villages are populated solely by Arabs in Israel, each with a population of less than 30,000. The only exception is Nazareth, which has over 60,000 inhabitants. Figure 11.1 outlines two of these concentrations in central and northern Israel.

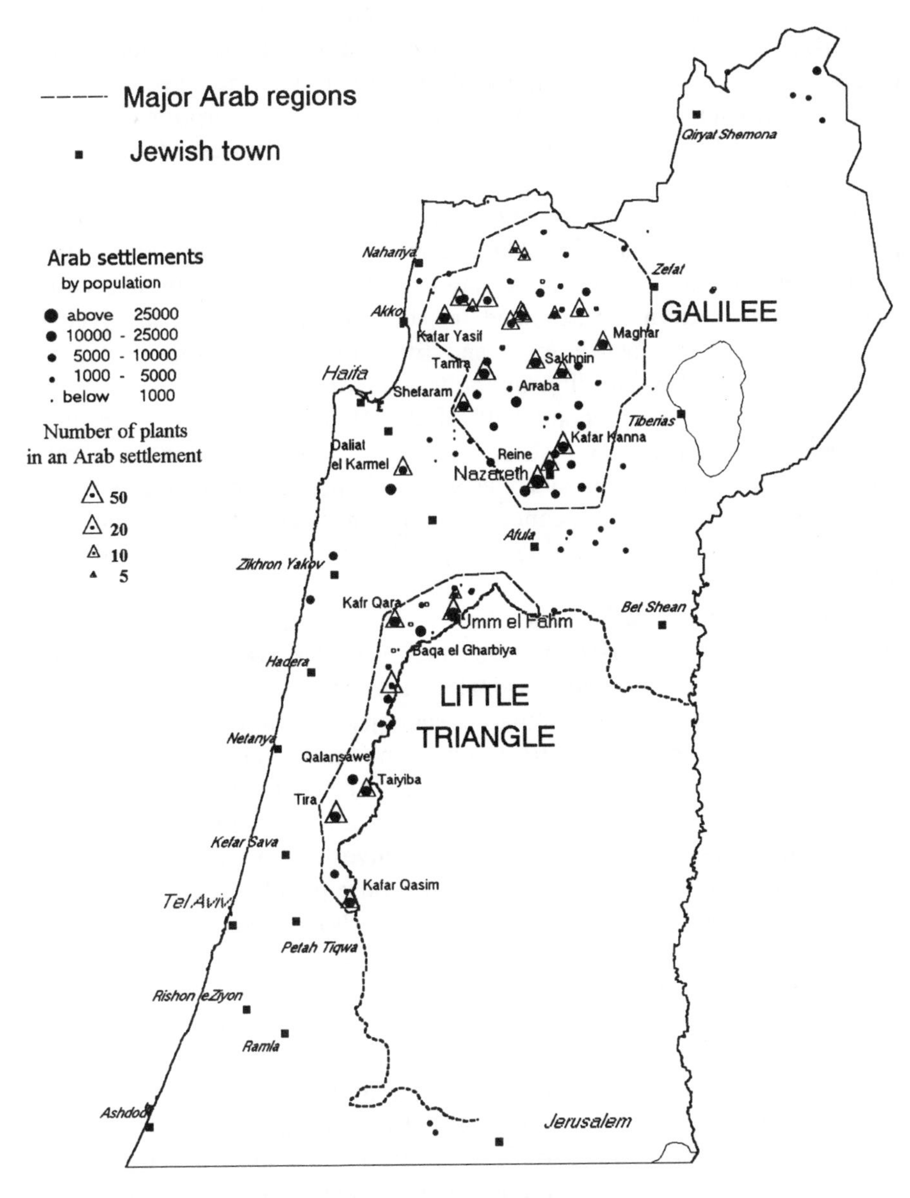

Figure 11.1 **The distribution of Arab owned enterprises by major settlements**

Source: field work

From Israel's independence in 1948 and until the mid 1990s, Arab industrialisation went through two periods of restructuring. Soon after independence the Israeli economy shifted to a state-managed capitalist system in which national capital was channelled to the absorption of Jewish immigrants and refugees from the Holocaust and Arab countries and to secure housing and employment in Jewish new towns. The Arab economy, mainly rural, was almost completely excluded from national development projects. Production facilities that did exist were characterised by their small size, low levels of capitalisation and intensive labour utilisation drawn primarily from the extended family. Most plants were involved in traditional branches such as coal and lime production, stone quarrying, olive oil pressing, flour milling and pottery making.

Only during the second half of the 1950s, after the government released some martial law restrictions, did a large proportion of labour-force begin to commute to work-places outside its place of residence. Arab labour was engaged in secondary labour market jobs, especially in industries such as construction, personal services and other low-pay occupations.

During this period, the first restructuring stage took place. Opportunities for industrial entrepreneurship in the Arab sector remained severely limited for several reasons: a lack of capital and professional skills; a lack of experience in a market economy; and most particularly, the absence of basic physical infrastructure in Arab communities (Haider, 1993). Risk minimisation strategies induced Arab entrepreneurs to imitate enterprises that had proven successful in the same economic environment in which they operated, and to refrain from investing in areas where others had failed (Czamanski and Taylor, 1986). Consequently, three industries – food, wood, and construction materials – showed significant growth (Khamaisi, 1984). The combination of an increased demand for housing, reliance on newly emerging local markets and capital accumulated by Arab manual workers, gave rise to the establishment of construction firms producing for the local housing market (Schnell, 1994).

A new period of development started after the 1967 Six-Day War, when the national economy restructured into a corporate dominated system (Hasson, 1981). During this period, large corporations located in the central region of the country developed production plants in the peripheral areas thereby creating jobs for Arab workers in neighbouring towns and villages (Gradus *et al.*, 1993). By the late 1970s most Arab settlements had acquired basic infrastructure. Thousands of Arab workers, through employment in Jewish-owned firms, had gained industrial and professional experience, and there was a rise in Arab educational levels. Under these conditions a new generation of entrepreneurs emerged and women began to join the labour force. Selective government policies, aimed at increasing Jewish population and economic activity, continued to ignore Arab industry, leaving the Arab periphery almost totally dependent either on growth from below or on sub-contracting for Jewish owned corporations.

The most significant change during this period was the introduction of textile and clothing plants. Managed by Arab sub-contractors, clothing plants employed the large Arab female labour force characterised by low geographic mobility. With these plants rapidly becoming large industrial employers, internal dependency was also created, with local labour largely dependent on these plants for employment.

The expansion of indigenous Arab industries during this period was based on regional markets and providing sub-contracting to Jewish corporations in the textile and clothing industries. In 1983, Khamaisi (1984) counted 415 plants in Arab settlements. Only one third of them employed more than 10 workers and most of these were sub-contractors to Jewish-owned plants. More than 40% were allied to the construction industry, about one third were in textile and clothing and most of the rest in food production. Arab plants (with at least three workers each) employed that year about 3,000 workers, with 70% of them women.

By 1992, over 900 firms existed in Arab settlements with about 13,000 employees (Atrash, 1993; Schnell *et al.*, 1995). Nevertheless, Arab-owned enterprises employing five workers or more represented only 5.5% of all such plants in Israel, and they employed only 3.2% of the industrial labour-force (Central Bureau of Statistics, 1995). The average size of Arab-owned plants grew from about seven workers in 1983 (Khamaisi, 1984), to around fifteen in 1992 (Schnell *et al.*, 1995). Almost all were owned jointly by family members, and only about one quarter were formally organised as corporations. Textiles became the leading industry with the largest number of plants and workers as well as sales per plant.

The ethnic structure of the Arab population is of considerable importance in understanding their industrialisation. The Christian Arabs and Druze communities each accounting for about 10% of the Israeli Arab population, have certain advantages that allow them to exploit economic opportunities and break out of traditional patterns of work and employment. The Christian community has a higher level of education, smaller families and thus more opportunity for saving and investment and training than the general Arab population. This allowed them easier entry into entrepreneurship (Schnell *et al.*, 1995).

Members of the Druze community are closer to the Moslems in terms of demographic behaviour and education. However, they have a certain preferential status regarding government aid. The Israeli government has assisted Druze villages and has provided basic physical infrastructure for industry. In addition, Druze youngsters serve in the Israeli army, exposing themselves to Jewish culture, and open social networks to Jewish society. As a result, the index of average entrepreneurship for Muslims lags somewhat behind the index for the Christians and Druze (Schnell *et al.*, 1995).

Conceptual Framework

The specific form of developing markets by Israeli Arabs may be explained by theories of ethnic entrepreneurship, in particular the ethnic enclave theory. This argues that ethnic economies are self-contained entities that generate a variety of inputs and outputs within the ethnic community. Profits and earnings remain internal, producing multiplier effects as firms purchase mainly from intra-ethnic suppliers (Waldinger, 1993). However, there are some debates concerning the advantages of enclave economies for ethnic communities. Some studies confirm the emergence of enclave economies and their benefits to ethnic communities (Portes and Jensen, 1989; Jensen and Portes, 1992; Zhou and Logan, 1989). Other studies question the benefits of limited and deprived enclave economies (Nee and Sanders, 1987; Sanders and Nee, 1992).

Studies of ethnic businesses rarely analyse the structural context in which ethnic industries develop (Barrett *et al.*, 1996). It is argued, however, that entrepreneurship is always embedded in socio-cultural structures (Grabher, 1993; Kloosterman *et al.*, 1999; Oinas, 1999; Taylor, 1999). Aldrich and Waldinger (1990) set the foundation for a theoretical framework which incorporates structural aspects in theorising ethnic economies. They stress three major factors. First, the importance of the opportunity structure as it is formed in a historical context and by political decisions. Forms of capital accumulation are an example of one key factor that may structure selective opportunities on both sides of the ethnic divide. Second, access to enterprise ownership may channel ethnic groups either to low rewarding enterprises, or to a segregated ethnic economy according to the degree of inter-ethnic competition. Third, ethnic group characteristics, such as their orientation toward entrepreneurship, may influence their forms of integration into the general economy.

Exposure to racism and differences in entrepreneurial cultures seem to affect ethnic entrepreneurs in their choice of markets. Ethnic entrepreneurs tend to operate in different business cultures. Therefore, members of marginal groups who seek to break out of their enclave economy are frequently forced to adapt to the dominant business culture. This may result in mounting difficulties, which must be overcome in order to establish business connections among bodies on the two sides of the ethnic divide (Camagni, 1991; Ratti, 1992). There is evidence that in many cases members of a minority group are marginalised and even excluded from certain markets as a result of racism (Miles, 1989). In response, ethnic minorities may enhance a spirit of mutual trust, co-operation and collective self-help, as a facilitator of ethnic enterprise development (Light, 1984; Light and Bonacich, 1988). Entrepreneurs may then use this sense of ethnic solidarity to enlist production factors, such as labour, capital, management, and markets, from more accessible intra-ethnic sources (Aldrich and Waldinger, 1990).

With respect to ethnic barriers, we can identify four marketing categories according to the size of the market hinterland and to ethnic boundaries (Jones *et al.*, 1992). First, there are local intra-ethnic markets, which follow the logic of the 'ethnic enclave theory' (Semyonov and Lewin-Epstein, 1993). Second, there are local inter-ethnic markets, which follow the 'middle man model' (Ward, 1985; Jones *et al.,* 1993). Third, there are non-local intra-ethnic markets, which may develop when ethnic barriers remain durable while economic opportunities are growing within ethnic markets elsewhere. In this case, an autonomous ethnic economy may emerge, including possibly high-order firms and wholesalers supplying the autonomous economy. Fourth, there are non-ethnic non-local markets, which may emerge when potential markets are unbounded (Jones *et al.*, 1993).

Location in peripheral regions may place an added economic burden on ethnic entrepreneurs. Distance in social and economic terms from large markets and complementary economic activities, upon which the functioning of the plant depends, obviously reduces opportunities. Moreover, remoteness from the economies of scale, from information on market conditions, and from the chance to evaluate the potential of various opportunities may block growth even further (Pred, 1977; Van Geenhuizen and Nijkamp, 1995). Peripheral industries in the early stages of growth tend to suffer from a lack of risk reducing mechanisms. As a consequence, entrepreneurs find it almost impossible to break the local barriers of economic networks. In many cases, further development may only be stimulated by public intervention (Grossman, 1984). Given the lack of risk reduction institutions, entrepreneurs may adopt a mimicking strategy which leads to the opening of a large number of similar small enterprises that are content to make do with the low profits afforded by limited local markets (Czamanski and Taylor, 1986).

Once such entrepreneurs gain some experience and they succeed in breaking local barriers, their routes of development may be channelled by large corporations. Industries in the peripheral regions tend to expand into the marketing chains in a clear spatial division of functions. The first type are the large plants, characterised by the labour-intensive mass production of inexpensive standardised products for national markets. Here, most sales are directed to the large markets located primarily in large metropolitan core areas. The second type includes small businesses existing alongside larger enterprises. These plants mainly manufacture intermediate products, which constitute inputs for the larger enterprises in their regions, or serve the demand created by local end-users. Most products of these plants are thus sold within the peripheral region itself (Felsenstein, 1992). Under such conditions, the factors of ethnicity and peripherality may place cumulative burdens on the processes of industrial growth and market expansion (Schnell *et al.*, 1995). Schnell, Benenson and Sofer (1999) show that both factors constitute an economic space that may be theoretically divided into sixteen possible sub-regions having markets on four different levels: local, regional, national and international, each of them sub-divided into either core or peripheral markets and intra- and inter-ethnic markets.

In applying this general process to the case of Arab industry in Israel, several patterns of business development emerge. Initially, Israeli Arabs relied on mimicking behaviour strategies which channelled them into local intra-ethnic markets. Only since the 1980s have an increasing number of Arab entrepreneurs attempted to break into new, more demanding, and riskier markets. At present, Israeli Arab entrepreneurs have expanded into the following five sub-markets (Schnell *et al.*, 1999):

- intra-settlement markets making up 25% of total sales;
- intra-ethnic markets, within the home region, counting for 26% of total sales;
- neighbouring Jewish markets making up about 19% of total sales;
- inter-regional, intra-ethnic markets counting for 3% of total sales; and
- core Jewish markets to which 27% of total sales are directed.

Most Arab plants expand into new markets using two major alternative strategies. Firstly, they expand from close markets to more distant ones, starting with local and home regional markets, then expanding into ethnic markets and only as a last choice expanding into distant Arab markets. Secondly, they adapt a strong orientation to metropolitan markets and expand from them to the other markets.

Research Method

Two major research questions inform our analysis of ethnic minorities' market expansion strategies. First, is it possible to understand the disparity in market expansion and the order of expansion or contraction of markets using variables such as the inter-generational gap, level of education and ethnic-religious groups? Based on earlier work we assume that until the 1970s Israeli Arab industry was solely based on local sales linkages and on intra-settlement markets. If so, any later expansion into additional markets may indicate a change in the pattern of sales linkages. Second, to what extent did entrepreneurs of different groups manage to break ethnic and peripheral barriers, thus enabling them to expand into inter-ethnic and inter-regional markets, and increasingly to exploit opportunities in the Israeli economy?

Sample

A sample of Arab settlements and enterprises was surveyed in 1992 for an extensive study of Arab industry and industrial entrepreneurship in Israel (Schnell *et al.*, 1995). An Arab enterprise was defined as any plant that is Arab-owned, acting as a production unit and employing at least three workers. Of the 900 Arab-owned enterprises operating in 1992, the managers and owners of 514 plants (57% of the total) were interviewed. These respondents were located in 35 of the 61 Arab town and villages that had industrial plants. This included 80% of the settlements in the mountainous Galilee and the Little Triangle regions (Figure 11.1). In most of the 35 settlements, all plants were surveyed. In the largest settlements (Nazareth, Shefaram, Taiyibe, and Um el Fahm) a random sample of plants was used. In Nazareth the survey covered about 25% of the plants, and about 70% in the other three settlements. For the aggregate analysis, data for these settlements were weighted by a factor determined by the size of the sample. The survey was comprehensive, examining among other questions, the origin of inputs, destination of outputs, and related characteristics such as sector and types of product (Schnell *et al.*, 1995).

Data Analysis

The data were analysed in two stages. First, groups of plants that participated in common configurations of the five sub-markets listed previously, were analysed, and the relative impact of each of the two structural forces of ethnicity and space economy on the emerging pattern was examined. Second, differences among the three generations of plants, religious origin, and educational groups of their owners were tested.

The investigation uses Partial Ordinal Scalogram Analysis - POSA (Guttman, 1965). POSA analyses the plant profile which describes its pattern of participation in the markets. The profile is a Boolean vector, with the number of components equal to the number of sub-markets discerned, and a given component of a profile equals one if the plant participates in a certain market and zero otherwise. We denote participation in a given sub-market using symbols. Thus, instead of using a binary rotation (1.0) to indicate sales to a metropolitan core region, we use M,0 where 'M' indicates sales to a metropolitan market and '0' indicates sales elsewhere.

POSA begins with a graph of market inclusions. The nodes of this graph correspond to profiles and are vertically arranged by levels, in accordance with the number of markets in which the plant participates. The lowest level consists of nodes corresponding to profiles of one market only, the next level those with two markets, and so on. The edges of a graph connect nodes of adjacent levels only: two nodes are connected if the profile on the upper level includes all markets of the profile on the lower one. The aim of POSA is to rearrange the nodes at each level in order to obtain a *planar* presentation of a graph, titled a scalogram. There are no available techniques for assessing the 'most planar' representation of a given graph. For the relatively simple graphs, considered below, this is done on the basis of trial and error.

Usually a planar scalogram can be constructed for a graph with a low number of edges only, but this is not the case for the all-market inclusions revealed for Arab industry. In this latter situation we consider the most planar presentation of a graph of inclusions, that is, we try to arrange the nodes in a way that minimises the number of intersecting edges. The planar or near planar scalogram displays the expansion or withdrawal of individual plants from markets (Lingoes, 1973), thus allowing an understanding of the evolution of the individual plants within the set of markets. The conclusions concerning markets development are based on the assumption that decisions are made in a pre-structured milieu, and, therefore, are repetitive. The POSA unravels common patterns of market expansion and contraction, as structured by past decisions.

In general, our interest is in the disparity between frequent and less frequent market profiles. We begin with constructing scalograms for the more frequent profiles, and subsequently extend the analysis by adding less frequent ones. This procedure allows the investigation of the major and marginal routes of plants' expansion into new markets or withdrawal from them. We are then able to compare across scalograms constructed for entrepreneurs who belong to different generations, ethnic groups and education levels.

According to our earlier study (Schnell *et al.*, 1999), the structure of developing markets for textile and clothing plants differs greatly from the plants belonging to other industries. Consequently, we always begin our analysis with all industries taken together (aside from textile and clothing), and test at a later stage whether the inclusion of the textile and clothing industry influences the results.

Results

Plants Forms of Participation in Markets

The number of all possible participation profiles for plants with five potential markets equals $2^5 - 1 = 31$. In practice, 22 forms of participation profiles are identified as shown in Table 11.1.

When comparing the number of plants for each profile within any specific level (Table 11.1), we immediately recognise that at each level there are either one (levels 1, 2, 4, 5) or two (level 3) modal profiles which contain most of the plants. At levels 1, 2, 4 and 5 the most frequent profile - the first for each level - contains at least 65% of the plants belonging to that level. At level 3, frequencies of two profiles – 0J0HS and M00HS - are high and together contain about 85% of the plants at that level. The scalogram of the six major profiles is presented in Figure 11.2, thus identifying the leading forms of market structures.

Table 11.1 The distribution of market profiles (including textiles and clothing in brackets)

Number of Markets (level)	Number of plants by profiles	Total
5	MJAHS – 9	9 (9)
4	MJ0HS – 43 (45), M0AHS - 7, MJA0S – 3, 0JAHS – 3	56 (58)
3	0J0HS – 76 (77), M00HS – 53(58), MJ00S – 17, 00AHS – 12, M0A0S – 7, 0JA0S - 2, MJA00 – 1	168 (174)
2	000HS – 112 (114), M000S – 23 (24), 0J00S - 18, MJ000 – 9(13), M0A00 – 3, 00A0S – 2	167 (174)
1	0000S – 32, M0000 – 5 (52) , 0J000 – 1 (4), 00A00 – 1 (1)	39 (89)

Profile components denote participation in the following method:
M = Metropolitan Jewish Market
J = Jewish neighbouring Market
A = Distant Arab Market
H = Home Region (Arab) Market
S = Local Village Market (Arab)

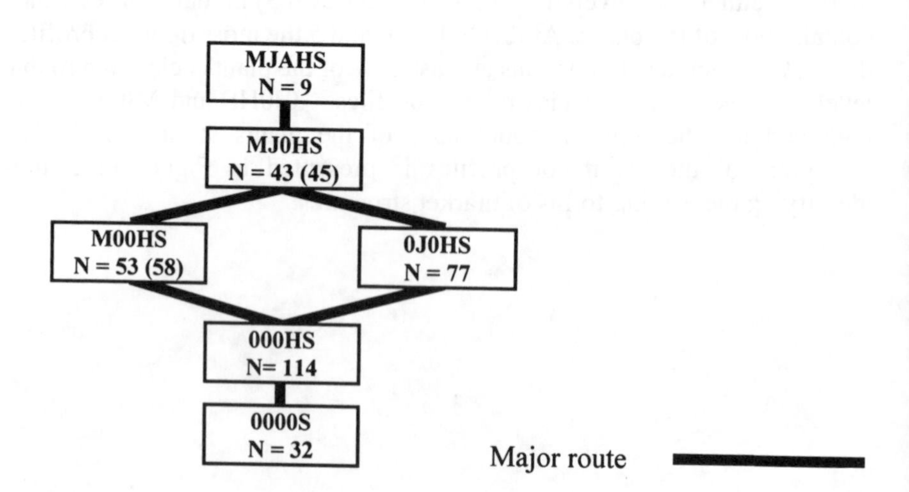

Figure 11.2 Scalograms of major market profiles

Main Forms of Market Expansion

Further analysis of the most frequent profiles is presented in Table 11.2 and Figure 11.2. Both show that at levels 1 - 4 about 70% of non-textiles plants are characterised by the five most frequent profiles only. The most frequent one 000HS (n = 112, 25.6%) consists of plants that sell within the Arab ethnic milieu, that is, within the settlement and the home regional market. The second, third and fourth most frequent profiles - 0J0HS, M00HS and MJ0HS (total n = 172, 39.3%) - differ from the first by adding Jewish markets – either neighbouring or metropolitan, or both. This large group breaks ethnic barriers either within their home region or at the national level.

Table 11.2 Characteristic of the major participation profiles (in brackets - including textiles and clothing)

Level, type of markets	Profile	Number of plants	Level Percentage*	Overall Percentage*
5 - All markets	MJAHS	9	100	2.1
4 - Inter-ethnic	MJ0HS	43 (45)	76.7 (77.6)	9.8 (8.9)
3 - Metro & Arab Home/Regional,	M00HS	53 (58)	31.4 (33.3)	12.1 (11.5)
- Local Jew & Arab Home/Regional	0J0HS	76 (77)	45.0 (44.2)	17.4 (15.2)
2 - Arab Home/Regional	000HS	112 (114)	66.7 (65.5)	25.6 (22.6)
1 - Arab Home	0000S	32	69.6	7.3
Total		325 (335)		74.2 (66.5)

Notes: * Percentage share of plants of a given profile, out of all plants excluding textiles plants
(In brackets - percentage share, including textile plants)

Three different forms of access into Jewish markets could be revealed by the analysis of the plants' characteristics and interviews with entrepreneurs. First, purchasing of inputs in the metropolitan core may encourage entrepreneurs to sell some of their products within their suppliers' vicinity. Comparisons of the purchasing modes of plants, which sell exclusively to Arab markets or to mixed markets, tends to confirm this hypothesis (Table 11.3).

Second, interviews with entrepreneurs belonging to the textile and clothing branch reveal that they started as sub-contractors to Jewish metropolitan-based industries and thus their dependence on Jewish markets is an inherent characteristic. Taking the textile and clothing industry into consideration enables one more high frequent profile, M0000, which consists of 47 textile plants and 5 plants belonging to other branches of industry (taken together, 10.3% of all plants). Combining all the surveyed plants, the proportion that belongs to the main profiles and breaks ethnic barrier is 45.8%.

Table 11.3 Market profile and purchasing rate from Jewish suppliers

Group of plants	Market profile	Purchases from Jews among all input purchases	Weighted mean
Plants that sell to Arab markets only	0000S	45%	52.2%
	000HS	55%	
	00AHS	45%	
Plants that sell to mixed markets	0J0HS	75%	75.3%
	M00HS	72%	
	MJ0HS	80%	

Interviews reveal a third form of access to Jewish markets which exists amongst entrepreneurs located in settlements closer to the metropolitan areas. This is characterised by sales linkages to occasional Jewish customers in trading linkages that take place in Arab towns and villages.

Secondary Forms of Market Expansion

By adding the second most frequent profiles to the scalogram of the major profiles, we are able to reveal alternative market strategies that are used by about 20% of Israeli Arab entrepreneurs, i.e. those using one or more marginal form of expansion when trying to access new markets (Figure 11.3). Some of these strategies may represent past and residual forms of market participation, while others may represent new forms of expansion into the market (Table 11.4).

Table 11.4 Secondary forms of participation in the market (textiles in brackets)

No. of markets	Type of markets	Profile	Number of plants	Level Percentage	Overall Percentage
4	Metro & All Arab markets	M0AHS	7	12.5	1.5
3	Metro & Local Jewish & Arab Home	MJ00S	17	10.1	3.9
	All Arab markets	00AHS	12	7.1	2.7
2	Metro & Arab Home	M000S	23 (24)	13.8	5.3 (4.8)
	Local Jewish & Arab Home	0J00S	18	10.8	4.1
	Metro & Local Jewish	MJ000	8 (13)	4.8	1.8 (2.6)
	Total		85 (91)		19.3 (19.6)

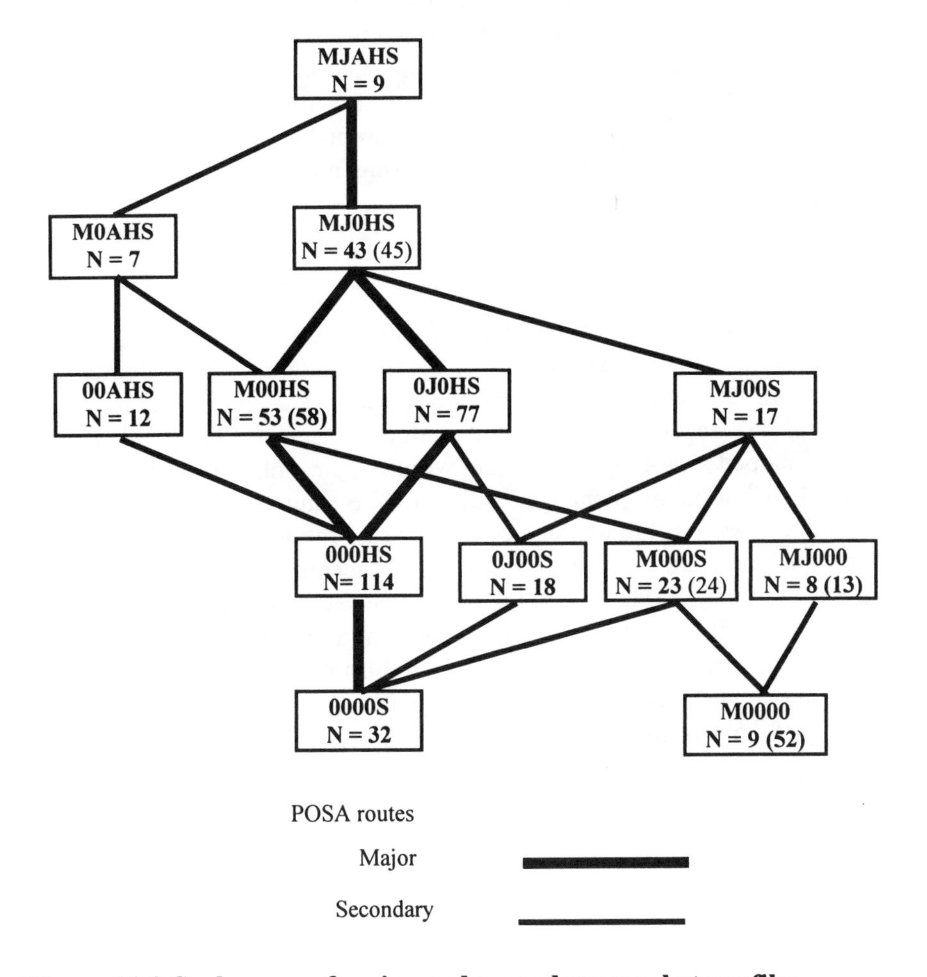

Figure 11.3 Scalogram of major and secondary market profiles

The sales patterns of plants involved in the secondary profiles (overall n = 85, 19.3%) concern access to both Arab and Jewish markets. Only one of these secondary profiles (00AHS, n = 12, 7.1%) represents entrepreneurs who have not expanded beyond the ethnic enclave. The two most frequent profiles are M000S and 0J00S which, besides local home markets, caters for either metropolitan cores or local Jewish markets. This means that combining major and secondary profiles results in 58.0% of plants selling to Jewish markets, and 54.6% when the textile and clothing industry is included.

Interviews with entrepreneurs belonging to this secondary group express a reluctance to take risks in their attempts to expand into more demanding markets free of kinship relations and mutual interests. One entrepreneur demonstrates this attitude by arguing that Jews refuse to buy from him because he is an Arab. When asked whether his products have the approval of the Israeli Standard Institute his answer was, "No! It costs too much money and people in the town know my products are of the best quality". This example emphasises entrepreneurs' perceptions of opportunities and constraints when exploring new opportunities outside traditional modes of action.

The Complete Scalogram

Eight of the existing 22 profiles contain four or less plants (including textile and clothing n = 19, 3.8%). These plants can be added to the scalogram of the main and secondary profiles without multiplying non-planar relations (Figure 11.4). According to this full scalogram, about 94% of all plants have developed markets other than within their home settlement. The overall distribution of plants according to the numbers of markets they accessed, shows that the tendencies revealed by the most frequent profiles remain unchanged, and most of them have concentrated their sales on two or three markets (Table 11.5). Of those who sell to two markets, 65% belong to the 000HS profile, which means intra-ethnic and intra-regional markets. Once an entrepreneur penetrates a third market, the tendency is to expand into Jewish markets either in the metropolitan core or in the Jewish periphery (87.5% of those selling to three markets). Thirteen percent of the plants expanded to four or five markets, thus presenting a relatively high degree of integration into the national economy.

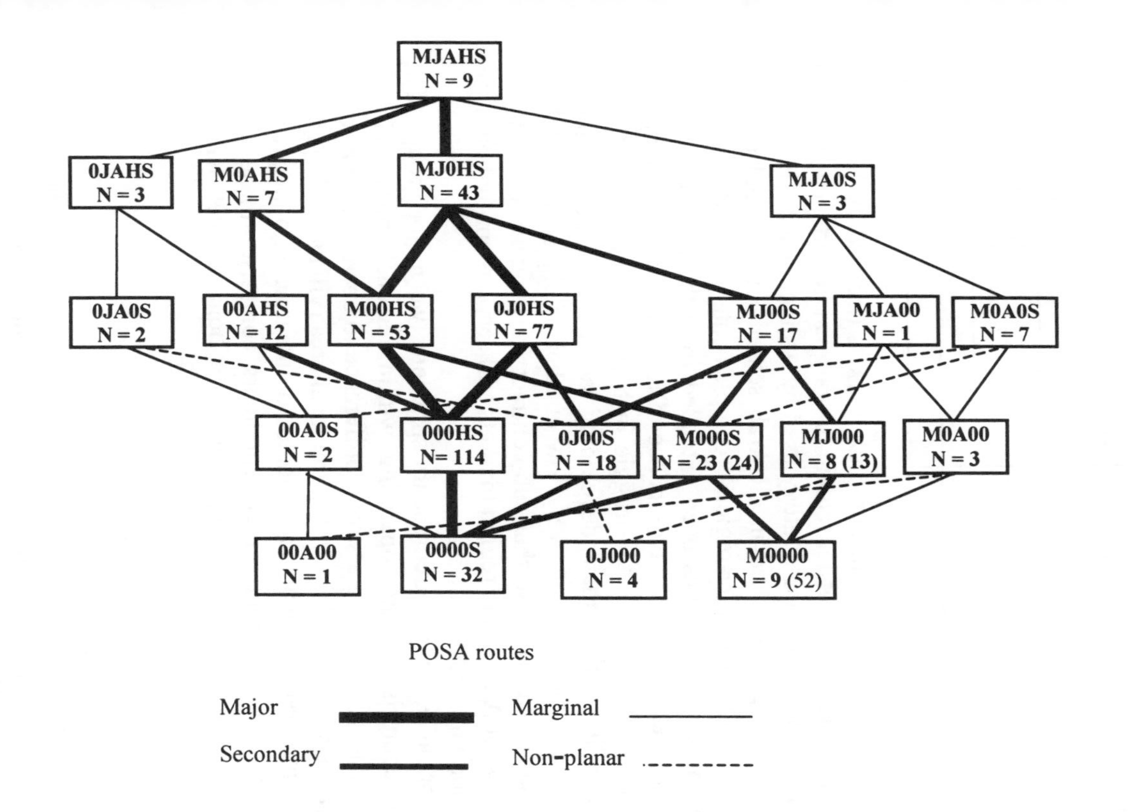

Figure 11.4 Complete scalogram of market profiles

Table 11.5 Number of plants by number of markets (in brackets – including textiles and clothing)

Number of markets	Number of plants	Percentage share of plants
5	9	2.1
4	56 (58)	12.6 (11.5)
3	168 (174)	38.4 (34.5)
2	167 (174)	38.1 (34.5)
1	39 (89)	8.8 (17.7)
Total	438 (504)	100.0

It is worth noting that distant Arab markets (marked by letter A for the third component of a profile) remain of marginal importance. This may be the result of the mimicking strategy used by Arab industry. The fact that similar plants were established in all Arab settlements blocked potential opportunities to develop inter-regional complementary relations. It seems that the mimicking model has been extended from the production to the marketing spheres through the emergence of regional home markets. Arab entrepreneurs seem to prefer to struggle in breaking barriers of ethnicity and peripherality rather than developing autonomous Arab markets. Nevertheless, the fact that relatively low number of plants rely solely on Jewish markets (except those that sub-contract for plants in metropolitan areas) shows that Arab home region markets still serve as spring boards for this expansion into wider markets.

To conclude, Arab entrepreneurs split into two major groups. The first group comprises plants operating within regional markets in the national periphery, exhibiting a strong preference to operate in intra-ethnic markets. Those who sell solely within the local area (village, town) may be viewed as entrepreneurs who are unable to integrate into the regional economy. The second group of entrepreneurs comprises those able to expand into Jewish markets once opportunities in home markets have been exploited.

Differences Among Groups of Entrepreneurs

Three entrepreneurial characteristics seem to be of importance when businesses choose markets: religious origin, age and education. None of these provides significant differences between the scalograms as a whole when all profiles are accounted for. However, stratifying by religious group and date of establishment yields some revealing results.

It can be suggested the routes of market expansion adopted by Moslem, Christian, and Druze communities may be different due to their different reactions to ethnic barriers. Moslems may rely on larger local and intra-ethnic markets, while Christian and Druze communities may suffer from disadvantages of being minorities within a minority. Druze may better integrate into the Jewish economy, due to the fact that they serve in the army and therefore gain more legitimacy by the Jewish society.

Partial Ordinal Scalogram Analysis for the three communities demonstrates considerable similarity in the forms of local market expansion they adopt. There are, nevertheless, differences in the forms they select for expansion beyond the settlement. More than 80% of the Moslem-owned plants expand their markets by adding the neighbouring settlements to progress from the 0000S market to 000HS market. By contrast, half of the Christian entrepreneurs choose the Jewish market as a second destination when expanding outside 0000S (Table 11.6, Figure 11.5). The number of Druze-owned plants is too low for comparison, though they tend to follow the Moslem pattern in this stage of market expansion.

It is interesting to investigate whether variables measuring the age (i.e. the generation of entrepreneurship) and education of entrepreneurs have an impact on firm's expansion into a third market, thus breaking barriers of peripherality and ethnicity. We hypothesise that less educated entrepreneurs are less qualified and have greater difficulties of integrating into a modern economy. In other words, plants founded during the last decade have more opportunities to integrate into the larger national economy than those established earlier.

Table 11.6 The distribution of second markets by communities (textiles and clothing data in brackets)

Religious Origin of Entrepreneur	Second market exists (number of plants having 000HS profile)	Jews as a second market (number of plants having M000S, 0J00S, MJ00S profiles)	Percentage of H as a second market
Moslems	85 (86)	29 (34)	74.5 (71.7)
Christians	17	16	51.5
Druze	10 (11)	5	66.7 (68.7)

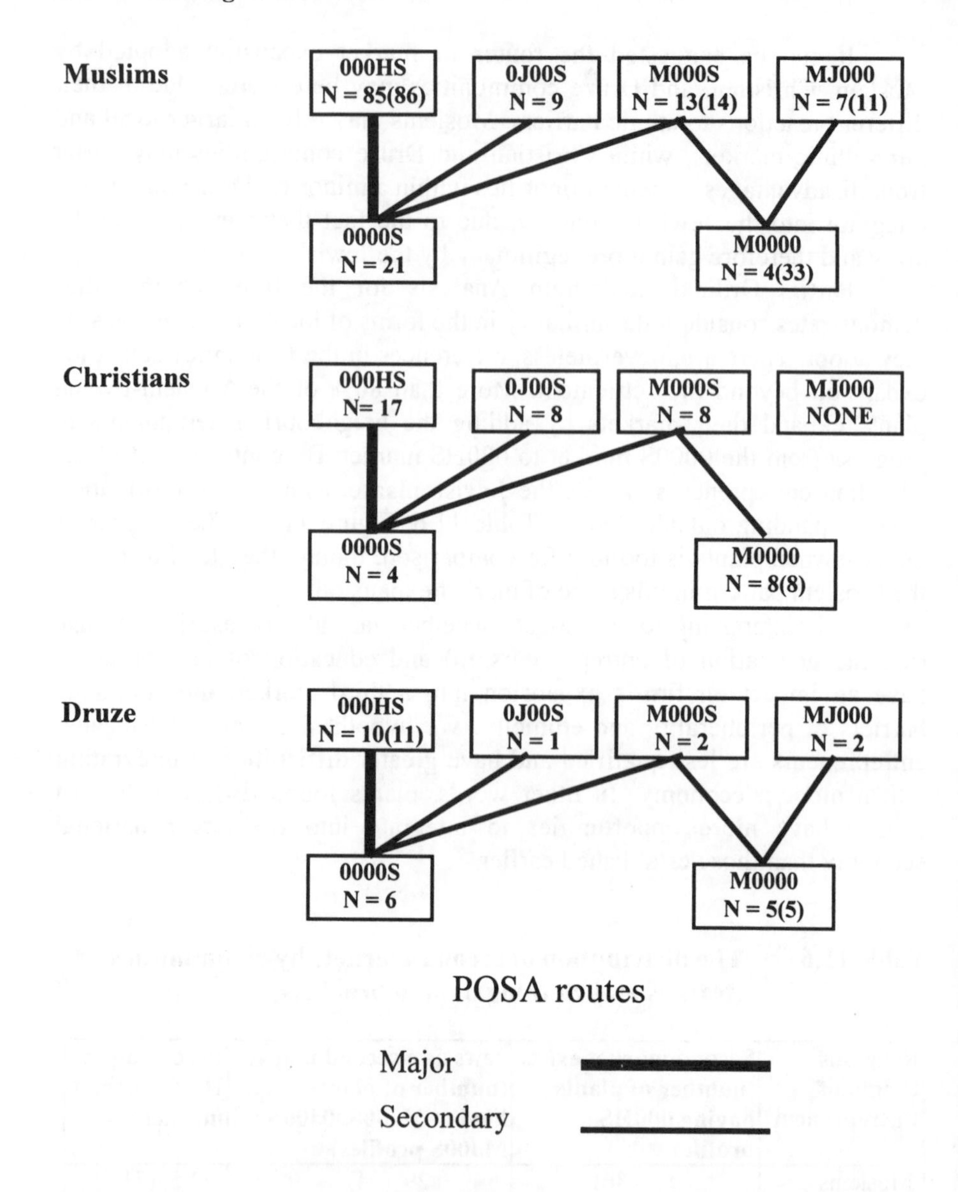

Figure 11.5 **Scalograms of expansion from local to metropolitan markets by religious origin**

Figure 11.6 shows a comparison of scalograms for three periods of plant openings, which serves as surrogate for different generations. The figure shows that different generations expand out of the local market using different entrepreneurial strategies. An innovative trend that started to appear in the 1990s is the tendency to expand from markets in the home settlement directly into neighbouring Jewish markets. The younger generation follows a pattern of expansion similar that of Christian entrepreneurs. Owners of recently established plants select the Jewish market as a second destination more frequently than plants established in earlier periods (Table 11.7, where for non-textile plants $\chi^2 = 4.61$, df = 2, $p \sim 0.1$). We hypothesise that this tendency characterises young and highly educated entrepreneurs, who face the low potential of local intra-ethnic markets, and who make deliberate efforts to break ethnic barriers.

Table 11.7 The distribution of second markets by period of establishment (textiles and clothing data in brackets)

Period of establishment	Second market exists (number of plants having 000HS profile)	Jews as a second market (numbers of plants having M000S, 0J00S, MJ00S profiles)	Percentage of H when A second market
Before 1976	20	5 (6)	80.0 (76.9)
1976 – 1985	57 (59)	21	73.1 (73.7)
After 1985	35	24 (28)	59.3 (55.5)

Conclusions

Arab industry in Israel has undergone significant restructuring. Traditionally, Arab entrepreneurs were almost completely excluded from the country's more privileged, large, institutionalised markets, being channelled either into their limited home region markets, occasional markets or secondary markets; some of which are in the vicinity of their suppliers in the national core. The majority of Arab entrepreneurs have made only limited efforts to expand into distant Arab markets, relying on home region markets as an anchor for further expansion into Jewish markets. Others have used secondary metropolitan markets as a springboard to expand into intra-regional Arab markets.

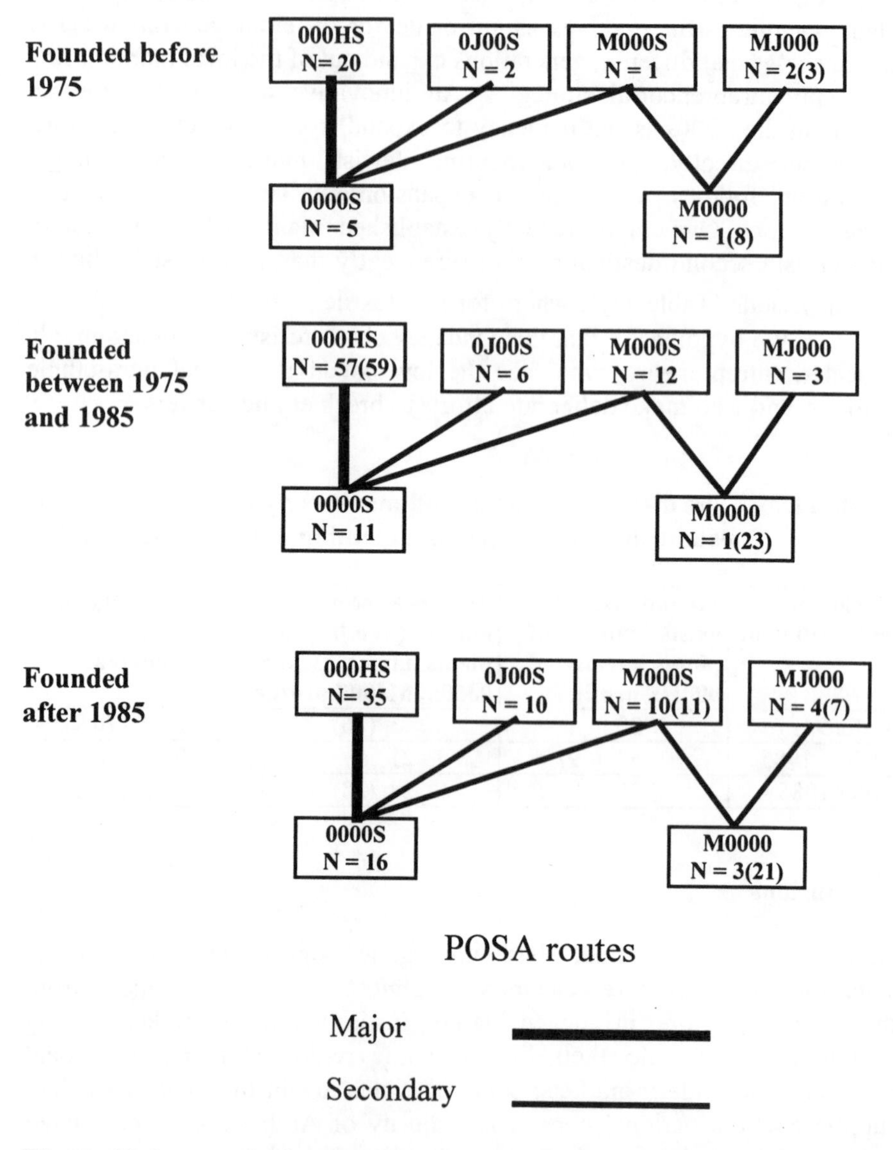

Figure 11.6 Scalograms of expansion from local and metropolitan markets by period of establishment

The analyses presented in this chapter demonstrate that ethnic minorities in Israel use two major routes to achieve market expansion; from the close to the distant, and from the metropolitan to the other markets. They represent two forms of marginal development. The dominant route is characteristic of entrepreneurship that grows from below (Falah, 1993). Only when entrepreneurs shift into a third type of market are the intra-regional and intra-ethnic barriers of the national periphery broken. The second route, expanding into ethnically mixed markets with emphasis on metropolitan markets, is executed under conditions of dependency, with Arab entrepreneurs sub-contracting for Jewish firms, or relying on occasional (Jewish) customers.

According to ethnic enclave theory, entrepreneurs can be expected to prefer intra-ethnic market outlets and to use them as an anchor for further expansion. The current case shows that the majority of ethnic minority entrepreneurs in Israel follow this line of development. However, small groups of entrepreneurs tend to search for alternative routes for market expansion. Investigation of these alternatives confirms our argument, presented in earlier research (Schnell *et al.,* 1995), that Arab entrepreneurs are highly motivated and willing to exploit any slight chances for market expansion. Two significant developments are of particular interest. First, these are those firms that anchor their market expansion on marginal metropolitan markets, and second, there is a small group of new entrepreneurs who, in the first instance, make an effort to expand into Jewish markets.

Our results further confirm that the majority of ethnic minority plants remain restricted to regional markets within peripheral areas, and that even those that expand beyond the region remain subject to other barriers of peripherality. Several results suggest the conclusion that these barriers of peripheriality are stronger than barriers of ethnicity. While more than two thirds of the entrepreneurs break ethnic barriers by selling to Jewish markets, less than half sell to core metropolitan markets, and a smaller proportion sell to large corporations. In addition, we found that Druze entrepreneurs show a tendency to expand in the earlier stages of their enterprise development into Jewish markets, but that their pattern of sales overall does not differ significantly from that of the Moslems. This is due to the fact that both face peripheriality constraints.

References

Aldrich, H.E. and Waldinger, R. (1990), 'Ethnicity and Entrepreneurship', *Annual Review of Sociology*, vol. 16, pp. 111-135.

Atrash, A. (1993), 'The Arab industry in Israel: Branch Structure, Employment and Plant Formation', *Economics Quarterly*, vol. 152, pp. 112-120 (Hebrew).

Barrett, G.A., Jones, T.V. and McEvoy, D. (1996), 'Ethnic Minority Business: Theoretical Discourse in Britain and North America', *Urban Studies*, vol. 33, no. 4-5, pp. 783-809.

Best, M. (1990), *The New Competition: Institutions of Industrial Restructuring*, Harvard University Press, Cambridge, MA.

Camagni, R. (1991), 'Local 'Milieu', Uncertainty and Innovation Network: Toward a New Dynamic Theory of Economic Space', in R. Camagni (ed.), *Innovation Networks: Spatial Perspectives*, Belhaven Press, London, pp. 121-144.

Central Bureau of Statistics (1995), *Statistical Abstracts of Israel*, Jerusalem.

Czamanski, D. T. and Taylor, T. K. (1986), 'Dynamic Aspects of Entrepreneurship in Arab Settlements in Israel', *Horizons*, vol. 17-18, pp. 125-144 (Hebrew).

Falah, G. (1993), 'Trends in the Urbanization of Arab Settlements in Galilee', *Urban Geography*, vol. 14, no. 2, pp. 145-164.

Felsenstein, D. (1992), 'Assessing the Effectiveness of Small Business Financing Schemes: Some Evidence From Israel', *Small Business Economics*, vol. 4, pp. 273-285.

Grabher, G. (1993), *The Embedded Firm*, Routledge, London.

Gradus, Y., Razin, E. and Krakover, S. (1993), *The Industrial Geography of Israel*, Routledge, London.

Grossman, G.M. (1984), 'International Trade, Foreign Investment, and the Formation of the Entrepreneurial Class', *The American Economic Review*, vol. 10, pp. 605-613.

Guttman, L. A. (1965), 'A General Nonmetric Technique for Finding the Smallest Euclidean Space for a Configuration of Points', *Psychometrika*, vol. 33, pp. 469-506.

Haidar, A. (1993), *Obstacles to Economic Development in the Arab Sector in Israel*, The Israeli Arab Centre for Economic Development, Tel-Aviv (Hebrew).

Harrison, B. (1992), 'Industrial Districts: Old Wine in a New Bottle?', *Regional Studies*, vol. 26, pp. 469-483.

Hasson, S. (1981), 'Social and Spatial Conflicts: The settlement Process in Israel During the 1950s and the 1960s', *Espace Geographique*, vol. 2, pp. 120-147.

Jensen, L. and Portes, A. (1992), 'Correction to the Enclave and the Entrants: Patterns of Economic Enterprise Before and After Mariel', *American Sociological Review*, vol. 57, pp. 411-414.

Jones, T., McEvoy, D. and Barrett, G. (1992), *Ethnic Identity and Entrepreneurial Predisposition: Business Entry, Motives of Asians, Afro-Caribbeans and Whites.* Paper to the URSC Small Business Initiative, University of Warwick.

Jones, T., McEvoy, D. and Barrett, G. (1993), 'Labour Intensive Practices in the Ethnic Minority Firms', in J. Atkinson and D. Storey (eds), *Finance and the Small Firm,* Routledge, London, pp. 145-181.

Khamaisi, R. (1984), *Arab Industry in Israel*, Unpublished Magister Thesis, Technion, Israel (Hebrew).

Kloosterman, R., Van Der Leun, J. and Rath, J. (1999), 'Mixed Embeddedness: In Formal Economic Activities and Immigrant Businesses in the Netherlands', *International Journal of Urban and Regional Research*, vol. 23, no. 2, pp. 252-266.

Lakshmanan, T.R. and Okumura, M. (1995), 'The Nature and Evolution of Knowledge Networks in Japanese Manufacturing', *Papers in Regional* Science, vol. 74, no. 1, pp. 63-86.

Light, I. (1984), 'Immigrant and Ethnic Enterprise in North America', *Ethnic and Racial Studies,* vol. 7, pp. 195-216.

Light, I. and Bonachich, E. (1988), *Immigrant entrepreneurs,* University of California Press, California.

Lingoes, J. C. (1973), *Guttman-Lingoes The Nonmetric Program Series*, Mathesis Press, Ann Arbor, MI.

Markusen, A. (1994), 'Studying Regions by Studying Firms', *Professional Geographer,* vol. 46, no. 4, pp. 477-490.

Miles, R. (1989), *Racism,* Routledge, London.

Nee, V. and Sanders, J. M. (1987), 'On Testing the Enclave-Economy Hypothesis', *American Sociological Review*, vol. 52, pp. 771-773.

Oinas, P. (1999), 'Voices and Silences: the Problem of Access to Embeddedness'. *Geoforum*, vol. 30, pp. 351-361.

Portes, A. and Jensen, L. (1989), 'What's an Ethnic Enclave? The Case for Conceptual Clarity', *American Sociological Review*, vol. 76, pp. 1029-1047.

Pred, A. (1977), *City Systems in Advanced Economies: Past Growth, Present Processes and Future Options,* Hutchinson, London.

Ratti, R. (1992), *Innovation Technologique et Development Regional*, Instituto di Recerche Economiche, Bellinzona.

Sanders, J. M. and Nee, V. (1992), 'Problems in Resolving the Enclave Economy Debate', *American Sociological Review*, vol. 57, pp. 415-418.

Schnell, I. (1994), 'Urban Restructuring in Israeli Arab Settlements', *Middle Eastern Studies*, vol. 30, no. 2, pp. 330-350.

Schnell, I., Sofer, M. and Drori, I. (1995), *Arab Industrialization in Israel*, Praeger, Westport.

Schnell, I., Benenson, I. and Sofer, M. (1999), 'The Spatial Pattern of Arab Industrial Markets in Israel', *Annals of the Association of American Geographers*, vol. 89, no. 2, pp. 311-336.

Scott, A. J. (1991), 'The Aerospace-Electronics Industrial Complex of Southern California: The Formative Years 1940-1960', *Research Policy*, vol. 20, pp. 439-456.

Semyonov, M. and Lewin-Epstein, N. (1993), *The Arab Minority in Israel's Economy.* Westview Press, Boulder.

Staber, U. and Schaefer, N. (eds) (1996), *Business Networks: Prospects for Regional Development*, Walter de Gruyer, Berlin.

Taylor, M. (1999), *Enterprise Embeddedness as a Mechanism of Exclusion: The Experience in Fiji.* Paper presented at the IGU Commission on the Organization of Industrial Space, Haifa and Beer Sheba, Israel.

Van Geenhuizen, M. and Nijkamp, P. (1995), 'Industrial Dynamics, Company Life Histories and Core-Periphery Dilemma', *Geography Research Forum*, vol. 15, pp. 49-69.

Waldinger, R. (1993), 'The Ethnic Enclave Debate Revisited', *International Journal of Urban and Regional Research*, vol. 17, pp. 444-452.

Ward, R. (1985), 'Minority Settlement and the Local Economy', in B. Roberts, R. Finnegan and D. Gallie (eds), *New Approaches to Economic Life*, Manchester University Press, Manchester, pp. 198-211.

Zhou, M. and Logan, J. R. (1989), 'New York City Chinatown', *American Sociological Review*, vol. 54, pp. 809-820.

12 Electronic Commerce: Global-Local Relationships in Financial Services

John Langdale

Introduction

Electronic commerce (e-commerce) has been widely predicted to revolutionise the way in which business is conducted and to have major social implications. Widespread adoption of computers and the Internet by households and by business has led to a surge in e-commerce applications in most industrialised countries. Unfortunately, it is difficult to separate the real impact of e-commerce from the hyperbole, since numerous vested interests, ranging from technology enthusiasts and consultants to companies that expect to gain business from e-commerce, have adopted an 'evangelical' attitude to the advent of e-commerce.

It is commonly argued that e-commerce is a relatively new development that has emerged largely as a result of the rapid adoption of the Internet by business. In fact, the use of information and communications technologies (ICTs) in business has been expanding rapidly since the 1960s, with a reduction in the cost of electronics components leading to cheaper computers and telecommunications. While intra-organisational IT expanded rapidly, it was difficult to integrate IT systems in different organisations and consequently industry-wide ICT networks were rare. Despite these problems, a number of global industry ICT networks emerged in the 1970s because of the compelling need within particular industries for linkages between participants. In particular, the airline and banking industries had well developed e-commerce systems prior to the advent of the Internet.

The two major areas of e-commerce are business-to-business (B2B) and business-to-consumer (B2C). It is not possible, however, to make a clear distinction between B2C and B2B, since many firms use e-commerce to link to final consumers as well as to industry suppliers. This chapter examines the B2B area, with a focus on banking and finance. In particular, the implications of e-commerce in the emergence of twenty-four hour trading in foreign exchange and securities markets are considered.

B2B e-commerce is difficult to neatly classify into categories because of the fluidity of relationships between different players and the extraordinarily rapid growth of the market. It is possible, however, to divide e-commerce firms into three groups: ICT firms, user firms and specialised B2B intermediaries.

The ICT industries include telecommunications, media (television, pay-television, film), 'new' media (e.g., multimedia, computer games), Internet (e.g., Internet Service Providers or ISPs) firms. The characteristics of these industries are that they provide either the carriage (e.g., telecommunications carriers and ISPs) or the content (media and multimedia) for e-commerce. The motives for these firms' entry into e-commerce vary, but most see it as a means of defending their existing positions and/or taking advantage of emerging opportunities. For example, telecommunications companies see e-commerce as a rapidly growing market given that their corporate customers are introducing these systems. More significantly, e-commerce is redefining the nature of the telecommunications industry by forcing carriers to be far more responsive to their customers' changing ICT needs. E-commerce and the Internet is requiring carriers to fundamentally change their way of doing business.

The second group of firms involved in B2B e-commerce, the user firms, are industries whose main business activity is not in e-commerce. E-commerce applications cover a wide range of user industries (e.g. automobiles, airline, retailing) and are likely to have a major impact on many of them, particularly in terms of the ease with which firms in different parts of the supply chain will be interconnected. There are, however, likely to be significant competitive problems in achieving such interconnection, given that firms will have to divulge confidential information to join e-commerce consortia. While most e-commerce systems in user industries are national in scope, it is likely that many will be extended globally in the near future.

The third group of firms are specialised B2B intermediaries who have either established their own B2B hubs, or provide hardware and/or software to facilitate e-commerce. It is difficult neatly to classify these firms since the pace of change is extraordinarily rapid. One classification divides such firms into enablers and B2B hubs (Reede, 2000). Enablers provide the software and/or hardware for B2B e-commerce, while B2B hubs provide portals that enable applications in a particular industry to facilitate B2B transactions by industry participants. Such hubs are emerging in the plastics, telecommunications, steel and energy industries.

It is likely that the distinction between these vertical hubs and the second category of user industries will blur over time as participants from each group merge into broader e-commerce systems. Many major user firms (e.g. automobile, airline, plastics and steel) are expanding into e-commerce and it is likely that they will shift more of their business onto their own systems and/or buy out the independent firms. E-commerce systems rely on economies of scale and it is likely that the system that gains the largest number of users will become the dominant e-commerce system in an industry.

The competition between different user groups and Internet firms is illustrated by the rivalry between hotel operators, airline companies and online travel agencies. The large hotel operators in the Asia-Pacific region have formed an Internet alliance to try to regain control over room bookings from airlines and online travel agencies. Similarly, major airlines in the Asia-Pacific region have also formed an e-commerce alliance to defend their positions from online travel agencies (Kermode, 2000). It is likely that, over time, mergers will take place between the various e-commerce exchanges, but it is difficult to predict how this will actually take place.

A second issue considered in the chapter is the role of e-commerce in shaping the competitive and cooperative relationships between firms and institutions. It is generally assumed that e-commerce will disintermediate many existing service functions, thus reducing transaction costs for customers. While this is taking place, past experience with ICTs suggests that the adoption process is quite complex and needs to be seen in the context of the roles of powerful firms and institutions that want to ensure favourable outcomes for their own interests. The locational implications of this competitive rivalry between firms and institutions in e-commerce are unclear, but it appears that the outcomes will shape the future role of major global cities that dominate the services industry.

Geographical Perspectives in E-Commerce

The recency and speed of the growth of e-commerce has meant that geographers and other social scientists have undertaken little research in this area. Some geographical research has examined the impact of the Internet on the city (Graham, 1999; Kitchin, 1998). Little attention, however, has been given to the role of e-commerce in the geographical restructuring of local, national and global economies, although research is emerging in this area (Brunn and Leinbach, 2001).

The changing interrelationships between the global and the local is a central issue in much geographical research and is a key theme in this chapter (Gertler, 1997). These global-local interrelationships may be conceptualised using Castells' global spaces of flows and places (Castells, 1996, 1998). *Spaces of flows* reflect the range and size of connections, which link the global economy together. They include flows of goods (agricultural, mining and manufactured goods), people (migration, tourism and business travel) as well as information (telecommunications and mail). The volume of electronic information is rising dramatically, particularly between major industrialised countries of the US, Europe and Japan.

Spaces of places are primarily a local perspective, although they are closely intertwined with the global spaces of flows. Unfortunately, little information is available on the global linkages of these cities (Beaverstock *et al.*, 2000). Global cities are critically important places in the global economy. They are global, national and local centres for the collection, processing and dissemination of information. This information role is a key reason for large corporations, business services and government agencies clustering in global cities. Thus, major global cities such as London, New York and, to a lesser extent, Tokyo are key information nodes in the global economy (Sassen, 1994). Other global cities such as Paris, Frankfurt, Los Angeles, Hong Kong and Singapore also play a significant role in global information flows, although their reach is weaker.

A central theme in the chapter is that e-commerce will have a major impact on the role of the global economy, although it is difficult to predict the long-term impact, since many e-commerce systems are in their infancy. E-commerce may shift power to end-users and away from intermediaries located in major cities, given that users are geographically dispersed. Thus, major cities may have a reduced role. Alternatively, and possibly more likely, these cities may perform new service intermediary functions and shift into skill-intensive functions such as risk management in financial services. E-commerce is also reducing the cost of overcoming distance, thus making it easier for firms to centralise functions in global cities. Consequently, it appears that a number of centralising and decentralising forces are associated with e-commerce, which will significantly alter the nature of global-local relationships.

A more fundamental issue relates to the global political economy implications of e-commerce. This issue has attracted very little research, but the emergence of e-commerce as a powerful force in domestic and global economies will mean that those organisations that control dominant B2B e-commerce systems will have enormous power and profitability in the future. A contrast may be drawn between e-commerce systems that have an open membership versus those that are closed in membership. Closed e-commerce systems owned by transnational corporations (TNCs) from major industrialised countries are of concern for small industrialised and developing countries given the future potential power and profitability of such systems. The geography of power in the global economy is likely to be shaped by the ownership of such e-commerce systems.

Global-Local Interdependencies and Electronic Commerce

Two themes are considered in this section. The first is the impact of e-commerce on the location and functions of specialised business and financial services intermediaries in major cities. The widespread adoption of e-commerce, especially in the B2B area, will profoundly affect the interrelationships between the global and the local. Global cities are particularly likely to be affected, given the very high geographical concentration of business and financial services in them (Sassen, 1994, 1999). The role of these services is likely to be significantly restructured by the disintermediating effects of B2B e-commerce. A second theme is related to the first and examines the role of e-commerce in the global spaces of flows and places (Castells, 1996, 1998). Castells' perspective recognises the centrality of key hubs or major cities in the global economy, as well as the importance of the movement of information, goods and people, which link the hubs together and to the rest of the world.

Spaces of Places

Global cities are the hubs of the global economy. While major cities have for centuries been the hubs for the collection, processing and dissemination of information (Braudel, 1984; Meyer, 1991), the volume and specialisation of business services intermediaries has mushroomed with recent globalisation trends.

The geographical agglomeration of specialised business services, corporate headquarters and government services in relatively close proximity is a key function of global cities. Clearly, place matters, since these firms and other institutions are paying very high rents to locate in the heart of a global city; any firm or part of a firm (e.g. a part performing back office functions) that does not need to be located in such high rental districts is likely to move out.

The role of business networks in the functioning of global cities has attracted a significant level of research interest, although few detailed studies have surveyed the nature of these networks. Amin and Thrift (1992) have pointed out that an important role of major cities is that they generate and disseminate collective beliefs about what is happening in an industry and in the national and global economy.

Thus, global cities are places that give structure to information, since it is vital in a highly uncertain environment for individuals and firms to understand the context of the information collected. These contextual frameworks are developed and reinforced through the extensive face-to-face contacts in informal business networks, although electronic communications are also used. Global cities are also centres for gathering information, establishing or maintaining coalitions and trust. These networks are particularly important in business and financial services, many of which are clustered in the CBDs or inner suburbs of major cities. Close geographical proximity is essential to initiate and maintain these networks in business and financial services industries.

While face-to-face contacts are a key means for forming and maintaining business networks, business and financial services intermediaries are increasingly interlinked via electronic communications to others at a local scale and on a global basis. These electronic communications networks, however, cannot completely substitute for geographical proximity, since most people use regular face-to-face contacts to initiate business networks and to build trust. Once trust is established, electronic communications often perform an important complementary role to face-to-face contact (Blanc and Sierra, 1999). Unfortunately, little research has examined the role of various forms of communications in the nature of business networks.

It is unclear what impact e-commerce is likely to have on the balance between face-to-face contacts and electronic communications. The e-commerce literature has emphasised the increasing electronic nature of business relationships. It appears, however, that personal business networks will have an on-going importance despite the rising significance of e-commerce. The head of a London-based metals broker commenting on the relationship between e-commerce and personal networks pointed out that "you can't reproduce over the Internet ... the interpersonal relationship between traders ... Trading still won't be done without personal relationships being built up" (Tait and Solman, 1999, p. 26). Personal business networks are of considerable importance for traders and sales/marketing staff in the foreign exchange market, which is the most highly globalised and heavily reliant on electronic trading (Langdale, 2001).

Spaces of Flows

The spaces of flows are becoming more complex as globalisation advances. This complexity reflects larger flows of goods, people and information and a growing range of countries connected to these global spaces of flows. Despite this trend, flows are still concentrated on major industrialised countries and global cities.

Global information flows are of particular importance in the emerging global economy and these are predominantly between major industrialised countries and, to a lesser extent, the Asian Newly Industrialising Economies. Most developing countries, on the other hand, are minimally connected in these global spaces of flows (International Telecommunications Union, 1998).

Global electronic information flows tend to be geographically centralised on major global cities, reflecting the information intensity of specialised business services and such high-technology manufacturing industries as aerospace and electronics, which are clustered in these cities. Unfortunately, little evidence is available on the actual origin and destination of the flows and the volume of flows generated by different industries (Langdale, 1989, 1991; Graham, 1999).

E-commerce will dramatically increase the range and intensity of these global spaces of flows. The volume of international telecommunications traffic has been rising at very rapid rates in recent years, particularly as a result of bandwidth-intensive applications on the Internet. A massive submarine fibre optic cable laying programme combined with very rapid change in fibre technology has taken place from the mid-1990s, but has barely been able to keep pace with the demand for Internet-based telecommunications (Gannon, 1999). The rate of adoption of e-commerce is much higher in major industrialised countries, whereas developing countries lag massively behind (International Telecommunication Union, 1999). In contrast, the chief policy issue in most developing countries is to provide basic telephone connections to all people; this is a particular problem in rural areas in developing countries.

Twenty-Four Hour Global Markets

The growing importance of globalisation is leading to a rise in the volume of information transmitted around the world on a twenty-four hour basis. These information flows are both facilitating and responding to the emergence of global markets.

Types of Twenty-Four Hour Global Networks

Regional hub-and-spoke networks are commonly used by firms and industries to facilitate twenty-four hour trading. These networks grew rapidly in the 1980s, with banks, airlines and high technology manufacturing firms (electronics, computer) using leased circuit networks to link their global operations (Langdale, 1989). They are characterised by high-speed broadband telecommunications networks providing the global backbone routes, with medium- and low-speed links for the spokes. Firms with more sophisticated networks have a greater degree of redundancy built into their networks and have a number of alternative high-speed backbone routes connecting global cities.

A common form of these networks is to have three regional hubs: London is often used as a hub for Europe, Africa and the Middle East; New York is generally the hub for the Americas; and Singapore or Hong Kong are the hubs for the Asia-Pacific region. In situations in which organisations are trading in financial or commodity markets, this triad network organisation allows traders in one centre to work for eight hours before passing on their positions to the next centre. Such a system allows trading to 'follow the sun' on a twenty-four hour basis (Langdale, 2000). This organisation is common in the case of the foreign exchange market, with the regional hubs undertaking trading in a number of major currencies and operating as centres for the collection and processing of information (Langdale, 2001). The spokes, on the other hand, are the firms' branch national offices, which concentrate on trading in their national currency. While these branch offices have access to the firms' global information network, they have a much more restricted role in firms' operations.

Different advantages and disadvantages are present for centralised versus decentralised global networks. The advantages of centralisation are that the expenses of operating centralised trading hubs are reduced, given that fewer electronic trading terminals and associated IT are needed, and office space requirements may be reduced. In addition, a firm may be able to employ fewer traders in one or two centralised hubs. In the past, such a centralised system would have led to more expensive telecommunications, but this is less of a problem with cheaper international telecommunications charges and e-commerce. For example, the global commodities trader Cargill has reduced the number of oil trading hubs from three to two (New York and Geneva), partly because of improvements in international telecommunications and e-commerce. Cargill, however, is continuing to operate other regional trading functions (palm oil and rubber) in Singapore ('Cargill to halt oil trading in Singapore', *Straits Times*, 9 May, 2000).

The disadvantages of centralising trading primarily relate to separating the location of trading from where business is taking place. Traders in centralised networks have access to excellent electronic information through information vendors such as Reuters and Bloomberg, as well as their own firms' internal information networks. They are, however, remote from the local forces driving financial markets in countries around the world. This is not much of a problem if a country contributes a small amount of business to the firm, but it can be significant for a large market.

Demand for Twenty-Four Hour Information Flows

Demand for twenty-four hour information flows reflects a number of factors. The first is that major TNCs have operations in a diverse range of countries and their subsidiaries around the world increasingly need to access corporate databases at any time of the day. The growing global integration of TNCs, particularly in high-technology industries and business and financial services, requires heavy information flows to support such operations. E-commerce is having a major impact on the intercorporate communications associated with globalisation. Prior to the advent of e-commerce, most global information flows on a twenty-four hour basis took place within firms, largely because of the difficulties of achieving inter-operability between computer systems in different organisations. The airline industry, with its SITA network, and banking and finance, with the SWIFT network, were notable exceptions (Langdale, 1985). E-commerce is allowing firms in a wide range of industries to procure supplies and reduce inefficiencies in the supply chain on a global basis.

Second, twenty-four hour information flows have also been driven by the globalisation of the banking and finance industry. Most global ICT applications in the banking and finance industry have either been for linking banks and financial institutions together (e.g. SWIFT), or for providing connections for large corporate customers (Langdale, 1985, 1989). Globalisation of economies has increased risks for banks and financial firms as well as for corporate clients. Growing use of ICTs in financial services has partly helped to create additional risk at the global scale, but has also led to the introduction of new financial products (e.g. derivatives) to manage that risk.

E-commerce is allowing more firms, ranging from large corporations to small firms, to operate on a global scale and thus increasing the need for financial services (e.g. risk management) to be available on a twenty-four hour basis. Electronic procurement and supply chains that are emerging in global B2B e-commerce are likely to operate on a twenty-four hour basis and thus expose participants to a wider range of risks.

Electronic Commerce and Financial Markets

Global Electronic Securities Markets

Stock markets and futures exchanges around the world have undergone major changes since the early1990s. An important development has been the shift away from floor-based to electronic trading. Also, alliances have been formed between exchanges with the aim of facilitating cross-border investment. Unlike the foreign exchange market, securities markets have not developed extensive twenty-four hour trading. A problem is, however, that the greatest liquidity for stocks is generally available only during the business hours of a company's home exchange. This reflects the fact that information about a company is greatest in its home market and most of this information is released during that country's business day. Furthermore, the lack of efficient clearing and settlement mechanisms on a global basis remains a significant problem for the emergence of twenty-four hour trading.

Nevertheless, a trend towards global twenty-four hour electronic trading in securities is emerging. For example, in the futures market the Chicago Mercantile Exchange and SIMEX have operated a mutual offset system since the mid-1980s, which allowed futures traders in eurodollar, euroyen and other contracts to trade over a thirteen-hour period between Chicago and Singapore. It traded 11.1 million contracts in 1998 (Cavaletti, 1999). This system was extended to include the Paris Bourse in 1999 and now provides a full twenty-four hour trading potential.

The rapid adoption of electronic futures trading, especially outside the US, is facilitating linkages between exchanges and the emergence of further twenty-four hour trading. European futures exchanges, in particular, have aggressively moved into electronic trading, as well as merging and/or forming alliances. These strategies have given them a greater ability to link up with exchanges elsewhere in the world to achieve twenty-four hour trading.

Considerable interest exists in twenty-four hour electronic trading in stock markets. Numerous proposals have been put forward since the late-1990s for twenty-four hour stock exchanges. The key proposals involve the major US (New York Stock Exchange and Nasdaq) and European (Paris, Frankfurt and London) exchanges. Such linkages would provide a larger global investor pool and greater liquidity for stocks and allow for global electronic trading of shares for retail and corporate clients.

One limited approach is to extend the operating hours of exchanges. For example, the Frankfurt Stock Exchange extended its opening hours by two-and-a-half hours to 8pm in June 2000 in order to attract trading from US investors and to allow European private investors to react to changes that are taking place during the US trading day, especially in the rapidly fluctuating technology stocks on the Nasdaq. The chief disadvantage is that trading firms in Frankfurt are faced with considerably higher staffing costs as a result of the longer operating hours in the exchange (Shorecki, 2000).

Exchanges have adopted a number of strategies to achieve twenty-four hour trading. One, which is followed by many futures exchanges, is to either adopt full electronic trading (e.g. Sydney Futures Exchange and Deutsche Terminbourse in Germany), or after-hours electronic trading and floor-based trading during normal business hours (e.g. Chicago Board of Trade's Project A, Chicago Mercantile Exchange's Globex and Liffe's Automated Pit Trading).

A second strategy to achieve twenty-four hour trading, which is often followed in conjunction with the first one, is to form alliances with exchanges in other locations. The make-up of exchanges and implementation of these alliances is complicated by the competitive rivalry between exchanges within countries (primarily New York Stock Exchange and Nasdaq) and between countries. Failure to form alliances or merge exchanges, however, is likely to lead to a number of exchanges being bypassed by Electronic Communications Networks (ECNs) such as Instinet and Archipelago, which have lowered costs for share trading. In 1999, ECNs accounted for approximately 30% of Nasdaq's volume of trading and this percentage is growing rapidly (Labate and Harris, 1999).

The large global investment banks are key players in the securities industry. The top ten banks account for about 55% to 60% of the order flow of the New York Stock Exchange. The investment banks have invested in ECNs, partly because they fear that they will be bypassed in the shift to electronic trading (Currie, 1999b).

The central complicating feature in the globalisation of securities exchanges is the conflict between the global and the local. Stock and futures exchanges are primarily based in a single place and are owned by local members, although several exchanges have listed on the stock market recently. Their inability to fully adopt electronic trading and form global alliances reflects the diverse interests of their local members who want to maximise profits on their exchange. Pitted against the locally-based exchanges are globally-oriented large investment banks and B2B e-commerce exchanges. At present, e-commerce exchanges are primarily nationally oriented, but they are likely to shift to global operations in the near future. Such e-commerce exchanges will reduce the middleman's role and consequently cut the number of staff and profitability of securities exchanges, investment banks and brokers.

The global investment banks tend not to care where their business is executed, but want lower trading costs and cheap and reliable clearing and settlement for their trades. Furthermore, these firms would prefer to deal with a smaller number of securities exchanges. Their bargaining strength is that they are generating a larger percentage of trading and are forming alternative trading exchanges (ECNs), or are placing their business with Internet-based B2B exchanges. Securities exchanges are thus caught in a dilemma: if they remain locally-based, they are likely to lose a large amount of business from investment banks to ECNs and B2B e-commerce exchanges. On the other hand, if they merge and/or form alliances with other exchanges and adopt global electronic trading, they will alienate their local members.

Twenty-Four Hour Markets and the 'New' E-Commerce

The recent growth of e-commerce throughout the world is expanding the demand for twenty-four hour markets. The emerging globalisation of economic activity is driving many of these developments. Twenty-four hour trading is likely to emerge in B2B e-commerce applications in industries which are time-sensitive. We have already seen that such markets have already emerged in banking and finance, but e-commerce will significantly widen the area of applications. Furthermore, the growing spread and depth of globalisation is also expanding the range of time-sensitive global trading.

It is difficult to forecast the impact of the emergence of twenty-four hour markets in these B2B e-commerce applications since they are still being established. For example, the automobile industry expects to make major savings from B2B e-commerce and a number of major automobile manufacturers have launched Covisint, a B2B e-commerce exchange linking manufacturers and suppliers. A major application is in global procurement, whereby firms in an industry will be able to globally procure supplies from firms at significantly reduced costs. It has been suggested, however, that while procurement savings will be important, the chief advantages will be faster and standardised communication in the supply chains, as well as the scope for eliminating unnecessary inventory and working capital (Tait, 2000).

It is likely that automobile firms around the world will use e-commerce on a twenty-four hour basis, thus speeding up time-sensitive operations. While automobile firms have operated intra-corporate information systems to produce global cars in the past, Covisint will allow growing global inter-corporate communications to take place on a twenty-four hour basis. This trend is likely to be also accelerated by the rapid expansion of international mergers and alliances between automobile companies. The automobile industry is being driven by massive economies of scale and scope; companies need to spread the heavy design costs for new automobiles over a much larger production base.

The development of global B2B e-commerce in a wide range of industries will have major implications for the banking and finance industry. Participants in B2B e-commerce exchanges will be faced with a higher level of financial risk, since they will be operating twenty-four hours a day on a global basis. For example, they will have a higher foreign exchange risk, since they will be exposed to fluctuations in a number of currencies. Thus, rapid expansion of global B2B e-commerce exchanges will lead to a growth in associated banking and financial systems to handle risk and the associated payments needed in such systems.

Disintermediation Impact of E-commerce on Global Cities

It is difficult to establish at this early stage in the adoption of e-commerce what the impact of e-commerce on major cities is likely to be. It is widely argued that e-commerce will increase the level of disintermediation taking place in different markets. Given that business and financial services intermediaries are disproportionately located in major cities, much of the impact will be felt in these cities.

One possible implication of e-commerce applications is that a shift of power to customers will take place away from the business and financial services hubs in major cities and particularly in New York and London. The Internet reduces the privileged information situation of business and financial services intermediaries. This issue has already arisen with respect to financial services firms. For example, a major concern for investment banks is that e-commerce will reduce their dominant role in information flows. "To the investment bank, the combination of technology, connectivity, availability and affordability is fundamentally shifting the balance of power: investment banks no longer have near-exclusive control over the flow of information" (Currie, 1999a, p. 56). Thus, corporate clients of investment banks are able to access the information themselves rather than using their broker. The Internet makes key information available more widely and, importantly, increases the transparency of the financial services market. Given that global investment banks' headquarters are heavily concentrated in London and New York, the size and range of their operations in these cities will be disproportionately affected unless they are able to shift into new areas of business.

The impact of e-commerce on global cities may take place in a number of different ways. Electronic trading is reducing the number of traders needed in those areas of financial services that have already adopted e-commerce such as foreign exchange and futures trading. Traders in these markets have traditionally been geographically concentrated in major global cities. The impact is particularly severe for 'plain vanilla' types of trading in financial markets (e.g. spot trading in foreign exchange and in government bonds), but more sophisticated transactions (e.g. options trading) still require the skills held by traders and deal makers. It is not clear, however, that e-commerce will have an impact on all areas of financial services. For example, even the electronic bond market will still require bond salesmen and research offered by the major investment banks.

Major financial services firms (e.g. global commercial banks such as Citigroup and Chase Manhattan and global investment banks such as Merrill Lynch and Goldman Sachs) have assigned a high priority to their e-commerce activities and are likely to be prominent players in the future (Currie, 1999a, 1999b). However, their profitability is likely to be severely eroded, since the Internet will reduce their former oligopolistic position as information providers to corporate clients.

Global investment banks acquired more power in the 1990s. The top ten investment banks accounted for almost 90% of cross-border securities business, compared with some 60% at the beginning of the 1990s (Luce, 1999). This dominance has given them greater leverage to force futures and stock exchanges to shift towards electronic trading and to consolidate clearing and settlement systems.

It is likely that financial services firms in global cities will shift into more sophisticated areas, as e-commerce disintermediates the 'plain vanilla' transactions and automates basic trading functions. For example, the risk management area in derivative trading has been a high growth one in the 1990s and rising globalisation of economies is likely to increase risk for corporate customers, thus ensuring a demand for the services of financial services firms.

Conclusions

It is difficult to forecast the locational impacts of B2B e-commerce given that widespread adoption of such systems is still taking place. While B2B e-commerce is attracting enormous interest from the business community, it is unknown how many proposals will be implemented or which ones will be successful. Despite the emergence of many competing B2B e-commerce exchanges, rapid consolidation of these exchanges will take place over the next few years given the economies of scale of dominant exchanges.

E-commerce is having a major impact on the nature of global-local relationships. At a macro-scale, it is leading to new opportunities and threats for countries and regions engaging with globalisation forces. In terms of opportunities, e-commerce is encouraging growth of twenty-four hour markets, which potentially allow countries throughout the world to participate in global industries. At the same time, it also poses significant threats in terms of locking smaller industrialised and developing countries into new forms of dependency.

While this chapter has focused on global twenty-four hour financial markets, similar developments are taking place in high technology industries. For example, Indian exports of computer services are expanding rapidly, partly as a result of shortages of IT skills in industrialised countries, but also because of the availability of skilled staff in India (Rao, 2000). A number of Indian software companies are using the time zone difference between the US and Europe with India to undertake work while the major industrialised countries are asleep.

E-commerce is also likely to facilitate the growth of twenty-four hour global design and engineering teams in high technology industries such as automobile and aircraft, but also in architecture. In the past, major automobile companies have had company design teams working on a global twenty-four hour basis to design cars. The shift to an open e-commerce environment is likely to increase the ability of these automobile companies to incorporate outside firms into their design teams. Twenty-four hour operations are particularly important in time-sensitive applications in which the speed of production is a key competitive issue.

E-commerce has major global political economy implications, since it is likely that it will give major industrialised countries a key advantage as compared with smaller industrialised and developing countries. Most B2B ventures have involved TNCs from major industrialised countries. While the ventures have been initially designed for either the US or Europe, it is clear that, if successful, they will be extended to cover the globe. Thus, these joint ventures between major firms could lead to new forms of dominance and dependency in the global economy.

Two broad outcomes are possible in global e-commerce. One is that barriers to entry are reduced for firms and for localities outside the existing dominant TNCs and global cities. In this scenario the Internet would facilitate geographical decentralisation of economic activity by allowing peripherally located firms to participate more fully in the global economy.

This scenario requires that the e-commerce consortia are open to entry to large and small firms from all countries.

An alternative scenario is that B2B e-commerce may reinforce existing dominance-dependency relationships amongst countries. In this scenario firms in major industrialised countries are the early adopters of B2B e-commerce and, because of their huge size in global markets, are able to dominate ownership of the emerging consortia in B2B industries. Thus, while firms in all countries would benefit from a lowering of costs, control and profits from the B2B consortia would be locationally centralised in major industrialised countries. Such a global economy would be comprised of a small number of hubs (global cities) and many spokes (smaller industrialised and developing countries).

Global e-commerce is thus likely to have a variety of impacts, some of which enhance locational centralisation of economic activity, while other increase locational decentralisation. Much depends on the way in which firms and institutions adopt the new technologies. This conclusion is hardly surprising, since it is one that has characterised ICT adoption processes in recent decades. The locational impacts of e-commerce may turn out to be an intensification of long-term processes and not the revolutionary impacts that many have imagined.

References

Amin, A. and Thrift, N. (1992), 'Neo-Marshallian Nodes in Global Networks', *International Journal of Urban and Regional Research*, vol. 16, no. 4, pp. 571-587.

Beaverstock, J.V., Smith, R.G. and Taylor, P.J. (2000), 'World-City Network: A New Metageography?', *Annals of the Association of American Geographers*, vol. 90, no. 1, pp. 123-134.

Blanc, H. and Sierra, C. (1999), 'The Internationalisation of R&D by Multinationals: A Trade-Off Between External and Internal Proximity', *Cambridge Journal of Economics*, vol. 23, pp. 187-206.

Braudel, F. (1984), *Civilization and Capitalism, 15th, 18th Century*, University of California Press, Berkeley, CA.

Brunn, S. and Leinbach, T. (eds) (2001), *The Wired Worlds of Electronic Commerce*, Wiley, forthcoming.

Castells, M. (1996), *The Rise of the Network Society. The Information Age*, Volume 1, Blackwell, Oxford.

Castells, M. (1998), *End of Millennium. The Information Age: Economy, Society and Culture*, Volume 3, Blackwell, Oxford.

Cavaletti, C. (1999), 'The Past, Present and Future of Exchange Alliances', *Futures*, vol. 28, no. 8, pp. 76-78.

Currie, A. (1999a), 'The New Battleground', *Euromoney*, September, pp. 53-66.

Currie, A. (1999b), 'Redefining Exchanges', *Euromoney*, December, pp. 48-49.

Gannon, P. (1999), '20,000 Leagues Under the Sea', *Communications International*, January, pp. 32-36.

Gertler, M. (1997), 'Globality and Locality: the Future of 'Geography' and the Nation-State', in P.J. Rimmer (ed.), *Pacific Rim Development: Integration and Globalisation in the Asia-Pacific Economy*, Allen and Unwin, Sydney, pp. 12-33.

Graham, S. (1999), 'Global Grids of Glass: On Global Cities, Telecommunications and Planetary Urban Networks', *Urban Studies*, vol. 36, no. 5-6, pp. 929-949.

International Telecommunication Union (1998), *World Telecommunication Development Report: Universal Access*, Geneva.

International Telecommunication Union (1999), *Challenges to the Network: Internet for Development*, Geneva.

Kermode, P. (2000), 'Hotels Form Internet Hub to Protect Trade', *Australian Financial Review*, 20 June, p. 35.

Kitchin, R. (1998), 'Towards Geographies of Cyberspace', *Progress in Human Geography*, vol. 22, no. 3, pp. 385-406.

Labate, J. and Harris, C. (1999), 'Fighting for a Share', *Financial Times*, 26 May, p. 14.

Langdale, J.V. (1985), 'Electronics funds transfer and the internationalisation of the banking and finance industry', *Geoforum*, vol. 16, pp. 1-13.

Langdale, J.V. (1989), 'The Geography of International Business Telecommunications: the Role of Leased Networks', *Annals of the Association of American Geographers*, vol. 79, no. 4, pp. 501-522.

Langdale, J.V. (1991), 'Telecommunications and International Transactions in Information Services', in S. Brunn and T.R. Leinbach (eds), *Collapsing Space and Time: Geographic Aspects of Communication and Information*, HarperCollins Academic, London, pp. 193-214.

Langdale, J.V. (2000), 'Telecommunications and Twenty-Four Hour Trading in the International Securities Industry', in K. Corey and M.I. Wilson (eds), *Information Tectonics: Space, Place, and Technology in an Information Age*, Wiley, Chichester, pp. 89-99.

Langdale, J.V. (2001), 'Global Electronic Spaces: Singapore's Role in the Foreign Exchange Market in the Asia-Pacific Region', in S. Brunn and T. Leinbach (eds), *The Wired Worlds of Electronic Commerce*, Wiley, Chichester, forthcoming.

Luce, E. (1999), 'Bonding Together', *Financial Times*, 6 August, p. 15.

Meyer, D.R. (1991), 'Change in the World System of Metropolises - the Role of Business Intermediaries, World Cities', *Urban Geography*, vol. 12, no. 5, pp. 393-416.

Rao, K. (2000), 'A Round-the-Clock Office for the West', *Euromoney*, June, pp. 51-60.

Reede, M. (2000), *B2B transaction hubs*, Unpublished paper, Gilbert and Tobin, Sydney, June.

Sassen, S. (1994), *Cities in a World Economy*, Sage, London.

Sassen, S. (1999), 'Global Financial Centers', *Foreign Affairs*, vol. 78, no. 1, pp. 75-87.

Shorecki, A. (2000), 'London Holds Back from Opening all Hours', *Financial Times*, 20 April, p. 27.

Tait, N. (2000), 'Technical Hitch Stalls 'Big Three' Trading Site', *Financial Times*, 14 June, Survey on FT Auto, p. II.

Tait, N. and Solman, P. (1999), 'Trading Metal in a Virtual Marketplace', *Financial Times*, 13 October, p. 26.

13 The Internet Age: Not the End of Geography

Edward J. Malecki

Introduction

The Internet is the defining technology at the dawn of the twenty-first century, changing much about how firms and people are connected. It also is built upon a large number of incremental changes to previous technologies. In particular, the Internet strongly depends on the telephone network, on satellites and on long-distance cables, all of which have formed the infrastructure for telecommunications. The geography of the Internet, then, is closely related to both the structures that link, at the highest level, world cities, and to the demand of large business customers that have always driven the demand for new telecommunications technology. This chapter reviews these themes and, with some evidence from the USA, concludes that, while some shifts are evident in the urban hierarchy, much stability remains.

> What, exactly, is the Internet? Basically it is a global network exchanging digitised data in such a way that any computer, anywhere, that is equipped with a device called a 'modem,' can make a noise like a duck choking on a kazoo. This is called 'logging on,' and once you are 'logged on,' you can move the 'pointer' of your 'mouse' to a 'hyperlink,' and simply by 'clicking' on it, change your 'pointer' to an 'hourglass.' Then you can go to 'lunch,' and when you get back, there, on your computer screen, as if by magic, will be at least 14 advertisements (Barry, 1999).

The quotation above depicts much of the story of the Internet to average users, who depend on the telephone network to connect to the network of networks. High-speed, or broadband, technology will change this, but not overnight, and maybe not at all for some people and some places. "Global services need local networks" (Staple, 1999, p. 3). How well linked to global networks a place is will have an increasingly important impact on local development prospects. This is particularly the case as new investments in high-speed (or high-bandwidth) infrastructure are made between some – but not all – places.

This chapter examines in a general way two distinct topics. First, the chapter examines what is new and what is not so new about the Internet as a technology. Second, it then turns to the layers of the infrastructure of the Internet, a 'network of networks': the backbone layer and the local layer. Between them, technology and infrastructure define several geographical patterns evident at this stage in the evolution of the Internet. While it is impossible to cite current data on all aspects of the Internet and its infrastructure, and use, Paltridge (1998a) provides several indicators and Web sites that track them.

The Internet: More of the Same or Something New?

It is valuable to examine what is new – and what is not so new – about the Internet. If the Internet is 'the world's next growth engine' it merely follows other historical precedents – earlier disruptive technologies – such as railroads, the telephone, the internal combustion engine, electric power, radio, television and movies, plastics, and microelectronics (Mandel, 1999). Two earlier technologies are most often cited as having had an impact like that of the Internet: the printing press and the telegraph.

The first of these, Gutenberg's printing press, ushered in dramatic changes by permitting one-to-many communications and new ways of learning that changed little until the possibilities of networked computers enabled many-to-many communications. Now, new ways of preserving, updating and disseminating knowledge are having similar disruptive impacts (Dewar, 1997). Another Internet parallel is the telegraph, which Standage (1998) calls 'the Victorian Internet'. The printing press had been unable to alter the tyranny of distance: "sending a message 100 miles took the best part of a day – the time it took a messenger on horseback to cover the distance. This unavoidable delay had remained constant for thousands of years" (Standage, 1998, p. 2). The telegraph lives on within the communications technologies that have built upon its foundations: the telephone, the fax machine and, more recently, the Internet, which "has the most in common with its telegraphic ancestor" (Standage, 1998, p. 205). The similarities are many and include:

- communication over distances using interconnected networks;
- common rules and protocols (which enable any machine to exchange messages);
- hops from server to server akin to the hops from one telegraph office to the next;
- a codebook of 8-bit codes (Morse and ASCII codes);
- the protocols used are decided by the ITU, which was founded in 1865 to regulate international telegraphy. Its name was subsequently changed from International Telegraph Union to International Telecommunications Union;
- worries about security and personal signatures;
- on-line weddings;
- businesses became the most enthusiastic adopters of both technologies; and
- both made possible new business practices (such as the large company with control at a head office).

"Today we are repeatedly told that we are in the midst of a communications revolution. But the electric telegraph was, in many ways, far more disconcerting for the inhabitants of the time than today's advances are for us" (Standage, 1998, pp. 205-213). Stratton (1996) also sees historical origins of cyberspace in the mid-1800s telegraph and attempts to speed up circulation time.

Much of the research about the Internet has centred around its transformation of social relations. Cyberspace is thought to have effects on 'place' and space, with 'virtual places' or cyberspace as utopias or ideal 'places' for human interaction (Adams, 1997; Healy, 1996; Kitchin, 1998a, 1998b; Mitchell, 1995). The following section identifies several other aspects of the Internet that are not so new but continue established practices. Then, the chapter turns to other, arguably unprecedented aspects of the 'network of networks' and its infrastructure.

What's Not So New?

The impact of telecommunications began in 1844 with the telegraph, the first effective mechanism for telecommunication (Hugill, 1999; Standage, 1998). Almost immediately, businesses – notably banks and other financial service firms – began to take advantage of the ability to exchange information without simultaneous human movement (Beniger, 1986; Gabel, 1996). For over 150 years, then, the technologies of knowledge and information have mediated and reflected relations of power (Robins and

Webster, 1988; Symons, 1997). Corporate needs have shaped telecommunications infrastructure – and its geography – and continue to influence the Internet (Schiller, 1999). Indeed, the need to develop ways in which trust can be formed has parallels with the ways in which large-scale migration westward in the USA during the 1800s disrupted established personal forms of trust. Consequently, banks and firms pushed for institutional-based trust to extend across distances. This was followed by markets, professions, regulation, laws and legislation as trust-producing mechanisms (Zucker, 1986). Internet retailers similarly need to form trust in cyberspace with customers located potentially world-wide (Schneider, 1999).

With few exceptions, improvements in telecommunications and transportation technologies have been conceptualised as similar – collapsing or shrinking time and space, time-space compression, and the annihilation of space through time (Brunn and Leinbach, 1991; Harvey, 1989). Space and time remain significant, however, for three reasons: (1) connections and bandwidth[1] are unequally distributed, (2) information is only useful at the locale, (3) cyberspace depends on real world spatial fixity. What Batty (1997) calls 'cyberplace' is the impact of the infrastructure of cyberspace – the physical wires, satellites and networks that enable interactivity among remote computers – that is embedded in real-world places and structures. Location is not going to become irrelevant because of cyberspace (Kitchin, 1998a). As Arnum and Conti (1998, p. 1) remind us, "the new superhighway follows the old wires, rails, and roads".

There are at least two central differences between telecommunications and transportation, which have been present since the telegraph. First, moving intangible, invisible information, however, is not the same as the transportation of goods (Hillis, 1998). The distinction is real: the big on-line or e-commerce sellers are intangibles: travel and ticketing services, software, entertainment (including adult, gambling, online games and music), and financial services (Wyckoff, 1997). Tradeable information has created a variety of new opportunities within all sectors as well as creating new ones (Goddard, 1991; Evans and Wurster, 1997). Second, although many technologies embody systemic characteristics, telecommunications is at "the extreme end of the systemness spectrum" because of its primary distinctive feature: to function as a network with simultaneous utilization by many users (Rosenberg, 1994, p. 208).

A further 'old hat' aspect of the Internet is the importance of 'private roads'. Private telecommunications networks are hardly a new phenomenon. The use of leased fibre-optic lines by global firms for their internal networks was already commonplace in the 1980s (Bakis, 1987;

Hagström, 1992; Langdale, 1989; Mansell, 1994), merely continuing a trend that began in the 1870s, when US banking firms assembled coast-to-coast private telephone networks. The early private networks were created to establish better, more reliable service, not only for banks but also for newspapers to transmit telephotographs and facsimiles. After the Second World War, complaints again grew about the quality of service on the public switched telephone network (PSTN) as well as the types of service offered. The slow pace of new technology deployment was a constant complaint, and new central office (CO) switches were needed to bring data, video and voice together. The PSTNs could not obtain economies of scope for PSTN data/video services, so their costs were higher than those of private networks (Gabel, 1996). These private networks of leased lines remain the core of the Internet, and collectively are far larger than the public Internet (Coffman and Odlyzko, 1998; Paltridge, 1999). The result is that the Internet is an unregulated system into which corporate networks have hooked (Schiller, 1999). "It is possible that eventually the only communication infrastructure will be a set of interfaces among myriad private networks" (Crandall, 1997, p. 168).

What's New About the Internet?

The Internet is a convergence or fusion of information types, information media, and information operators (Kellerman, 1997). The convergence of industries has changed the products, functions, technologies, and services provided by a soft-network of applications that remain dependent on access to a hard-network infrastructure (Antonelli, 1997). Since the 1960s, new technologies and applications across several fields (including computer-aided design, remote sensing, management information systems, and data bases) have led to a convergence of computing or information technology and communications into a single phenomenon (Arnold and Guy, 1989; Hall and Preston, 1988) During the 1990s, it has become apparent that at least four formerly distinct industries – computers, communications, software, and entertainment – have merged, through substitutions and complementarities, into a single sector (Yoffie, 1997). Technological convergence has increased speed of access to and processing of information and increased control over decentralised systems, not so much abolishing geography as assisting its more efficient exploitation (Charles, 1996; Capello, 1994). Rather than shrinking space, telecommunications enables 'human extensibility' (Abler *et al.*, 1975), a concept paralleled in Melody's (1991) view that new technologies permit markets to be extended to the global level.

The Internet has had remarkable impact in its short life, already having been identified as a *general-purpose technology*, joining writing, printing, electricity, and a handful of more recent technologies, including lasers, the factory system, mass production, and flexible manufacturing (Lipsey *et al.*, 1998). Mandel's (1999) list of 'disruptive technologies' is similar: railroads, the telephone, internal combustion engine, electric power, radio, television and movies, plastics, and microelectronics. Drucker (1999, p. 50) sees e-commerce as "a totally new, totally unprecedented, totally unexpected development." Its only historical parallel is the railroad: "In the new mental geography created by the railroad, business mastered distance. In the mental geography of e-commerce, distance has been eliminated. There is only one economy and only one market" rather than local markets constrained by distance (Drucker, 1999, p. 50).

Commercialisation has been the principal driver of growth and change during the Internet's evolution. The commercial Internet is the latest of four stages in the Internet's development. The first was as a testbed or scientists' playground (most of the 1970s), followed by emergence of an Internet community (the late 1970s to 1987). The third stage was as a general academic resource (1987-1993), followed by the current period, in which we are seeing its transformation into a general, commercial information infrastructure (Thomas and Wyatt, 1999). The key date in the switch of the Internet from an academically oriented network to a commercially oriented one was 1991 when the US National Science Foundation (NSF) decided to amend the 'acceptable use policy' to allow commercial traffic. Growth of the commercial part of the Internet was pushed after WWW protocol (one of the last innovations from the 'academic' era) became accessible via a graphical user interface. One key problem for the new commercial actors was how to capture value from the Internet, i.e. how to make customers pay for Internet content. The new commercial practices are tending to undermine the attributes of the Internet, which made it successful in the first place (Thomas and Wyatt, 1999).

Nguyen and Phan (1998) suggest that both the personal computer and the Internet were technological surprises – technologies to which firms in the computer, telecommunications, and information services sectors have had to adapt and learn. Most firms' strategies looked toward the past – and were wrong, as things turned out. Convergence has not occurred at all as expected, but the Internet, and its learning by using, looks like self-organization. Six industries are feeling the greatest impact of the Internet: computing and electronics, telecommunications, financial services, retailing, energy, and travel. All of these benefit from the fact that the Internet "puts the customer in charge as never before. Until the Net, buyers faced huge obstacles to extracting the best prices and service. Research was time-consuming, and everyone from producer to retailer guarded information like the crown jewels – which it was. Quoting Gary Hamel: 'For many companies, customer ignorance was a profit center'" (Hof, 1999, p. 86). In most industries, however, the Internet "has wrought flashy but mainly superficial change" (Schlender, 1999, p. 138).

Recent dramatic improvements in the computational power and networkability of computers have created new demands for data transmission, with the result that data transfer has surpassed voice in the quantity of traffic on the public network (Coffman and Odlyzko, 1998). Many new technologies focus on *data* rather than *voice* transmission; a sign of convergence is the growing use of the Internet for IP telephony, or voice calls over the Internet. Access to the World Wide Web and its graphical interface has meant that high-speed (broadband) capability, low error rates (packet loss), and ready connectivity are the hallmarks of high-quality communications from the point of view of today's consumer.

Even as the converged telecommunications infrastructure moves toward IP in response to shifts in traffic from predominantly voice to predominantly data, it is unlikely that telecommunications infrastructure will ever handle only data. As in the past, the needs of large corporations suggest what the future will look like. For example, the call centres that large firms operate for both incoming and outgoing telephone calls increasingly require a mix of voice and data capability, so that operators can access Internet data, or so that Web customers can receive customer service with a human being. Thus, Internet (IP) networks are only partially independent of the telephone network, and even firms that focus on providing by-pass service for large businesses must be connected to the telephone network.

What remains true is that real places must be connected and the network of 'world cities' is reflected in the network of telecommunications links. Has the Internet altered the global urban hierarchy? Perhaps. London and New York rank 1-2 in international bandwidth, but Tokyo ranks only 15th, behind Amsterdam, Frankfurt, Paris, Brussels, Geneva, Stockholm, Washington, San Francisco, Toronto, Chicago, Seattle, and Vancouver (TeleGeography, 1999). Europe's major cities are well-connected on the Internet. Most of them, however, remain more expensive places for business, mainly because competition is less, and therefore fewer fibre-optic networks are available. Greater competition leads to lower prices and to newer technology. New York has more providers with networks (9), followed by San Francisco/San Jose, Los Angeles, and London (6), and Atlanta, Chicago, and Kuala Lumpur (5), Hong Kong and Toronto (4). Amsterdam, Paris, and Frankfurt, along with Mexico City follow with three networks each (Finnie, 1998). The result is predictable. Bandwidth prices in Europe are up to 10 times greater than those in the US (Shetty, 1999; Paltridge, 1999).

The critical importance of access to new technologies has highlighted the characteristic diffusion pattern: hierarchical – beginning first in large cities, where the largest markets are found, and then to progressively smaller places. Connecting major cities has been a priority of all – and especially new – communications providers. In the USA, this has meant that the large-city business routes, connecting markets in New York, the Boston-Washington corridor, and then Chicago and (via Texas) Los Angeles, are the consistent priority of telecommunications providers, both in conventional telephone networks (Langdale, 1983) and in Internet backbones (Gorman and Malecki, 2000; Moss and Townsend, 1998; Wheeler and O'Kelly, 1999). The higher level of telecommunications technology available in the largest cities has been a consistent feature, one reinforced by the concentrated location of service firms (Moss, 1991; Salomon, 1996; Sassen, 1994, 1995; Warf, 1995). The result for most cities is 'cherry-picking' by the private networks, resulting in patchworks and intensification of social and spatial inequalities (Graham and Marvin, 1995).

Castells' (1989; 1996) term, *the space of flows*, consists of three layers, which acknowledge the significance of infrastructure as well as of corporate structures:

- the first layer, the material support for the space of flows, constituted by a circuit of electronic impulses. It is largely the technological infrastructure of telecommunications networks;
- the second layer of the space of flows, constituted by its nodes and hubs, which are hierarchically organised; and
- the third layer, which refers to the spatial organisation of the dominant, managerial elites (Castells, 1996).

Nowhere is the space of flows more evident than in financial services, which have been transformed by the electronic networks that enable transactions of 'international money' to take place non-stop (Corbridge *et al.*, 1994; Leyshon and Thrift, 1997; Thrift, 1996a). In finance as well as in other producer services and knowledge-related sectors, large cities are the main locations (Sassen, 2000). These patterns are examples of the "dual causality of telecommunications and development" articulated by Qvortrup (1994, p. 163).

Both the dominance of large cities and the significance of the Internet on commerce are evident in Zook's (2000) research in mid-1998 on the geography of domain names in the US. He finds the largest numbers registered by firms in the New York (142,375), San Francisco (122,970), and Los Angeles (118,000) CMSAs. Chicago is a distant fourth with 50,222. This ranking reflects both the location of established businesses and of new firm formation. However, both rely on the availability of broadband connectivity, which has favoured cities at the top of the urban hierarchy.

The attraction of large-city markets results in several competitors, and network redundancy, in the Internet backbone on some city-pairs. The San Francisco-Los Angeles, New York-Washington, and Washington-Atlanta routes, for example, each has 20 or more providers operating direct fibre-optic service (Moss and Townsend, 1998). So much transmission capacity is available on high-volume routes that a spot market has emerged (Cavanaugh, 1999; Paltridge, 1999). However, not every city is served by every firm, with the result that broadband access is disproportionately available in large urban areas. Recent analyses of Internet backbones suggest that Internet accessibility is responding to demand beyond that measured by population alone. Rankings of US cities or metropolitan areas according to measures of their Internet connectivity show that San Francisco, Washington, and Dallas generally outrank the much larger areas of New York and Los Angeles. This finding is especially strong when bandwidth-weighted links are analysed (Malecki and Gorman, 2000; Moss and Townsend, 1998; Wheeler and O'Kelly, 1999). High-speed bandwidth is concentrated in large urban areas even more than slower forms of Internet access, a fact that constrains companies as well as individual users, and works against those in small towns and rural areas most (Conte, 1999).

There are inequalities, then, in the infrastructure that permits access to the Internet. Access can be obtained through a 'direct connection' or by dial-up access through the telephone network. Direct connections are common for large employers, who have installed the necessary hardware. Small businesses and organizations, and individuals using the Internet from home, generally rely on the infrastructure in place in the telephone network, including central office (CO) switching equipment, and on Internet service providers (ISPs). Inadequate infrastructure is a key reason for the much lower levels of Internet use outside the developed countries (Arnum and Conti, 1998; Hargittai, 1998); nowhere is the situation more grim than in Africa (Butterly, 1997).

Deregulation and Competition

The technological changes would seem evolutionary rather than revolutionary if it were not for another aspect of recent change. Occurring simultaneously with technological convergence and the emergence of the Internet as a phenomenon, deregulation and competition in telecommunications has been taking place. Privatisation of telecommunications on the US model (including mergers and acquisitions) has spread to many other countries (Schiller, 1999). The abrupt neoliberal push has brought both old trends and new trends together, yet differently in different countries (D'Haenens, 1999).

It has been evident for several years that telecommunications was never quite like transportation, using only public infrastructure. Instead, "the new 'electronic highways' of the information society are not ... public thoroughfares but are more akin to a myriad of private roads" Gillespie and Robins (1989, p. 12). Even when a firm's computer network incorporates part of the public telecommunications infrastructure, "each computer network is essentially private and proprietary" (Gillespie and Williams, 1988, p. 1317). These private roads have created 'new monopolies and empires' and new forms of global subordination and domination between cities and regions (Gillespie and Robins, 1989). The push to commercialise, privatise, and deregulate actually results in monopolization, not increased competition, and in catering to wealthy corporate customers.

Despite this warning, telecommunications is becoming less regulated. The US Federal Communications Commission (FCC), while continuing to have a strong influence on voice telephone service, has treated data services as a separate category, virtually free of regulation (Kennard, 1999; Leo and Huber, 1997). The deregulated context in the US has created many new competitors in telecommunications, including wireless carriers, competitive access providers of advanced fibre networks, and personal communications services (PCS). Many of the fibre-optic networks – now Internet backbones as well as private leased lines – predated the federal Telecommunications Act of 1996.

A new oligopoly is evolving in US telecommunications. Soon, 171 million households will be served by one of the 'big four': AT&T/TCI, Bell Atlantic/GTE, SBC/Ameritech, and MCI WorldCom. These giant firms are unlikely to serve rural areas, small cities, and even large cities where state-level regulation thwarts competition (Conte, 1999). In long-distance, the three largest networks (AT&T, MCI WorldCom, and Sprint) have been joined by Qwest/GTE, IXC, Williams, Frontier, and Level 3 (King, 1999). These new firms are pushing broadband technologies that are in high demand by Internet users. The combined data networks of UUNet (another WorldCom subsidiary), and MCI carry more than 50% of the world's Internet backbone traffic (Perrin, 1998). A portion of that has gone to Cable & Wireless as a condition for MCI WorldCom's acquisition of Sprint. There are global impacts as well. Most ISPs outside the USA lease large amounts of bandwidth to the US because it is so much cheaper. Traffic from France that passes through the US amounts to more than 50% of total French Net traffic (Cukier, 1999).

Will the Internet Eliminate Geography?

Large (increasingly global) corporations initially implemented telecommunications for control and coordination of distant operations. Corporate spatial systems, interconnected today by the Internet, give firms – large and small – the ability to transcend space, interconnecting their operations and partners with broadband, high-speed links that often are barely utilized but are there when they are needed (Odlyzko, 1998). These links, formerly dedicated lines connecting only intracorporate sites, began during the 1990s to link with collaborators and alliance partners, and this necessitated standardizing around Internet (TCP/IP) data transmission protocols (Schiller, 1999).

For over a decade, geographers have investigated information flows and activities within organizations. Hepworth (1989), for example, detailed how information systems centralised the structure of several Canadian firms. Li (1995a, b) and Symons (1997) have shown that telecommunications permits the automation of management functions, and that both centralisation and decentralisation within the firm take place. A consistent key finding is the complementarity of telecommunications and face-to-face interaction. This has two sets of implications.

First, large cities will not be replaced by cyberspace and 'telemediated' services, but will continue to be essential (Goddard and Richardson, 1996; Graham, 1998). Telecommunications is both centralising and decentralising – not just decentralising, as many early accounts predicted, because of the importance of face-to-face contact for nonroutine activities (Gaspar and Glaeser, 1998; Kellerman, 1993; Moss, 1998). The result is that some cities have become the nodes or switching centres of this network-based economy (Goddard, 1991). These cities are the nodes in the network of world cities (Knox and Taylor, 1995; Sassen, 2000).

Second, travel is increasing, not declining, among the 'islands' of competence within corporate networks, because telecommunications and travel are not substitutes (Lorentzon, 1995). Predictions that telecommunications will substitute for transportation "simply ignore the synergetic effects of improved communications on the need for face-to-face contacts that, for institutional or cultural reasons, cannot be handled on-line. The point is that better telecommunications services are likely to both encourage substitution away from transportation and induce new transportation demands" (Nicol, 1985, p. 195). Although telecommunications may reduce some travel for shopping trips, it necessitates greater use of transport for delivery of goods.

While some observers continue to predict the 'death of distance' and the 'end of geography' because of telecommunications, overall there is little sign that this is actually taking place. Cyberspace represents continuity more than discontinuity – a continuation, with minor changes, of existing practices. Companies take advantage of opportunities afforded by telecommunications for enormous locational flexibility (Gertler, 1989; Li, 1995a, b).

The Role of Infrastructure: The Layered Telecommunications Network

The Internet, the 'network of networks', is typically depicted as a 'cloud' as are the individual networks or *autonomous systems* that comprise it. In essence, users transfer data to another network without concern with the details of how the other networks operates (Bailey, 1997). Telecommunications networks are a non-standard infrastructure, made up of juxtaposed sub-networks, based on different hardware, software, and standards (Bar *et al.*, 1989). Equally importantly, interconnection involves financial settlements between firms (Huston, 1999a; Kahin and Keller, 1997; Paltridge, 1998b).

The components of the Internet's infrastructure are typically described as 'layers' (Gong and Srinagesh, 1997; Heldman, 1997). There are (at least) five layers, each of which has potential bottlenecks and problems of access:

- Internet backbones – these are the fibre-optic lines that also serve as the leased lines on which large businesses depend for internal and external communications;
- Internet service providers – these firms, both local and national, provide Internet access (i.e. connection to Internet backbones) to dial-up users via 'points of presence' (POPs) located in most local areas;

- local direct connections – these fibre links by-pass the local switched network to provide direct connections to Internet routers for large customers;
- local telephone services – households and small businesses use dial-up service through the nearest PSTN central office switch - this layer can itself comprise several layers; and
- intra-building wiring – intra-building private networks for which plumbing, not highways, is the better metaphor, because the configuration is up to the user in the building. Public carriers have to provide an increasingly heterogeneous set of services to hook up to users' diversely equipped sites (Mueller, 1996).

At the macroscale, Internet backbones, the fibre-optic lines that also serve as the leased lines on which large businesses depend for internal and external communications, determine an urban hierarchy of accessibility. Global telecom firms have concentrated their network investments in world cities, connecting them in order to serve large corporate customers that operate in many global cities (Graham, 1999; Warf, 1998). Internet service providers, both local and national, provide Internet access (i.e. connection to Internet backbones) to dial-up users via a local POP. POPs are the meso level or 'junction points' of the local infrastructure with the Internet backbone.

At the microscale, the nearest central office switch through which households and small businesses use dial-up service are characterised by different degrees of capability, and not all are able to send data at even the relatively slow 56 Kbps (bits per second) of which current modems are capable. Generally, only a few types of areas – office-based activities, such as producer services and call centres, and high-density, high-income residential areas – have been targeted as early markets for broadband, and it is unclear if innovations will trickle down to rural areas at all. A wealth of infrastructure is available, for example, in the City of London (Graham, 1999; Thrift, 1996b) and in and around New York City's Financial District (Longcore and Rees, 1996).

Internet Backbones

The core of the Internet resides in its 'backbones', long-haul fibre-optic lines that originated in NSFNET's system for interconnecting its supercomputer centres (MacKie-Mason and Varian, 1997). Indirectly, everyone who accesses the Internet uses a backbone, except (and sometimes even) for local connections. Internet service providers (ISPs) must connect with transit backbone providers (some of which are ISPs themselves). Alternatively, ISPs can attempt to save money by buying bandwidth that is increasingly available on the spot market rather than entering into a long-term contract (Cavanaugh, 1999). The backbone providers provide the infrastructure – the network of networks – in the Internet. Although there are 48 national backbone operators in the US, three firms – MCI WorldCom (which includes UUNET, ANS, and CompuServe), Cable & Wireless (which acquired MCI's backbone), and Sprint – together provide backbone service for 86% of all ISPs in the US. ISPs frequently have redundant links (with more than one backbone), but 58% of ISPs link with MCI WorldCom.

One of the more obscure but, at the same time, one of the most critical aspects of the Internet is *peering*. Only through interconnection or peering do the individual backbone networks become interconnected to form the Internet. Peering is more than simply transferring data; it also allows a peer access to a network's topology and routing tables, integrating the two networks so that data can be transited back and forth seamlessly. Originally, peering was to be done mainly at network access points (NAPs), but the growth of the Internet caused the NAPs to be overburdened, resulting in a 20-30% packet loss rate (Cukier, 1998). Lost packets, which must be sent again, cause further congestion of both the network and the NAPs as well as delay for the users. Although the NAPs and Internet Exchange (IX) Points theoretically provide complete interconnectivity of the Internet, they are both public and congested.

Table 13.1 shows the leading Internet Exchange (IX) Points globally, listed by the number of ISPs connected at each. Although the top three, and four of the top five, IXs are in the USA, only 8 of the top 25 are located there. IXs in Europe are connection points for the massive – and growing – amount of bandwidth there. Although the leading Internet route globally in 1999, measured in bandwidth available, was the London-New York link, the next six were intra-European: London-Paris, Amsterdam-Frankfurt, Amsterdam-Brussels, Brussels-London, Geneva-Paris, and Frankfurt-Geneva (TeleGeography, 1999). The numbers of ISPs connected are large, and reflect the availability of choice and competition, which, in turn, lowers the cost of service to customers.

The Internet Exchange (IX) points are central to the inner workings of the Internet, not only to the overall hierarchy of cities, because of peering. To provide the quality of service demanded by their business customers, large backbone providers started peering privately – sharing routing tables – with each other. Peering and financial settlements are the core of interconnection. An ISP must pay for knowledge of the routes that can take data onward or upstream in the Internet. Without payment, routing information is not uniformly available (Huston, 1999a). *Peer-to-peer bilateral* interconnections are private peering points established between large firms that see themselves as equals (thus the term peers) (Bailey, 1997). However, only the top tier of operators (five or six carriers globally) exchanges traffic without charge for destinations on the network. Other carriers must pay to have their traffic delivered (Staple, 1999).

Table 13.1 Leading Internet Exchange (IX) points

Internet Exchange	Location	Number of ISPs connected
MAE-East	Washington, DC	92
Chicago NAP	Chicago	83
MAE-West	San Jose, California	83
LINX (London Internet Exchange)	London	82
PAIX (Palo Alto Internet Exchange)	Palo Alto, California	80
AMS-IX (Amsterdam IX)	Amsterdam	71
M9-IX (M9 Telehouse IX)	Moscow	54
DeCIX (Deutsche Commercial IX)	Frankfurt	51
PacBell NAP	San Francisco and Los Angeles	50
HKIX (Hong Kong IX)	Hong Kong	48
SFINX (Service for French Internet Exchange)	Paris	47
VIX (Vienna IX)	Vienna	43
JPIX (Japan IX)	Tokyo	38
IIX (Indonesia IX)	Jakarta	35
New York NAP	Pennsauken, New Jersey	32
L2IX (Layer 2 Internet Exchange/DACOM-IX)	Seoul	32
BNIX (BelNet IX)	Brussels	30
CIXP (Cern IX Point)	Geneva	28
THIX (ThaiIX)	Bangkok	27
SIX (Seattle IX)	Seattle	24
NYIIX (New York International IX)	New York	23
TWIX (Taiwan IX)	Taipei	23
STIX (SingTel IX)	Singapore	22
DIX (Danish IX)	Lyngby (near Copenhagen)	21
SIX (Slovak IX)	Bratislava	21

Source: Adapted from TeleGeography, 2000, pp. 120-121.

Private peering has become so common that many backbone providers are leaving the NAPs and IXs entirely and are refusing to peer with smaller network providers. In order for small companies to get their data to a non-peering provider, they must pay transit fees to stay connected. The two-party contracts define these *hierarchical bilateral* interconnections, currently the most pervasive interconnection model in today's Internet. In general, however, the large networks do not make public their peering criteria under non-disclosure agreements, keeping smaller ISPs at a disadvantage (Bailey, 1997). Hierarchical peering acknowledges the great power wielded by the backbone providers.

The Mesoscale: Points of Presence (POPs)

The mesoscale is represented by the POP, which is typically a junction point of the high-bandwidth backbone network (Huston, 1999a). Downes and Greenstein (1998) and Greenstein (1998) have identified nearly 12,000 POPs for 3,531 ISPs, most of which were concentrated in the counties of large metropolitan statistical areas (MSAs). Over 50% of US counties do not have a single POP; at the other end of the hierarchy, nearly 10% (307) counties are well-served with alternatives and have over ten ISPs with local POPs.

Some fibre-optic networks are unrelated to backbone providers, such as the metropolitan fibre rings and other local loops installed by large LECs (Greenstein *et al.*, 1995). Many of these fibre-optic lines have been installed to serve as by-pass conduits for corporate data and, more recently, to provide virtual private network (VPN) capability to multi-locational firms.

Central Office (CO) Switches

While there are several reasons why an individual's 'download' might be slow, including modem and processor speed, the most likely reason is the capability of the local central office (CO) switch. The local telephone network relies both on the 'loop' of copper wires and fibre-optic cables and on the technology of the CO switch. The CO switch, a specialised mainframe computer that typically costs about $1 million and routes phones calls toward their destination, has grown well beyond being a simple replacement for manual operators. Sophisticated software permits caller identification, call waiting, toll-free services, custom billing and other information services (Heldman, 1997). The quality of *local* service – and of an individual customer's service – depends critically on the capabilities of the local CO switch, and the customer's distance to it, rather than on the availability of fibre-optic trunk lines. The first (or last) 100 feet to the customer remains among the most unresolved aspects of the future of telecommunications (Akimaru *et al.*, 1997; Clark, 1999; Hurley and Keller, 1999). A highly capable CO switch increasingly is part of office complexes and office parks, and room for wires and cables must be part of the buildings' architecture. The infrastructure of office buildings and complexes has become an extension of the computer and computer networks (Longcore and Rees, 1996; Moss, 1991).

The radically different method of networking represented by packet-switching poses huge challenges for the telephone network, within which the Internet largely developed. A switch is needed even in wireless networks. In this regard, the infrastructure of the telephone system, particularly the CO switch, is a key piece in widespread Internet access (Schoen *et al.*, 1998). The integration of IP into next-generation telecommunication networks, and of IP and telecommunications generally, is needed because remote access (via modem) will continue to be needed for many users. New forms of access (cable, wireless, cellular, direct satellite) also must be meshed into both existing and new networks.

Also utilising the copper infrastructure is a family of digital subscriber line (DSL) technologies that provide higher-speed data access. DSL services are considered the best technology at present for residential and small business customers who are unable to get or unable to afford the monthly cost of a fibre-optic line that by-passes the local telephone network. The advantage of asymmetric DSL (ADSL) is its correspondence to Internet usage: it transmits downstream (e.g. a Web page or video to the user) at much higher speeds (up to 8 Mbps) than upstream (e.g. a request for a link to Web page). DSL requires certain types of switching equipment, and a physical constraint imposes geographical limits as well. The highest speeds are confined to a short distance from the CO switch: 12,000-15,000 feet (less than 3 miles) from the CO. Very High Rate DSL (VDSL) provides even faster speed (51.84 Mbps), but only within 1,000 feet of a switch, declining to 12.96 Mbps at 4,500 feet (i.e. less than one mile) from the switch (Northern Telecom, 1998). Such distance-constrained technologies will be located at first in high-density neighbourhoods, where early adopters are found (Carey, 1999), and where a high density of business customers willing to pay for premium service are clustered.

CO switches are frequently installed with other switches at the same location, rather than spread evenly throughout an urban area. Colocation permits a firm to upgrade older switches for existing lines, and colocation with competitive firms is a common, if grudging, practice permitted in the USA since the 1996 Telecommunications Act, and increasingly encouraged. New technology tends to be found at these 'super switches' rather than equally across an urban area.

The Future Internet

The Internet of tomorrow has a few features detectible today. The transition to commercialism has caused several 'pressure points' for the Internet (Thomas and Wyatt, 1999). First, interconnection, which was a non-problem when no one was trying to make money, and there was no need to calculate payments for traffic flow. New actors from the telephony system have introduced 'settlements' which are a growing issue as Internet content is increasingly in the form of graphics, videos, and sound clips. A relatively simple Web page can contain 1 megabyte (Mb) or more of data. "The congestion cause by video use is pernicious; it destroys some valuable mechanisms that are part of the Internet's discipline and efficiency" (Srinagesh, 1997, p. 136). As a result, most backbone providers are upgrading to high-bandwidth fibre links and developing new mechanisms to charge not only for traffic in general but for quality of service (QoS) (Huston, 1999b; Kahin and Keller, 1997).

Second, Internet telephony continues to challenge traditional pricing models for long-distance and international service (Staple, 1999). Third, new protocols such as Ipv6 and bandwidth charging will increase inequalities in Internet service provision, but increase the possibility of providing end-to-end bandwidth guarantees (rather than just to the nearest interconnection point) that will be a valuable commodity for some customers. Although far removed from the scientific and engineering origins of the Internet, among the greatest pushes toward the future Internet are audio and video, for which quality of service transmission is being devised instead of only 'best effort' (Clark, 1999). Fourth, domain names have become hot properties, sparking lawsuits and controversy. Among the developments is that Moldova, a former Soviet republic with few exports, has taken to selling rights to its domain to medical doctors who want the instant recognition that an 'md' domain name provides.

Fifth, 'push' technology and portals make deals with ISPs to be the default browser page, creating the perception that not everyone's door to the Internet is the same. Sixth, intellectual property rights are called into question. 'Users' become 'producers' easily with HTML, but this poses a problem for people who are trying to make a living from the provision of content. Seventh, an international standard for encryption will be needed to make e-commerce truly secure. Eighth, taxation has not been important while e-commerce represents a small portion of all commerce, but the avoidance of local (and sometimes national) taxes in online commerce persists (Soete and ter Weel, 1998). Finally, censorship issues remain, whether related to cybersex, free speech, or political information by opposition groups.

Those playing significant roles in the evolution of the Internet include: the old academic and technical Internet community; Internet hardware, software, access and consultancy providers; telecommunications companies; 'traditional' media companies; the commercial sector generally; governments; individual users. Different metaphors are used by each – e.g. highways, railroads, webs, and frontiers. Tidal waves, matrices, libraries, shopping malls, village squares and town halls. A 'metaphor war' is taking place between those who use engineering metaphors (e.g. information superhighway) and those who use evolutionary metaphors, e.g. wired. Californian ideology is more into organic metaphors (growth and change) (Thomas and Wyatt, 1999). "The planet earth will don an electronic skin" (Gross, 1999).

Internet speed is increasing because of wavelength division multiplexing (WDM) on fibre. Quality of service is being devised instead of 'best effort' because of music and video. The Next Generation Internet (NGI) is being funded by the US government with two goals: (1) make sure universities have research infrastructure, and (2) make sure that students are stimulated to be creative (Clark, 1999). At the same time, Internet2 is a university-led collaboration to develop and test new technologies, applications and capabilities in a pre-commercial setting (UCAID, 1998). The greatest remaining barrier is high-speed connections to homes and small businesses; only 1% have broadband (mostly cable or DSL); rewiring homes for broadband will cost $100 billion (Clark, 1999).

As both technological change and market demand push further utilization of Internet-based business, geography and uneven local development are reaffirmed rather than eliminated. The major concentrations of corporations, particularly world cities, are served first – and most – by new infrastructure. Although Melody (1999) sees stability of existing structures, a few shifts are evident. In the US, New York and Los Angeles are perhaps sharing dominance with the San Francisco-Silicon Valley area and with Washington, DC. In Europe and Asia, however, the hierarchy seems more intact, but with some focus on a single city in each country. While the primacy of London and Paris are not new, the position of Frankfurt is more firmly established as Germany's economic capital.

Therefore, geography remains significant even as distance has become more insignificant in flows of information, money and digital content. The geography that persists is the urban hierarchy as well as the local network that connects homes and small businesses to the information highway. Here, distance is not dead and its importance has increased as new technologies and their distance constraints become more consequential. Many of the links to the information highway are unpaved paths, providing only slow access to the Internet.

Note

1 Bandwidth is the term commonly used to designate transmission speed, measured in bits per second. A simple "rule of thumb is that good video requires about a thousand times as much bandwidth as speech. A picture is truly worth a thousand words" (Mitchell (1995, p. 180, note 28). *Broadband* generally refers to transmission speeds above 64kbps, the base normal speed of a voice call (Huston 1999a). Higher bandwidths are made possible by multiplexing the base line.

References

Abler, R., Janelle, D., Philbrick, A. and Sommer, J. (1975), 'Introduction: The Study of Spatial Futures', in R. Abler, D. Janelle, A. Philbrick and J. Sommer (eds), *Human Geography in a Shrinking World*, Duxbury, North Scituate, MA, pp. 3-16.

Adams P.C. (1997), 'Cyberspace and Virtual Places', *Geographical Review*, vol. 87, pp. 155-171.

Akimaru, H., Finley, M.R. and Niu, Z. (1997), 'Elements of the Emerging Broadband Information Highway', *IEEE Communications Magazine*, vol. 35, no. 6, pp. 84-92.

Antonelli, C. (1997), 'A Regulatory Regime for Innovation in the Communications Industries', *Telecommunications Policy*, vol. 21, pp. 35-45.

Arnold, E. and Guy, K. (1989), *Policy Options for Promoting Growth through Information Technology*, Organisation for Economic Co-operation and Development, Paris, pp. 133-201.

Arnum, E. and Conti, S. (1998), 'Internet Deployment Worldwide: the New Superhighway Follows the Old Wires, Rails, and Roads'. Available at: http://www.isoc.org/inet98/proceedings/5c/5c_5.htm

Bailey J.P. (1997), 'The Economics of Internet Interconnection Agreements', in L.W. McKnight and J.P. Bailey (eds), *Internet Economics,* MIT Press, Cambridge, MA, pp. 155-168.

Bakis, H. (1987), 'Telecommunications and the Global Firm', in F.E.I. Hamilton (ed.), *Industrial Change in Advanced Economies*, Croom Helm, London, pp. 130-160.

Bakis, H., Abler, A. and Roche, E.M. (eds), (1993), *Corporate Networks, International Telecommunications and Interdependence*, Belhaven, London.

Bar, F., Borrus, M., Cohen, S. and Zysman, J. (1989), 'The Evolution and Growth Potential of Electronics-Based Technologies', *STI Review*, vol. 5, pp. 7-58.

Barry, D. (1999), 'Be an Internet Millionaire, and Be Happy', *Gainesville Sun*, August 15, vol. 1D, 7D.

Batty, M. (1997), 'Virtual Geography', *Futures*, vol. 29, pp. 337-352.

Beniger, J.R. (1986), *The Control Revolution: Technological and Economic Origins of the Information Society*, Harvard University Press, Cambridge, MA.

Brunn, S.D. and Leinbach, T.R. (eds), *Collapsing Space and Time: Geographic Aspects of Communication and Information*, Harper Collins Academic, New York.

Butterly, T. (1997), 'Constraints to the Development of the 'Wired' Economy in Africa', *Nua Internet Surveys*, Available at: http://www.nua.net/surveys/analysis/african_analysis.html

Capello, R. (1994), 'Towards New Industrial and Spatial Systems: The Role of New Technologies', *Papers in Regional Science*, vol. 73, pp. 189-208.

Carey, J. (1999), 'The First 100 Feet for Households: Consumer Adoption Patterns', in D. Hurley and J.H. Keller (eds), *The First 100 Feet: Options for Internet and Broadband Access*, Cambridge, MIT Press, MA, pp. 39-58.

Castells, M. (1989), *The Informational City*, Blackwell, Oxford.

Castells, M. (1996), *The Rise of the Network Society*, Blackwell, Oxford.

Cavanaugh, K. (1999), 'Bandwidth's New Bargaineers', *Technology Review*, vol. 101, no. 6, pp. 62-65.

Charles, D.R. (1996), 'Information Technology and Production Systems', in P.W. Daniels and W.F. Lever (eds), *The Global Economy in Transition*, Longman, London, pp. 83-102.

Clark, D. (1999), 'The Internet of Tomorrow', *Science*, vol. 285 (16 July), pp. 353.

Coffman, K.G. and Odlyzko, A.M. (1998), 'The Size and Growth Rate of the Internet', *First Monday*, vol. 3, pp. 10. Available at: http://www.firstmonday.dk/issues/issue3_10/coffman/index.html

Conte, C. (1999), 'The Telecom Disconnect', *Governing*, vol. 12, no. 10, July, pp. 20-25.

Corbridge, S., Martin, R. and Thrift, N. (eds) (1994), *Money, Power and Space*, Blackwell, Oxford.

Crandall, R.W. (1997), 'Are Telecommunications Facilities 'Infrastructure'? If They Are, So What?', *Regional Science and Urban Economics*, vol. 27, pp. 161-179.

Cukier, K.N. (1998), 'The Global Internet: A Primer', *TeleGeography 1999*, TeleGeography, Washington, pp. 112-145.

Cukier K.N. (1999), 'Bandwidth Colonialism? The Implications of Internet Infrastructure on International E-commerce', *INET99 Proceedings*, Internet Society, Reston, VA. Available at: http//www.isoc.org/inet99/proceedings/1e/1e_2.htm

D'Haenens, L. (1999), 'Beyond Infrastructure: Europe, the USA, and Canada on the Information Highway', in J. Downey and J. McGuigan (eds), *Technocities*, Sage, London, pp. 139-152.

Dewar, J.A. (1997), *The Information Age and the Printing Press: Looking Backward to See Ahead*, Rand Paper P-8014, Rand Corporation, Santa Monica, CA. Available at: http://www.rand.org/ publications/P/P8014/

Downes, T. and Greenstein, S. (1998), 'Universal Access and Local Commercial Internet Markets'. Available at: http://www.kellogg.nwu.edu/faculty/greenstein/images/research.html

Drucker, P. (1999), 'Beyond the Information Revolution', *The Atlantic Monthly*, vol. 284, no. 4, October, pp. 47-57.

Evans, P.B. and Wurster, T.S. (1997), 'Strategy and the New Economics of Information', *Harvard Business Review*, vol. 75, no. 5. pp.71-82.

Finnie, G. (1998), 'Wired Cities', *Communications Week International*, vol. 18, May, pp. 19-23.

Gabel, D. (1996), 'Private Telecommunications Networks: An Historical Perspective', in E. Noam and A. Níshúilleabháin (eds), *Private Networks Public Objectives*, Elsevier, Amsterdam, pp. 35-49.

Gaspar, J. and Glaeser, E.L. (1998), 'Information Technology and the Future of Cities', *Journal of Urban Economics*, vol. 43, pp. 136-156.

Gertler, L. (1989), 'Telecommunication and the Changing Global Context of Urban Settlements', in R.V. Knight and G. Gappert (eds), *Cities in a Global Society*, Sage, Newbury Park, CA, pp. 272-284.

Gillespie, A. and Robins, K. (1989), 'Geographical Inequalities: The Spatial Bias of the New Communications Technologies', *Journal of Communication*, vol. 39, no. 3, pp. 7-18.

Gillespie, A. and Williams, H. (1988), 'Telecommunications and the Reconstruction of Regional Comparative Advantage', *Environment and Planning A*, vol. 20, pp. 1311-1321.

Goddard, J.B. (1991), 'New technology and the Geography of the UK Information Economy', in J. Brotchie, M. Batty, P. Hall and P. Newton (eds), *Cities of the 21st Century: New Technologies and Spatial Systems*, Longman Cheshire, Melbourne, pp. 191-213.

Goddard, J.B. and Richardson, R. (1996), 'Why Geography Will Still Matter: What Jobs Go Where?', in W.H. Dutton (ed.), *Information and Communication Technologies: Visions and Realities*, Oxford University Press, Oxford, pp. 197-214.

Gong, J. and Srinagesh, P. (1997), 'The Economics of Layered Networks', in L.W. McKnight and J.P. Bailey (eds), *Internet Economics*, MIT Press, Cambridge, MA, pp. 63-75.

Gorman, S.P. and Malecki, E.J. (2000), 'The Networks of the Internet: An Analysis of Provider Networks in the USA', *Telecommunications Policy*, vol. 24, pp. 113-134.

Graham, S. (1998), 'The End of Geography or the Explosion of Place? Conceptualizing Space, Place and Information Technology', *Progress in Human Geography*, vol. 22, pp. 165-185.

Graham, S. (1999), 'Global Grids of Glass: On Global Cities, Telecommunications and Planetary Urban Networks', *Urban Studies*, vol. 36, pp. 929-949.

Graham, S. and Marvin, S. (1995), 'More than Ducts and Wires: Post-Fordism, Cities and Utility Networks', in P. Healey *et al.* (eds), *Managing Cities: The New Urban Context*, John Wiley, Chichester, pp. 169-189.

Greenstein, S. (1998), 'Universal Service in the Digital Age: The Commercialization and Geography of US Internet Access'. Available at: http://www.kellogg.nwu.edu/faculty/greenstein/images/research.html

Greenstein, S., McMaster, S. and Spiller, P.T. (1995), 'The Effect of Incentive Regulation on Infrastructure Modernization: Local Exchange Companies' Deployment of Digital Technology', *Journal of Economics and Management Strategy*, vol. 4, no. 2, Summer, pp. 187-236.

Gross, N. (1999), 'The Earth Will Don an Electronic Skin'. *Business Week*, August 30, pp. 132-134.

Hagström, P. (1992), 'Inside the 'Wired' MNC', in C. Antonelli (ed.), *The Economics of Information Networks*, North-Holland, Amsterdam, pp. 325-345.

Hall, P. and Preston, P. (1988), *The Carrier Wave: New Information Technology and the Geography of Innovation 1846-2003*, Unwin Hyman, London.

Hargittai, E. (1998), 'Holes in the Net: The Internet and International Stratification', *Internet Society INET'98 Proceedings*. Available at: http://www.isoc.org/inet98/proceedings/5d/5d_1.htm

Harvey, D. (1989), *The Condition of Postmodernity*, Blackwell, Oxford.

Healy, D. (1996), 'Cyberspace and Place: The Internet as Middle Landscape on the Electronic Frontier', in D. Porter (ed.), *Internet Culture*, Routledge, New York, pp. 55-68.

Heldman, P.K. (1997), *Competitive Telecommunications*, McGraw-Hill, New York.

Hepworth, M. (1989), *Geography of the Information Economy*, Belhaven, London.

Hillis, K. (1998), 'On the Margins: The Invisibility of Communications in Geography', *Progress in Human Geography*, vol. 22, pp. 543-566.

Hof, R.D. (1999), 'A New Era of Bright Hopes and Terrible Fears', *Business Week*, October 4, pp. 84-98.

Hugill, P.J. (1999), *Global Communications since 1844: Geopolitics and Technology*, Johns Hopkins University Press, Baltimore.

Hurley, D. and Keller, J.H. (eds) (1999), *The First 100 Feet: Options for Internet and Broadband Access*, MIT Press, Cambridge, MA.

Huston, G. (1999a), *ISP Survival Guide*, John Wiley, New York.

Huston, G. (1999b), 'Interconnection, Peering, and Settlements', in *INET99 Proceedings*, Internet Society, Reston, VA. Available at: http//www.isoc.org.inet99/proceedings/1e/1e_1.htm

Kahin, B. and Keller, J.H. (eds) (1997), *Coordinating the Internet*, MIT Press, Cambridge, MA.

Kellerman, A. (1993), *Telecommunications and Geography*, Belhaven, London.

Kennard, W.E. (1999), Remarks before the Federal Communications Bar, Northern California Chapter, San Francisco, July 20. Available at: http://www.fcc.gov

King, R. (1999), 'Too Much Long Distance', *Fortune*, vol. 139, no. 5, March 15, pp. 106-110.

Kitchin R. (1998a), *Cyberspace: The World in the Wires,* John Wiley, Chichester.

Kitchin, R. (1998b), 'Towards Geographies of Cyberspace', *Progress in Human Geography*, vol. 22, pp. 385-406.

Knox, P. and Taylor, P.J. (eds) (1995), *World Cities in a World-System*, Cambridge University Press, Cambridge.

Langdale, J.V. (1983), 'Competition in the United States Long Distance Telecommunications Industry', *Regional Studies*, vol. 17, pp. 393-409.

Langdale, J.V. (1989), 'The Geography of International Business Telecommunications: The Role of Leased Networks', *Annals of the Association of American Geographers*, vol. 79, pp. 501-522.

Leo, E. and Huber, P. (1997), 'The Incidental, Accidental Deregulation of Data … and Everything Else', *Industrial and Corporate Change*, vol. 6, pp. 807-828.

Leyshon, A. and Thrift, N. (1997), *Money/Space: Geographies of Monetary Transformation*, Routledge, London.

Li, F. (1995a), *The Geography of Business Information*, John Wiley, Chichester.

Li, F. (1995b), 'Corporate Networks and the Spatial and Functional Reorganization of Large Firms', *Environment and Planning A*, vol. 27, pp. 1627-1645.

Lipsey, R.G., Bekar, C. and Carlaw, K. (1998), 'What Requires Explanation?', in E. Helpman (ed.), *General Purpose Technologies and Economic Growth*, MIT Press, Cambridge, MA, pp. 15-54.

Longcore, T. and Rees, P. (1996), 'Information Technology and Downtown Restructuring: the Case of New York's Financial District', *Urban Geography*, vol. 17, pp. 354-372.

Lorentzon, S. (1995), 'The Use of ICT in TNCs: A Swedish Perspective on the Location of Corporate Functions', *Regional Studies*, vol. 29, pp. 673-685.

Lyne, J. (1999), 'The Call Center Location Curve: Sophistication, Site Selection Complexity Increase', *Site Selection*, vol. 44 (May), pp. 533-543.

MacKie-Mason, J.K. and Varian, H.R. (1997), 'Economic FAQs about the Internet', in L.W. McKnight and J.P. Bailey (eds), *Internet Economics*, MIT Press, Cambridge, MA, pp. 27-62.

Malecki, E.J. and Gorman, S.P. (2000), 'Maybe the Death of Distance, But Not the End of Geography: The Internet as a Network', in S.D. Brunn and T.R. Leinbach (eds), *The Worlds of Electronic Commerce*, John Wiley, Chichester (in press).

Mandel, M.J. (1999), 'The Internet Economy: The World's Next Growth Engine', *Business Week*, October 4, pp. 72-77.

Mansell, R. (1994), 'Multinational Organizations and International Private Networks: Opportunities and Constraints', in C. Steinfield, J.M. Bauer and L. Caby (eds), *Telecommunication in Transition: Policies, Services and Technologies in the European Community*, Sage, Thousand Oaks, CA, pp. 204-222.

Melody, W. (1991), 'New Telecommunications Networks and the Spatial Characteristics of Markets', in J. Brotchie, M. Batty, P. Hall and P. Newton (eds), *Cities of the 21st Century: New Technologies and Spatial Systems*, Longman Cheshire, Melbourne, pp. 65-72.

Melody, W.H. (1999), 'Mapping Information Societies', *Telecommunications Policy*, vol. 23, pp. 681-682.

Mitchell, W. (1995), *City of Bits: Space, Place and the Infobahn*, MIT Press, Cambridge, MA.

Moss, M.L. (1991), 'The Information City in the Global Economy', in J. Brotchie, M. Batty, P. Hall and P. Newton (eds), *Cities of the 21st Century: New Technologies and Spatial Systems*, Longman Cheshire, Melbourne, pp. 181-189.

Moss, M. (1998), 'Technologies and Cities', *Cityscape*, vol. 3, pp. 107-127.

Moss, M.L. and A.M. Townsend (1998), 'Spatial Analysis of the Internet in U.S. Cities and States', Paper prepared for the "Technological Futures – Urban Futures" Conference, Durham, England, April 1998. Available at: http://urban.nyu.edu/ research/newcastle/newcastle.html

Mueller, M. (1996), 'The User-Driven Network: The Present Extent of Private Networking in the United States', in E. Noam and A. Níshúilleabháin (eds), *Private Networks Public Objectives*, Elsevier, Amsterdam, pp. 65-82.

Nguyen, G.D. and Phan, D. (1998), 'Learning and the Diffusion of the Digital Paradigm in Information and Communication Technology', in S. Macdonald and G. Madden (eds), *Telecommunications and Socio-Economic Development*, Elsevier, Amsterdam, pp. 275-293.

Nicol, L. (1985), 'Communications Technology: Economic and Spatial Impacts', in M. Castells (ed.), *High Technology, Space, and Society*, Sage, Beverly Hills, CA, pp. 191-209.

Northern Telecom (1998), *Nortel Networks ISP Partner Program: High-Speed Access: xDSL and Other Alternatives for Last Mile Access*. Available at: http://www.nortelnetworks.com/prd/isppp/ gls_hispeed.html

Odlyzko, A. (1998), 'The Internet and Other Networks: Utilization Rates and Their Implications'. Available at: http://www.research.att.com/~amo/doc/internet.rates.pdf

Paltridge, S. (1998a), *Internet Infrastructure Indicators*, Organisation for Economic Co-operation and Development, Paris. Available at: http://www.oecd.org/dsti/sti

Paltridge S. (1998b), *Internet Traffic Exchange: Developments and Policy*, Organisation for Economic Co-operation and Development, Paris. Available at: http://www.oecd.org/dsti/sti

Paltridge, S. (1999), *Building Infrastructure Capacity for Electronic Commerce: Leased Line Developments and Pricing*, Organisation for Economic Co-operation and Development, Paris. Available at: http://www.oecd.org/dsti/sti

Perrin, S. (1998), 'The CLEC Market: Prospects, Problems, and Opportunities', *Telecommunications Online*, November. Available at: http://www.telecommagazine.com/issues/199811/tci/perrin.html

Qvortrup, L. (1994), 'Telematics and Regional Development: A Research Literature Review', *Prometheus*, vol. 12, pp. 152-172.

Robins, K. and Webster, F. (1988), 'Cybernetic Capitalism: Information, Technology, Everyday Life', in V. Mosco and J. Wasko (eds), *The Political Economy of Information*, University of Wisconsin Press, Madison, pp. 44-75.

Rosenberg, N. (1994), *Exploring the Black Box: Technology, Economics, and History*, Cambridge University Press, Cambridge.

Salomon, I. (1996), 'Telecommunications, Cities and Technological Opportunism', *Annals of Regional Science*, vol. 30, pp. 75-90.

Sassen, S. (1995), 'Urban Impacts of Economic Globalisation', in J. Brotchie *et al.* (eds), *Cities in Competition: Productive and Sustainable Cities for the 21st Century*, Longman Australia, Melbourne, pp. 36-57.

Sassen, S. (2000), *Cities in a World Economy*, second edition, Pine Forge Press, Thousand Oaks, CA.

Schiller, D. (1999), *Digital Capitalism: Networking the Global Market System*, MIT Press, Cambridge, MA.

Schlender, B. (1999), 'The *Real* Road Ahead', *Fortune*, vol. 140, no. 8, October 25, pp. 138-152.

Schneider, F.B. (ed.) (1999), *Trust in Cyberspace*, National Academy Press, Washington.

Schoen U., Hamann, J., Jugel, A., Kurzawa H. and Schmidt, C. (1998), 'Convergence Between Public Switching and the Internet', *IEEE Communications Magazine*, vol. 36, no. 1, pp. 50-65.

Shetty, V. (1999), 'Broadband Networks: Europe's Barons of Bandwidth'. Communications International 01 December. Available at: http://www.totaltele.com/secure/view.asp

Soete, L. and ter Weel, B. (1998), 'Cybertax', *Futures*, vol. 30, pp. 853-871.

Srinagesh, P. (1997), 'Internet Cost Structures and Interconnection Agreements', in L.W. McKnight and J.P. Bailey (eds), *Internet Economics*, MIT Press, Cambridge, MA, pp. 121-154.

Standage, T. (1998) *The Victorian Internet*, Walker, New York.

Staple, G. (1999), 'The Soft Network: Code is a Compass', in *TeleGeography 2000*, TeleGeography, Inc., Washington, DC. Available at: http://www.telegeography.com/Publications/tg00_intro.html

Stratton, J. (1996), 'Cyberspace and the Globalization of Culture', in D. Porter (ed.), *Internet Culture*, Routledge, New York, pp. 253-275.

Symons, F. (1997), 'Virtual Departments, Power, and Location in Different Organizational Settings', *Economic Geography*, vol. 73, pp. 427-444.

TeleGeography, Inc. (1999), *TeleGeography 2000*, TeleGeography, Inc., Washington, DC.

Thomas, G. and Wyatt, S. (1999), 'Shaping Cyberspace – Interpreting and Transforming the Internet', *Research Policy*, vol. 28, pp. 681-698.

Thrift, N. (1996a), 'A Phantom State? International Money, Electronic Networks and Global Cities', in N.J. Thrift, *Spatial Formations*, Sage, Thousand Oaks, CA, pp. 213-255.

Thrift, N. (1996b), 'New Urban Eras and Old Technological Fears: Reconfiguring the Goodwill of Electronic Things', *Urban Studies*, vol. 33, pp. 1463-1493.

UCAID [University Corporation for Advanced Internet Development] (1998), *Internet2: Frequently Asked Questions*. Available at: http://www.internet2.edu/html/faqs.html

Warf, B. (1995), 'Telecommunications and the Changing Geographies of Knowledge Transmission in the Late 20th Century', *Urban Studies*, vol. 32, pp. 361-378.

Warf, B. (1998), 'Reach Out and Touch Someone: AT&T's Global Operations in the 1990s', *Professional Geographer*, vol. 50, pp. 255-267.

Wheeler, D.C. and O'Kelly, M.E. (1999), 'Network Topology and City Accessibility of the Commercial Internet', *Professional Geographer*, vol. 51, pp. 327-339.

Wyckoff, A. (1997), 'Imagining the Impact of Electronic Commerce', *OECD Observer*, vol. 208 (Oct/Nov), pp. 5-8.

Yoffie, D.B. (ed.) (1997), *Competing in the Age of Digital Convergence*, Harvard Business School Press, Boston.

Zook, M (2000), 'The Web of Consumption: The Spatial Organization of the Internet Industry in the United States', *Environment and Planning A*, vol. 32, pp. 411-426.

Zucker, L.G. (1986), 'Production of Trust: Institutional Sources of Economic Structure, 1840-1920', in B.M. Staw and L.L. Cummings (eds), *Research in Organizational Behavior*, vol. 8., JAI Press, Greenwich, CT, pp. 53-111.

14 Patterns of Suburban Office Development: The Spatial Reach of Office Firms in Metropolitan Tel Aviv

Baruch A. Kipnis and Oded Borenstein

Introduction

Globally oriented post-industrial service industry, viewed as a propulsive sector in contemporary urban economies, tends to spatially agglomerate in a few urban centres equipped with rich and diversified social, cultural, and economic opportunities and with quality-of-life assets. In these agglomerated environments, office-based service can maintain strong backward and forward economic linkages both as suppliers of intermediate and final demand goods and services, and as consumers of inputs from joint ventures, sub-contractors, and service suppliers.

Until the early 1960s, offices in the major urban centres of advanced economies tended to agglomerate in CBDs, enjoying superior accessibility and benefiting from proximity to other economic activities and services. For the last three decades, metropolitan office firms in these urban agglomerations have sprawled into suburban locations (Daniels, 1974; Foley, 1956, 1957; Fulton, 1986; Pivo, 1988), in response growth stimuli, and building booms that have led to the large scale construction of modern Class A offices (Daniels, 1993).[1]

These processes began in metropolitan Tel Aviv, Israel's largest urban agglomeration, in the late 1980s. The first consequence was the emergence of the *Bursa* office cluster situated on the *Ayalon* expressway, Tel Aviv's only freeway, by the Diamond Exchange complex. The *Bursa* was built on land previously occupied by traditional manufacturing plants as part of a process of invasion and succession. It reached its peak in the mid 1990s, when suburban office space in metropolitan Tel Aviv was estimated at 316,000 square meters, and office construction projects under way and to be completed by 2000 were to add an additional 275,000 square meters of space (Borenstein, 1997).

Early studies on suburban office development (Foley, 1956; Pivo, 1988) focused on the numbers of suburban offices and the reasons for their development. As part of this work, Pivo (1990) suggested that suburban office expansion would follow two paths; the emergence of small, low-intensity office clusters built along freeways, and the emergence of a few larger clusters, at freeway intersections, that would contain a major share of a region's total office stock.

This chapter examines Pivo's hypothesis in the context of metropolitan Tel Aviv, using data on 105 office buildings in suburban metropolitan Tel Aviv compiled in 1995. In addition, using sample data for 415 office firms in these buildings, the chapter explores the significance they attach to a range of locational attributes. The chapter also examines the spatial pattern of firms' linkages with clients, suppliers and professional partners in joint ventures and sub-contracting. The nature and magnitude of suburban office development is interpreted in terms of Israel's recent entry into the post-industrial age and its integration into the global economy (Felsenstein and Razin, 1993; Kipnis, 1998a, b, c; Kipnis and Noam, 1998). These processes have been accelerated by the peace process in the Middle East, and by the absorption of over 1,000,000 immigrants between 1990 and the year 2000, mostly from the former Soviet Union. They have resulted in the rapid restructuring and upgrading of employment and the mix of industries in the Israeli economy, increased R&D expenditures, higher levels of foreign direct investment (FDI), and enhanced exports in terms of quantities, mix, and destinations (Bank Hapoalim, 1996, 1997). These changes have been attributed by Arazi (1998) to Israel's high-tech revolution, which has led some foreign investors to call Israel 'the new Silicon Valley' (Karp, 1998; Sagee and Dagoni, 1997).

During this period, metropolitan Tel Aviv, Israel's principal urban agglomeration, attracted extensive domestic and foreign investment and large-scale infrastructure developments, making it the national hub of Israel's 'post-industrial' globally-orientated economy. This role in the country's economy is expected to continue into the future, as implied in Israel's national planning efforts of the 1990s (Kipnis, 1996; Shachar, 1996): National Master Plan 31 (approved in 1992), Long-range Outline Plan for Israel 2020 (undertaken between 1992-1997) and the recently completed National Master Plan 35.

Patterns of Suburban Office Space

From his research on intra-metropolitan suburban office development in six North America cities (Los Angeles, San Francisco, Seattle, Houson, Denver and Toronto), Pivo (1990) identified four basic patterns of non-CBD office activity. These can be used as a model to explore the emerging pattern of non-CBD office agglomeration that has begun to emerge in post-industrial Israel. Pivo's four theoretical patterns of non-CBD office location are (Pivo, 1990):

- *Scattered*, low-density office developments spreading randomly across the suburban fabric of a 'monocentric metropolis'. This pattern is not associated with any particular suburban focal point or activity centre.
- *Suburban office clusters*, also known as 'urban villages', 'outer cities', 'suburban activity centres', or 'suburban downtown'. They are described as high density, prestige concentrations of urban jobs and services, usually amid lower density or near declining developments. Their location is explained by accessibility, short commuting distance, agglomeration economies, and so on.
- *Freeway corridors* are office developments along freeway routes. Their locations are determined by their visibility as well as by their accessibility to labour, to clients, and to their suppliers of goods and services.
- *A combination of the above* illustrates a situation in which office firms of different functions reveal their individual unique location preferences. Improved communication allows these firms to seek their optimal location, while at the same time to maintain adequate linkages with other firms and their subordinate departments.

Pivo (1990) concluded that spatial patterns of non-CBD office clusters resemble a '*net of mixed beads*', in most cases not in association with shopping centres. The 'nets' consist of a large number of office clusters, composed of one or two very big agglomerations (110,000-200,000 square meters), a few medium-size, and a large number of smaller clusters. The clusters were usually dispersed at irregular intervals along freeway corridors, the largest being located where freeways intersect.

Portraying a close-to-maturity stage of suburban office development, Pivo (1990) estimated that between 1960 and the late 1980s, US suburban office space increased between 2 and 10-fold depending on the region, while CBD office space declined. By the late 1980s, most of the non-CBD office space was in clusters, and each of the six metropolises he studies had between 33 and 273 clusters. Most clusters were small, but each region had a single 'primate' cluster, 2 to 4 times larger than the next largest cluster, containing some 36% of non-CBD office space. Pivo predicted that in the future new office clusters would be added at a slower rate than the addition of new office space, and that the existing clusters would grow larger. This would occur despite the fact that the very large clusters would no longer increase their share of new office space, and that a more even distribution of office space would be created (Pivo, 1990).

Data Base

Between October and December 1995, three field surveys were carried out in the urban settlements of metropolitan Tel Aviv (Figure 14.1). The metropolis extends over the Central and Tel Aviv districts and the port city of Ashdod, the northernmost part of the Southern district. It covers 1,520 square kilometers, 7% of Israel's land territory; and its 1997 population was 2.6 million, 44% of the nation's total. The *Bursa* office cluster, in the suburban city of Ramat Gan on the boundary of Tel Aviv, is considered in this study to be a suburban office cluster, though it has been said to be part of the metropolitan CBD, most of which is located in Tel Aviv city (Greicer, 1991; Kipnis, 1998b). The surveys covered all 105 office buildings whose principal tenants were business or market-orientated office firms. The three surveys were as follows.

- A survey of all 105 office buildings designed to record their physical attributes and the nature of their daily operations, and to register and classify all their 1,791 tenant office firms according to the 1993 Standard Classification of Economic Activities (Central Bureau of Statistics, 1993).
- A stratified sample of 416 office firms, 23% of the previously identified population of 1,791 firms. In each firm a key informant (the owner, the director, or another key official) was asked about the firm's organisational affiliations, the importance of each of 11 locational attributes for the firm, and to indicate where (locally, in Tel Aviv city, in the rest of the metropolis, in the rest of Israel, and abroad) most of its clients, its suppliers, and the firms it maintained strong professional linkages with (joint venturers and sub-contractors) were located.
- A survey of all office building projects under construction in 1995 whose intended year of occupancy was not later than 2000.

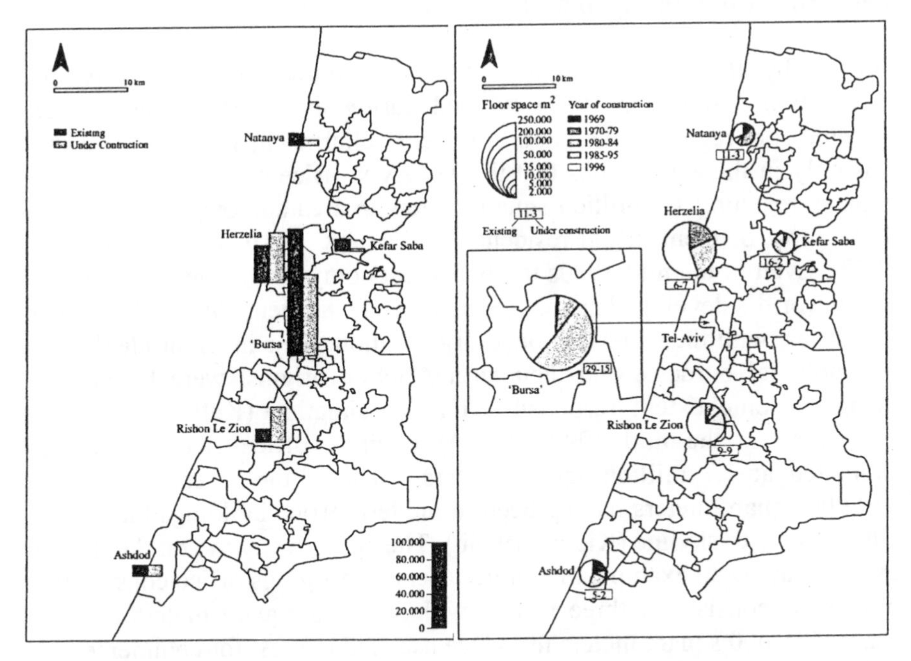

A. Office floor space existing and under construction by Cluster

B. Number of office buildings and office floor space by year of construction

Figure 14.1 Suburban office clusters in metropolitan Tel Aviv

Source: Field survey, 1995

To reveal the most important attributes of the suburban office clusters and of the firms' most important spatial linkage patterns, *coefficients of concentration* C_i were calculated,

C_i = the raw percentage of i / the raw percentage of Σ

where 'i' is the attribute under consideration. Note that the value of C_i denotes the relative importance of a given attribute variable comparable with the distribution of that variable in the entire sample. To highlight only the most important concentrations, only C_i> 1.20 are considered in this analysis.

The Office Community of Metropolitan Tel Aviv

Fuelled by the capital imports of Jewish immigrants, Tel Aviv city (population 350,000 in 1997) evolved during the 1920s as the service-oriented core region of the Land of Israel (Kellermann, 1986, 1993). But it was only in the early 1960s that the Tel Aviv office community started to expand, reaching 1.3 million square meters in area and over 10,000 office units, most of them in old residential structures, in 1974 (Har-Paz *et al.*, 1977). Even in the mid-1980s the Israeli economy still lagged behind most of the world's developed nations (Shachar and Choshen, 1993). The service industries, with their old-fashioned organisation, their inherent ideological and political burdens, and their strong public origins, were blamed for being to some extent responsible for the situation (Kellerman, 1986). However, by the mid 1990s Tel Aviv city's office floor space was estimated at 1.7 million square meters, most of the recent increase of 410,000 square meters having been in modern office buildings that totally altered the city skyline (Kipnis, 1998b). The *Azrielli Centre,* on the *Ayalon* expressway, is an example of a more recent, prestigious office construction project. It consists of three towers with 150,000 square meters of floor space, 125,000 square meters for office use, and the rest for commerce, and underground parking for 3,600 vehicles.

Although the office construction cycle that began in Tel Aviv in the late 1980s seems to have reached its peak (*Haaretz* reporter, 1995; Maor, 1996), an additional 3.3 million square meters of office space are projected to be built by 2020 (Pasternak, 1993). Most of the new construction will take place along the *Ayalon* expressway and its newly constructed southern and northern expansions, providing easy access to the rest of the metropolis, the rest of Israel, and the rest of the global economy via Ben-Gurion Airport.

De-concentration of office functions in Tel Aviv City started in the 1980s with the development of the *Bursa* office cluster. In the early 1990s, the *Bursa* cluster of office buildings, all class A, matured to become the northern extension of the metropolitan CBD (Graicer, 1991; Kipnis, 1998b). Table 14.1 summarises the overall de-concentration process. Two thirds of the 105 surveyed office buildings had been built since 1985, accounting for 72% of the 319,350 square meters of total floor space in 1995. Of these buildings and floor space, 45% had been added between 1990 and 1995. The number of structures under construction to be occupied by 2000 was 59 in 1995, having close to 275,000 square meters of floor space. The newly built structures were larger than the existing ones. They made up 56% of all the buildings, and 86% of all the floor space in 1995. Twenty five percent of the buildings had more than 5,000 square meters of floor space, compared with only 18% in the existing office building stock. A Chi square test revealed that the distribution of the newly built office structures is significantly different from the distribution of the existing ones, at the $p<0.001$ level.

Table 14.1 Office buildings by size and year of construction

	Existing buildings (percentages)				Buildings under construction %
	All office Buildings	Year of construction: Up to1984	1985-1989	1990+	
Less than 500 m^2	20	14	9	25	2
501 - 1,000 m^2	28	25	27	30	17
1,001 - 5,000 m^2	34	50	41	26	54
5,001 - 10,000 m^2	7	6	9	6	12
10,001 m^2 and more	11	5	14	13	15
Total	100	100	100	100	100
Total number of buildings	105	36	22	47	59
Percent of total	100	34	21	45	-- --
Total floor space (m^2) (sq.m)	319,350	91,485	85,400	142,465	274,970
Percent of total	100	28	27	45	-- --

Source: Field survey, 1995

The 'net of mixed beads' comprising the young, non-CBD office milieu of metropolitan Tel Aviv in the mid 1990s consists six spatially separate office clusters. The clusters are associated with processes of gentrification in the form of 'invasion and succession' in suburban industrial areas, and processes of suburban downtown revival. In addition, 32 office buildings, 35% of the total number of buildings, with 81,000 square meters or 25% of the suburban office floor space, are scattered randomly through Tel Aviv's suburban fabric. The 'net' has two major office clusters, both of which are gentrified industrial parks (Figure 14.1): the *Bursa*, with close to 250,000 square meters of office floor space, either existing or under construction, and Herzelia, with close to 100,000 square meters of office floor space.

Figure 14.1 shows that four other suburban office clusters - *Rishon* (Rishon Le-Zion), *Natanya*, *Kefar Saba,* and *Ashdod* - each have the same amount of existing floor space, some 12,500 - 14,000 square meters. Among these, the suburban cluster of *Rishon,* located at the edge of the inner belt of the metropolis, is notable for the amount of office space under construction. At the end of the 1990s, Rishon le Zion, Israel's fourth largest city with excellent accessibility to Tel Aviv city via the southern extension of the *Ayalon,* and plenty of potential land for development, became the third largest suburban office cluster of the metropolis. The other three clusters are '*edge cities*' of the metropolis: Natanya in the north with a population of 152,000 in 1997, Ashdod the metropolis's port city to the south with 147,000 people, and Kefar Saba to the northeast with 71,000 inhabitants in 1997. Significant, too, is the amount of office floor space under construction in small, dispersed office buildings.

In terms of their location relative to the metropolis's expressway system, the three larger clusters are located along the *Ayalon*. Among the three, the *Bursa* has the prime location at the municipal edge of Tel Aviv City, close to two major interchanges of the *Ayalon*. Herzelia is situated at the planned northern end of the *Ayalon* and its future junction with National Highway 2, connecting Tel Aviv through Natanya to Haifa, Israel's second largest metropolis. Rishon is on the southern extension of the *Ayalon,* not far from National Highway 4, connecting Tel Aviv via Ashdod with the south and metropolitan Beer Sheva.

Table 14.2 shows, in terms of coefficients of concentration (C_i > 1.20), the similarities and differences between Tel Aviv's suburban office clusters on a range of attributes. Very clearly, the *Bursa* and Herzelia, the two largest clusters, stand out in relation to the quality of their office buildings, the types of activity they accommodate (with high concentrations of business services and computer software and hardware firms), and the high concentration of firms' head offices within them. They are also notable for the emphasis their firms place on 'building image' and 'operational and convenient in-building services' as their most important location factors.

Table 14.2 Attributes of suburban office clusters in metropolitan Tel Aviv (Coefficients $C_i > 1.20$)

Attributes	Office clusters in metropolitan Tel Aviv						
	Bursa	Herzelia	Rishon	Natanya	Ashdod	Kefar Saba	Scattered
Quality grade of office buildings							
Very high (N=14)	2.60	5.00					
High (N=15)	1.35				1.40		
Medium (N=27)	1.35		1.30				1.34
Low (N=36)			1.30	1.86	4.85	1.82	
Very low (N= 13)							1.51
Average grade (5=very high-1=very low)	3.75	4.00	2.45	2.30	1.80	2.25	2.50
Type of aggregated activity:							
Business Services (N=202)	1.24	1.30					
FIRE* (N= 99)					1.26		
Consumer services (N=57)			2.04	1.46		1.75	
Government (N=11)			1.51	1.51	2.27		
Software and Hardware (N=24)	1.24	3.01					1.39
Others (N= 23)				1.45			1.45
Type of organization:							
Head office (N=13)	1.83	4.17					
Branch office (N=47)				1.42	1.59		1.33
Independent firm (N=337)							
Others (N=19)				1.75	1.75		
Location factors (rank order of N=416)							
Accessibility (N=356)							
Available parking (N=332)							
Rent payments (N= 312)							
Image of Building (N= 309)	1.21	1.20					
In-building services (N= 291)	1.29	1.22					
Accessibility to clients (N= 255)			1.24		1.44		
Proximity to services (N= 212)		1.53					
Proximity to institutions (N= 165)			1.31	1.51			
Near associated firms (N= 119)	1.40						
Convenient in-building services (N= 119)	1.50	1.44			1.61		
Near similar firms (N= 82)							

Note: FIRE= Finance, insurance and real estate
Source: Field survey, 1995

The Spatial Reach of Suburban Office Firms

To examine the spatial behaviour of the suburban office firms, C coefficients have been calculated to reveal the most significant spatial dimensions of the client, supplier and professional linkages of suburban office firms (C_i> 1.20). Values of C_i have been calculated separately for each of the 18 office activity types that the sampled firms can be divided into. The objective is to identify the most significant spatial category for each type of linkage and each type of office. Five categories of linkage length and destination have been recognised in the analysis, depicting the location of the counterpart in any linkage:

- the local area – the immediate locality of the office firm itself (its suburban cluster);
- Tel Aviv City;
- the rest of the Tel Aviv Metropolis;
- the rest of Israel; and
- abroad – the rest of the world.

Separately, for each type of linkage, the values of $C_i > 1.20$ for these spatial categories have been ranked, first, for the 18 categories of office activity and, second, for the suburban office clusters that firms, from the most localised to the least localised linkages. The results of this process are reported for linkage with clients in Table 14.3.

Firms' Linkages with Clients

Table 14.3 illustrates a hierarchy of office activities in terms of the spatial orientation of their market linkages with clients. Four activities have strong local market linkages: education, medical services, government, and manpower and security services. Real-estate dealing, management, and assessment firms serve clients located both locally and in Tel Aviv City. However, accounting and insurance firms are inclined to find clients locally and throughout metropolitan Tel Aviv. Suburban law firms, in contrast, tend to provide their services to clients located mostly in Tel Aviv City. Communication and advertising, and import and export (commerce) activities find clients everywhere other than in their own locality. Significant, too, are the suburban office activities whose firms are orientated towards markets abroad. These include, in descending order of significance, firms in computer software and hardware, commerce, finance and economic consulting, and communication and advertising.

Table 14.3 Firms spatial reach for clients - firms ranked according to their spatial reach and by suburban cluster (coefficients C_i> 1.20)

Activity / cluster	Location of main clients				
	Locally	In Tel Aviv city	In the rest of the metropolis	In the rest of Israel	Abroad
Economic activities					
Education	2.35				
Medical services – public	2.35				
Medical services – private	1.85				
Government – national, local	1.71				
Real estate - dealing, management, assessments	1.63	4.57			
Manpower, Security	1.33				
Accounting	1.24		1.22		
Insurance	1.21		1.21		
Law		3.64			
Communication and advertising		2.83	1.32	1.21	1.32
Miscellaneous		2.58		1.37	
Commerce - imports and exports		1.38		2.20	3.54
Finance and economic consulting			1.81	1.48	1.63
Finance - institutions, banks			1.54	3.15	
Travel agents, Tourist services			1.54		
Real estate – investment, development			1.30	1.45	
Engineering, Architecture				3.00	
Computers – software, hardware				1.31	6.36
Suburban office cluster					
Ashdod	2.16				
Natanya	1.64				
Rishon	1.55				
Kefar Saba	1.32				
Bursa		4.24	1.76	1.44	1.38
Scattered			1.36		
Herzelia				1.64	4.82

Source: Field survey, 1995

Among the client linkages of firms in different suburban clusters in Tel Aviv, there are quite distinctive spatial linkage patterns (Table 14.3).

Firms in Ashdod, Natanya, Rishon, and Kefar Saba tend to have principally local client linkages and to service the local market. The *Bursa*, in contrast, serves all markets, but primarily the market of Tel Aviv City. However, firms in Herzelia extend their services mainly to the rest of Israel and abroad, to the world at large.

Aside from the linkages with markets described above, similar patterns emerge when examining linkages with sub-contractors or suppliers of inputs. In the case of the former, accounting, legal and travel services have strong local links that extend to Tel Aviv City and to the rest of the metropolis. For many economic activities, Israel at large is the arena of sub-contracting. But, for some activities, such as finance and economic consulting, computer software and hardware, and commercial import and export, professional linkages extend to the world at large. For the last two of these activities, the values of C_i are very large (respectively 5.00 and 4.65), indicating the relative strength of these global-scale linkages.

When the spatial reach of firms' professional linkages is examined by suburban office cluster, Ashdod and Kefar Saba are primarily locally orientated. Herzelia and the *Bursa* have a stronger metropolitan orientation, and are linked significantly to Tel Aviv City and the rest of the metropolis. The only cluster to maintain significant professional linkages abroad is the *Bursa*, with C_i value of 5.94. The suburban clusters of Natanya and Rishon do not show a C_i value large enough to indicate that they maintain any strong (C_i>1.20) professional linkages with any particular type of destination.

In contrast to these patterns, links with suppliers appear to be spatially constrained to the immediate locality of the office or the Tel Aviv metropolitan area. Most of services are supplied within the local area or the city of Tel Aviv. Some office activities do draw the services they need from the rest of Israel and abroad (for example, computers, financial consulting, real estate and manpower services). Government is strongly nationally orientated in the suppliers it uses (C_i value of 3.79). Computer software and hardware firms and commercial import and export firms. Interestingly, communication and advertising firms, usually thought of as being locally orientated to satisfy local taste and values, draw the services they need principally from Tel Aviv City or the rest of the world, but not suburban Tel Aviv or the rest of Israel.

A distinctive pattern of input linkage is evident at the level of the individual suburban office cluster in Tel Aviv. In Ashdod, the spatial reach of office firms for inputs is very limited. The same hold true to some extent for *Bursa*, adjacent to the metropolitan CBD, whose spatial reach for services inputs favours Tel Aviv City and the rest of the metropolitan area. Office firms in Rishon, Natanya, and Kefar Saba, however, draw their services either from local sources or from the rest of Israel. Only Herzelia has strong global links for the supply of the services it uses while it still has significant links with suppliers in Tel Aviv.

Conclusions

This analysis has demonstrated the massive growth of office activity in Tel Aviv since the mid 1980s. This growth has, in turn, created new suburban office centres, whose spatial pattern closely resembles the 'net of mixed beads' identified in the US by Pivo (1990). More than 72% of Tel Aviv's office buildings have been built since 1985, and 45% since 1990. Half of the metropolitan area's office firms have been established since 1985 and over a third since 1990. Close to 70% of the firms have been in their present office building since the mid 1980s, and over 20% since the first half of the 1990s.

What this analysis has shown, however, is that the suburban office centres that have emerged in Tel Aviv are differentiated in the functions they contain and in the spatial reach of the firms that operate within them. The *Bursa* and Herzelia, are the principal suburban office centres in term of size, functional mix, and in their firms' spatial reach. Four clusters, Rishon, Natanya, Ashdod, and Kefar Saba, are mostly locally orientated, and this is also true for office firms scattered across the metropolis.

Different office activities also display different patterns of spatial linkage, reflecting the very different ways they fit into the Israeli economy. Education, medical services, government, real-estate dealing, management and assessment, and manpower and security services are mostly locally orientated for their acquisition of inputs and in terms of the clients they serve. The activities that are linked strongly to the rest of Israel and the global economy include commercial import and export, computer software and hardware, finance and economic consulting, and communication and advertising. Some activities, however are strongly linked to the city of Tel Aviv, even though they have suburban locations. These activities embrace firms engaged in the law, travel and tourist services, communication and advertising, accounting, commerce, and finance institutions. Some of these firms also have commercial links with other places inside and outside Israel.

The growth of office activities associated with the emergence of post-industrial society is, therefore, bringing new pressures to bear as the urban agglomerations within which they are found expand: pressures of building; pressures of employment concentration and spatial differentiation; and pressures on infrastructure. It is important to understand these pressures to begin to appreciate the wider impact that the tertiarisation of the economy is having in countries such as Israel.

Note

1 Class A edifices are medium-height to high office structures, charging high rents and maintenance fees, and occupied by 'primary users' such as corporation headquarters and professional service firms (Birch, 1986; Hartshorn, 1992). Class A are distinguished from class B, usually old buildings used by government and non-profit organizations, and by labor-intensive service outlets (Birch, 1986; Hartshorn, 1992).

References

Arazi, D. (1998), 'Three Spokes in the Wheel', *Arena*, Jan. 22.

Bank Hapoalim (1996, 1997), *Economic Review*, Bank Hapoalim Research Department, Tel Aviv (Hebrew).

Birch, D.L. (1986), *American Office Needs: 1985-1995.* Arthur Andersen & Co. and MIT Center for Real Estate Development, Boston.

Borenstein, O. (1997), *Sub-urbanization of Offices in a Metropolitan Region: A Case Study in Metropolitan Tel Aviv*, Master's Thesis, Faculty of Architecture, Urban and Regional Planning Program, The Technion, Haifa.

Daniels, P.W. (1974), 'New Offices in the Suburbs', in J.H. Johnson (ed.), *Suburban Growth: Geographical Processes at the Edge of the Western City*, Wiley, London, pp. 177-200.

Daniels, P.W. (1993), *Service Industries in the World Economy*, Blackwell, Oxford.

Felsenstein, D. and Razin, E. (1993), 'Post-industrial Processes in Israel and Their Impact on Spatial and Socio-Economic Organization of Israel', in A. Mazor (team leader), *Israel 2020, A General Plan for Israel in the 21st Century*, The Technion, Haifa, pp. 89-119 (Hebrew).

Foley, D.L. (1956), 'Factors in the Location of Administrative Offices', *Papers and Proceedings of the Regional Science Association* 2, pp. 318-326.

Foley, D.L. (1957), *The Suburbanization of Offices in the San Francisco Bay Area*, the Real Estate Research Program, University of California, Berkeley.

Fulton, W. (1986), 'Offices in the Dell', *Planning*, vol. 52, no. 7, pp. 13-17.

Greicer, I. (1991), 'The Central Business District of Tel Aviv in the 1980s: Structural Changes of Land Use', *Merhavim*, vol. 4, pp. 5-28 (Hebrew).

Haaretz reporter (1995), 'By 2000 Greater Tel Aviv will be Flooded with Office Space', *Haaretz*, January 1, 1995 (Hebrew).

Har-Paz, H., Shachar, A., Ganani, S., and Cohen, M. (1977), *Offices in Tel Aviv Yaffo: Development, Distribution and Characteristics of the Activities*, Tel Aviv Municipality, Tel Aviv, and the Department of Geography, the Hebrew University, Jerusalem (Hebrew).

Hartshorn, T.A. (1992), *Interpreting the City: An Urban Geography*, Wiley, New York.

Karp, M. (1998), 'Not a Powerhouse Yet'. *Israel's Business Arena - Globes,* Jan. 22.

Kellerman, A. (1986), 'Characteristics and Trends in the Israeli Service Economy', *The Service Industry Journal*, vol. 6, no. 2, pp. 205-226.

Kellerman, A. (1993), *Society and Settlement: Jewish Land of Israel in the Twentieth Century*, State University of New York Press, Albany.

Kipnis, B.A. (1996), 'From Dispersal to Concentration: Alternating Spatial Strategies in Israel', in Y. Gradus and G. Lipshitz (eds), *The Mosaic of Israeli Geography*, Ben Gurion University of the Negev Press, Beer Sheva pp. 13-28.

Kipnis, B.A. (1998a), 'Technology and Industrial Policy for a Metropolis at the Threshold of the Global Economy: The Case of Haifa, Israel', *Urban Studies*, vol. 35, no. 4, pp. 649-662.

Kipnis, B.A. (1998b), 'Location and Relocation of Class A Office Users: Case Study in the Metropolitan CBD of Tel Aviv, Israel', *Geographical Research Forum*, vol. 18, pp. 64-82.

Kipnis, B.A. (1998c), 'Spatial Reach of Office Firms: Case Study in the Metropolitan CBD of Tel Aviv, Israel', *Geografiska Annaler* 80 B, 1998, pp. 17-28.

Kipnis, B.A. and Noam, T. (1998), 'Restructuring of a Metropolitan Suburban Industrial Park: Case Study in Metropolitan Tel Aviv, Israel', *Geografiska Annaler* B, pp. 217-227.

Maor, Z. (1996), 'An Estimate: In the Central Region of Israel There Are 205,000 Square Meters of Vacant Space for Offices, Commerce and Industry'. *Ha'aretz Nadlan*, vol. 10, May 26, 1996, p. 7 (Hebrew).

Pasternak, E. (1993), *Tel Aviv - Industrial Zones - An Extension of Land-Use Definitions*, Tel Aviv Municipality, Tel-Aviv (Hebrew).

Pivo, G. (1988), 'The Intrasuburban Pattern of Office Suburbanization: Form and Impact in the San Francisco Bay Area'. In M.P. Tempe (ed.), *The City of the 21st Century,* Department of Planning, Arizona State University, Phoenix, AZ.

Pivo, G. (1990), 'The Net of Mixed Beads: Suburban Office Development in Six Metropolitan Regions', *Journal of the American Planners Association,* Autumn, pp. 457-469.

Sagee, M. and Dagoni, R. (1997), 'Business Week: Israel Is the New Silicon Valley', *Israel's Business Arena - Globes*, Jan. 2, 1997.

Shachar, A. (1996), 'National Planning at a Crossroads: The Evolution of a New Planning Doctrine for Israel', In Y. Gradus and G. Lipshitz (eds), *The Mosaic of Israeli Geography*, Ben Gurion University of the Negev Press, Beer Sheva, pp. 3-12.

Shachar, A. and Choshen, M. (1993), 'Israel Among the Nations: Comparative Evaluation of Israel's Status Between Developed and Developing Countries', *Studies in the Geography of Israel*, vol. 14, pp. 312-24 (Hebrew).

15 Hyper-Footloose Business Services: The Case of Swedish Distance Workers in the Mediterranean Sun Belt

Claes G. Alvstam and Annelie Jönsson

Introduction

One of the most spectacular phenomena of the spatial organisation of the industrial firm in the 1990s has been the rapid change of the staff functions surrounding manufacturing. After having been located in close contact with physical production, and organized as wholly-integrated and completely internalised parts of the business over the last decade, these functions have become separate, specialised, externalised and outsourced activities. They are no longer related to the principal company than any independent supplier or subcontractor. Outsourcing and the creation of an independent business services sector has thus been a means to implement improved cost-control, from the manufacturer's point of view.

The internationalisation process of conventional professional business services has also accelerated over the 1990s, as a consequence of a major restructuring. Small-scale, family-owned and domestically-oriented firms have become partners in global strategic alliances or consolidated business groups. Until recently, however, the face-to-face contact between producer and client was considered to be vital for the location of these kinds of services, generally in urban areas.

Another crucial factor influencing the dispersion of various activities within the firm has been the development of information and communication technology (ICT). The launching of the internet has already reshaped the world marketplace, and will doubtless alter the way producers of goods and services deal with customers and suppliers. This poses a challenge for the research community to find new ways of explaining the 'extended enterprise' and the virtual teams of producers and consumers that this implies.

This chapter develops a framework for the understanding of the spatial behaviour of extremely mobile forms of business services. In particular, we focus especially on outsourcing of those branches of business services that to a limited extent are dependent on frequent physical proximity to commissioners and customers, e.g. management consultancy, legal and financial advice, architectural design, construction and engineering consultants, and computer software programming development. These have all been early adaptors of new information and communication technology in order to locate their physical basing-point outside their business home-base, often in peripherical areas from a manufacturing point of view. The field-work for this study was carried out in the tourist resorts along the Spanish, French and Italian coastline, as well as in the cities of Barcelona, Toulouse, Lyon and Milan. In these locations, recent rapid growth of small enterprises, dependent on cooperation with remote clients, has occurred (Jönsson, 1998, 1999; Garvi-Kullenberg, 2000).

Our pre-understanding of this phenomenon was the observation that Swedish entrepreneurs, having grown up in the European periphery, in a country characterised by a large number of internationally active industrial companies, high taxation of personal income, as well as a harsh winter climate, have pioneered in this development, and that their experiences may well spread to larger northern European countries, e.g. U.K. and Germany. The choice of the Mediterranean coastline on one hand and Barcelona/Toulouse/Lyon/Milan on the other is based on the assumption that the pull-factors in these contrasting *milieux* are entirely different from each other. At the same time they are also complementary in that they have different sets of attractive attributes that match the push-factors behind the move of small companies from Sweden. The large tourist resorts along the Mediterranean coastline have gradually been transformed into centres with a wide variety of economic activities, and with emphasis on attracting temporary visitors to more permanent living. The successful transformation of these towns from being tourism centres to sites for the establishment of advanced business services firms, is not a clear-cut process. We can discern a natural stepwise process beginning with visiting an area as a tourist, through purchasing an apartment for semi-permanent living, to emigrating and establishing a new company on site, operating together with former clients in the home country, finding new clients in the host country, and at the same time establishing new business contacts in third countries. The cities investigated here represent various degrees of vibrant, dynamic, cosmopolitan environments attractive for new businesses. Barcelona was ranked as the sixth most attractive city in Europe after London, Paris, Frankfurt, Amsterdam and Brussels in Healey & Baker's yearly report on European cities (quoted by Garvi and Kullenberg, 2000, p.70).

Basic Definitions

The term 'hyper-footloose company' in this study is used to denote companies with high value added per employee, and an almost total lack of physical infrastructure around the added value process. The physical and technical barriers to mobility, for these firms, are thus, extremely low. By using little more than a portable computer, an inbuilt modem and a mobile phone, the manager of the company becomes totally 'mobile' or 'footloose'. The company may also be formally registered separately from the owners' residential location or from the location of its main market. The company can change its judicial domicile several times during its life-cycle. The absence of a physical context is seen as a necessary, although not sufficient condition for actual mobility. It is moreover assumed that new information and communication technologies have paved the way for the opportunity to work under hyper-footloose conditions. A distinction should also be made between physical and judicial persons. The individual, managing the company, may be more mobile than the company, and *vice versa.* As will be observed in the case-study, a hyper-footloose company may not be a one-person business only. Hyper-footloose business services are treated as a broader concept than pure 'teleworking'. They incorporate a more advanced form of service production, even though 'teleworking', (home use of telecommunication services, often a substitute to commuting), is an essential element in the firms within this study.

Whatever the case, the precondition for teleworking is usually to give the individual an opportunity to stay closer to home and to avoid time-consuming daily commuting or the need to follow the moves of the employer. This study however focuses on the opposite situation, in which the individual represents a high propensity to move, while the labour market is more immobile.

Theoretical Framework

The theoretical starting-points of this analysis are to be found in different contexts. The first context is that of theories on the emergence of independent producer services and their internationalisation process. The second, relates to theories of the impact of information and communication technology on the spatial pattern of work organization while the third deals with theories of firms in networks, in contrast to the local embeddedness of firms.

Externalisation of Business Services

The spatial behaviour of independent producer services was neglected for a long time in economic geography. Their footloose character, late arrival as a major force in economic life, and technical problems in their measurement, may all have contributed to this lack of interest. Only in the early 1980s did broader literature start to grow around service sector location. Research that has analysed producer services as an independent sector with a location pattern distinct from that of manufacturing has crystallised around the fact that independent business services are more mobile, flexible and small-scale than services that are internalised in manufacturing itself. (Bailly, 1995; Beyers, 1991; Bryson, 1997; Bryson and Daniels, 1998; Coffey, 1995; Daniels, 1985, 1991, 1993, 1995a, 1995b; Daniels and Moulaert, 1991; Gummesson, 1983; Harrington, 1995; Illeris, 1989, 1994; Lindahl-Beyers, 1999).

Information and Communication Technology

The impact of new information and communication technologies on human behaviour and societal organization is indeed one of the greatest challenges in current social science research. In this respect, a clear distinction exists between theories of new opportunities resulting from the introduction of new technologies, and the longer term indirect effects of ICT (Graham, 1998, Kitchin, 1998; Nunn *et al.*, 1998). Thus, a distinction needs to be made between the potential opportunities created by ICT, and the realised changes, directly caused by the new technologies. There is inconclusive evidence as to whether new ICT results in a decline in business travelling and commuting patterns, or not (Mokhtarian and Salomon, 1997; Skåmedal, 1999). However a high degree of proximity to clients is considered important in the producer services, due to the need for repeated and close contacts between supplier and client (Castells, 1996; Cornish, 1997; Gertler, 1995; Illeris, 1996).

Törnqvist (1996) suggests that even in the presence of ICT, there will be a continuous need for personal contact, and comes to the conclusion that well-structured, routinised information can be transmitted quickly and effectively with new ICT. Information is mostly a one-way flow, and follows well established formal channels with a low degree of insecurity. Technical accessories are not, however, sufficient when treating insecurity, unpredictability and surprise. Information associated with negotiations, orientation and investigation constitutes a main element in knowledge accumulation and renewal, diffused through direct personal contacts, dialogues and group discussions. Consequently ICT development also calls for investment in transport infrastructure, enabling people to be integrated into different networks (Lorentzon, 1997, 1998, 2000). So In practice, the introduction of new ICT has not resulted in declining travel volumes and physical contact, but rather has provided a complement for face-to-face contact, generally increasing the intensity of personal communication.

A related issue is ICT's effect on local communities, i.e. is community-enhancing rather than community-disrupting. The latter is the role usually assigned to ICT. This study is concerned with highlighting the opportunities for finding new forms of community survival.

Firms in Networks - Local Embeddedness

The view of the firm as a network of numerous relations ultimately coordinated and simultaneously bound together (Burt, 1992; Dicken-Thrift, 1992; Håkansson, 1989; Malmberg *et al.*, 1996) is a fruitful starting-point for an analysis of the anatomy of the hyper-footloose producer of professional business services. Externalised service activity includes numerous integrated agents, mutually interdependent on each other for the functioning of their business activities, but still free to make other commitments and form new alliances. The firm that is embedded in terms of social relations (Granovetter, 1985) can also be entrenched in a given locality through the loyalties, trust and norms that define its organisation (Markusen, 1994). However, the impact of new information and communication technology and the process of globalisation may very well contribute to the dissolving this spatial embeddedness, while maintaining all other forms of relationships necessary to continue the business relationship.

Data Sources

Data was collected on the basis of three rounds of field surveys conducted at the sunbelt regions along the Spanish, French and Italian coasts (the Costa del Sol in the South, along the Spanish Mediterranean coast, further to French Roussillon, Languedoc and Provence to the Ligurian coast in Italy), and in major cities in those countries (Valencia, Barcelona, Toulouse, Montpellier, Lyon, Nice and Milan). Surveys were conducted over three periods in 1998-1999, complemented by a few telephone interviews and visits to golf clubs and schools. The initial focus in the first field study was on individuals over the age of 45 with developed networks in the Swedish business community in Sweden, as well as in Spain. The second and third rounds of interviews were extended to include young people based on the pre-understanding that younger people usually have different motives for the move abroad and different reasons for their locational choice. Obviously, the younger group is over-represented in utilising new technology, and operating ICT-based business. Therefore, two major sub-populations characterise the second and third interview rounds. The first group consists of people working at a distance in one way or another, while the second is made up of young people working actively in self-owned companies. The focus, however, is on people working at a distance from a location in Spain, France or Italy. In total, 137 personal interviews were undertaken. Conducting research of this kind is a delicate matter due to a widespread concern among the interviewees that highlighting this type of economic activity may cause negative publicity in the Swedish media, and furthermore nourish populist opinions of tax evasion as being a main motive for their emigration from Sweden. Thus, there was initially a widespread hesitation among potential respondents, although such concerns were relatively easily overcome after a couple of personal contacts.

Results

The service companies studied are classified into two major groups. The first consists of Swedish companies established in the Mediterranean countries with the majority of their customers living in the same region. These companies have mainly customers of the same nationality and would probably not exist if there was not a big Swedish community, such as the one existing in the tourist resorts along the Costa del Sol and Torrevieja-Alicante. Figure 15.1 gives examples of different service companies that cater for the local market. The second group consists of companies that conduct business at a distance (Figure 15.2). It is assumed that the regions' tourist character has had an impact on the establishment of these businesses. Business owners have usually had their first acquaintance with the location while visiting as tourists. This first contact has led to permanent residence in the region.

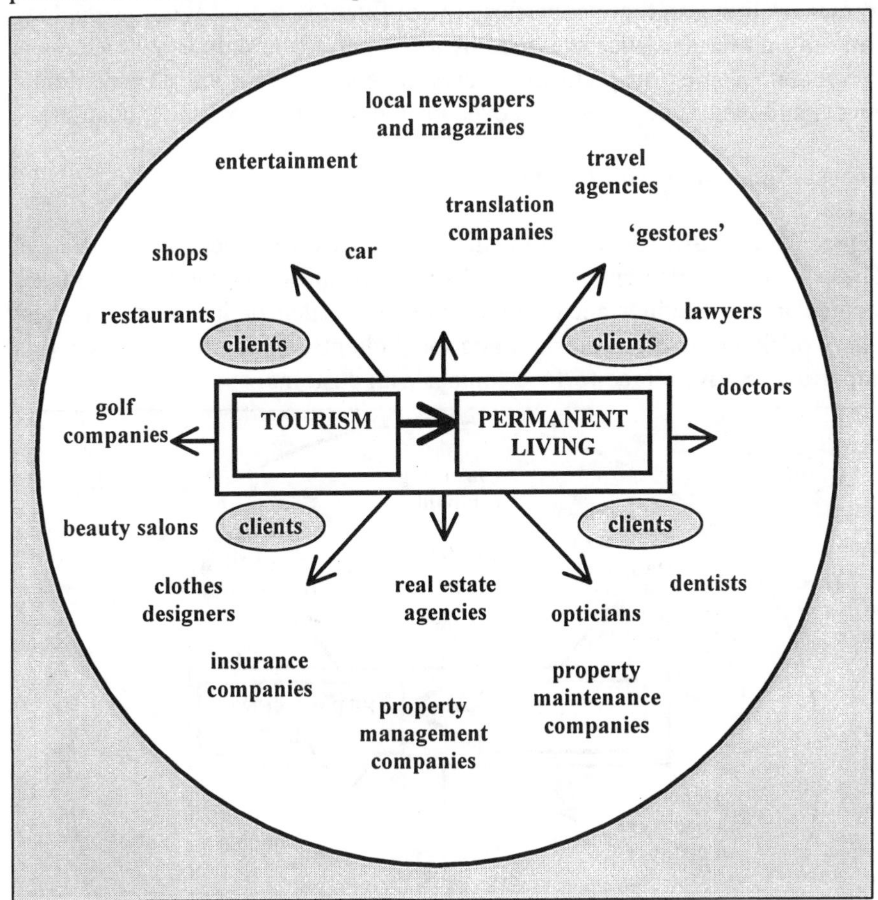

Figure 15.1 Service companies with local customers

Source: Jönsson (1999)

Service Companies with Local Customers

Service companies serving local customers are not normally seen as part of the context of hyper-footloose companies operating at a distance. Our research, however, indicates otherwise. New local companies act as complementary agents and provide a basic infrastructure that enhances and supports hyper-mobility. Local service companies are usually found in traditional service sectors. They concentrate on customers from the same country, and the customers mainly find the companies through adverts in local magazines and through personal recommendations. Since many foreigners living in these countries are not fluent in the local language, they naturally prefer a service provided by a company where their own language is spoken. Some local customer service companies such as real estate agents and insurance companies even provide services to remote customers.

Service Companies with Distance Business

Figure 15.2 shows those companies that provide services at a distance. These conduct different kinds of distance business. Not all of them are located in the Mediterranean; many are in Sweden or in other countries. Many of them have both distant and local clients, since in some cases it is important to have a face-to-face contact with customers.

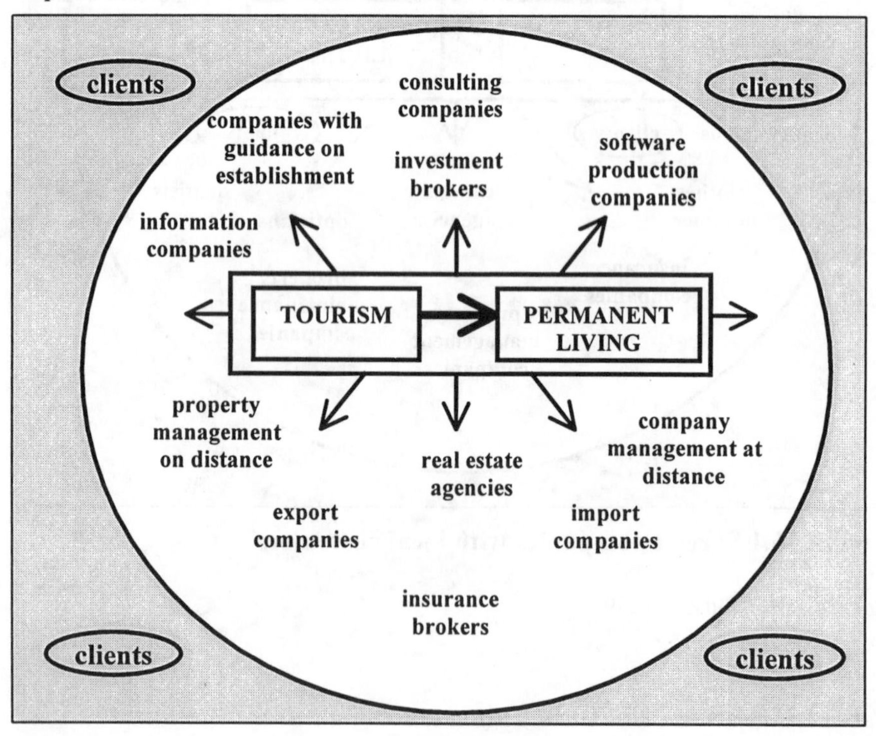

Figure 15.2 Service companies with distance business

Source: Jönsson (1999); Garvi-Kullenberg (2000)

There are many different types of companies engaged in distance business. In some instances the company is local while clients are distant, while in others the opposite is true, for example, an investment broker investing money for a local client in a trustee fund in another country.

Companies that can easily conduct business at a distance are for example consulting firms of different kinds and companies that produce software for other firms. These companies do not usually need direct contact with their clients. Other companies like real estate agents and investment and insurance brokers, can work at a distance most of the time, but periodically need face-to-face contact.

The Clients

Figure 15.3 summarises companies and clients across three categories; first those located in the Sun-Belt field study areas, second, those located in other parts of Spain, France or Italy, and those found in the rest of the world, primarily in Sweden. This classification generates nine possible groups, of which six are considered to be relevant for this study. It should be noted that this is a static picture, that does not take into consideration potential evolutionary dynamics, in which the geographical orientation of the clients will gradually change.

Company location in				
The rest of Europe and primarily in Sweden	Software producers, investment brokers		Building material company, insurance brokers	
Other areas in Spain, France, Italy	Golf material distribution companies			
The Sun Belt	Traditional companies in the service sector	Import companies	Real estate agents, investment brokers	
	The Sun Belt	Other areas in Spain, France, Italy	The rest of Europe and primarily in Sweden	Principal clientele lives in

Figure 15.3 The relationship between the location of the service company and the nationality of their principal clients; examples of typical companies within each relevant sector

Source: Adapted version of Jönsson (1999)

The Impact of New Information Technologies

The rise of the Internet has meant that the possibility of establishing a company in another area is easier now than in the past. It also makes people see the possibility to be in direct contact with clients, providers and other parts involved in the actual business, without being in face-to-face contact. Our field survey shows that some forms of communications technology are more important than are others. Among the 76 responses in the first round of field work, the most important technologies that emerged were the mobile phone and the computer with an Internet-connection. Table 15.1 shows the use of different ICT instruments amongst the responses. While very different businesses were represented in the survey, the results show a consistent use of certain forms of ICT.

Table 15.1 The frequent professional use of communication and information technologies

	Int. airport*	Mobile phone	Phone	Fax	ISDN	E-mail	Home-page	Internet	TOTAL
Real estate agents	1	8	7	2	0	8	6	8	8
Investment brokers	4	4	2	2	3	4	2	4	4
Insurance brokers	2	2	2	1	1	2	1	2	2
Software producers	1	0	0	0	1	2	1	2	2
Management of company at distance	1	2	1	0	0	1	0	0	2
Magazines & newspapers	2	2	1	0	1	2	2	2	2
Import/ export	2	2	3	3	1	2	2	3	4
TOTAL	13	20	16	8	7	21	14	21	

* The frequent use of the international airport is defined as at least once a month, either for the customer or for the producer of the service

Source: Jönsson (1999)

In almost all of the cases, extreme importance is attached to cellular communications, sending e-mail and using the Internet on a frequent basis. The ordinary telephone is also of importance, but not to the same extent. ISDN-connections are becoming more important, and will probably continue to be so in the future. The presence of an international airport is also seen as essential and a homepage on the web would seem important, both as a means of marketing the company and as a way of communicating with customers and co-workers.

Narratives of Different Types of Distance Work

One form of distance work is that conducted by the manager of the company who decides to move abroad while keeping the customers he had before moving. This sub-group is illustrated in Figure 15.4. Distance workers make the use of telephone, Internet and modem, without having to be in constant face-to-face contact with the customer.

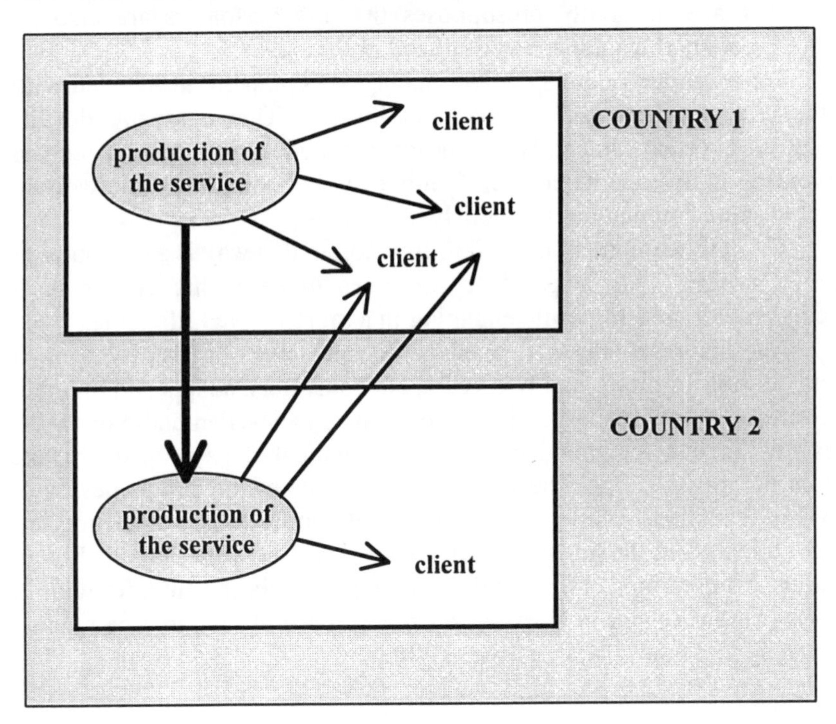

Figure 15.4 The relocation of a company in the service sector

Source: Jönsson (1999)

It is possible to move the company in the service sector from Country 1 to Country 2 while keeping the same clients. It is also possible that the same company will find clients in Country 2. Not all types of service companies however can be moved and function at a distance. This specific form of distance work will usually develop in a company that pre-exists the decision to move abroad.

Another form of distance work arises when a new company is started as a consequence of the move abroad. A third possibility is that the mover decides to close the company, existing in the origin country and to start a new company in Country 2, in order to provide a totally new service, with an entirely new clientele.

Distance Work with the Company Located in Another Area

Those people, working at a distance with their parent company located in another area, are usually providing services for large companies, especially in the area of programming and software. Distance work with the company in another area normally presupposes that the customers are also in a different geographical area, as visualized in Figure 15.5.

For example, one respondent manages a company in Sweden with a partner living where the company is located. This company develops CAD/CAM systems and software for technical education. After three years of working in Sweden he decided to move abroad doing the same work and with the same company as previously.

The only equipment needed to function in this way is a computer and Internet access. This respondent continues to serve his clients in the Swedish metal industry while engaging in a more relaxed life-style.

Another respondent engaged in distance working has lived on the Costa del Sol for more than two years. He constructs, adapts and maintains software programs for large airline companies in Sweden and Norway. He does not have a registered company in Spain, but pays Spanish taxes. Before the move to Spain he was working in one of the companies that he now has as a client. While employed by that company, he started servicing distant clients, but the geographic distances involved were obviously much smaller. All communications with customers are via the Internet with one week per month spent in Norway, for the purpose of face-to-face meetings over designing new custom software systems.

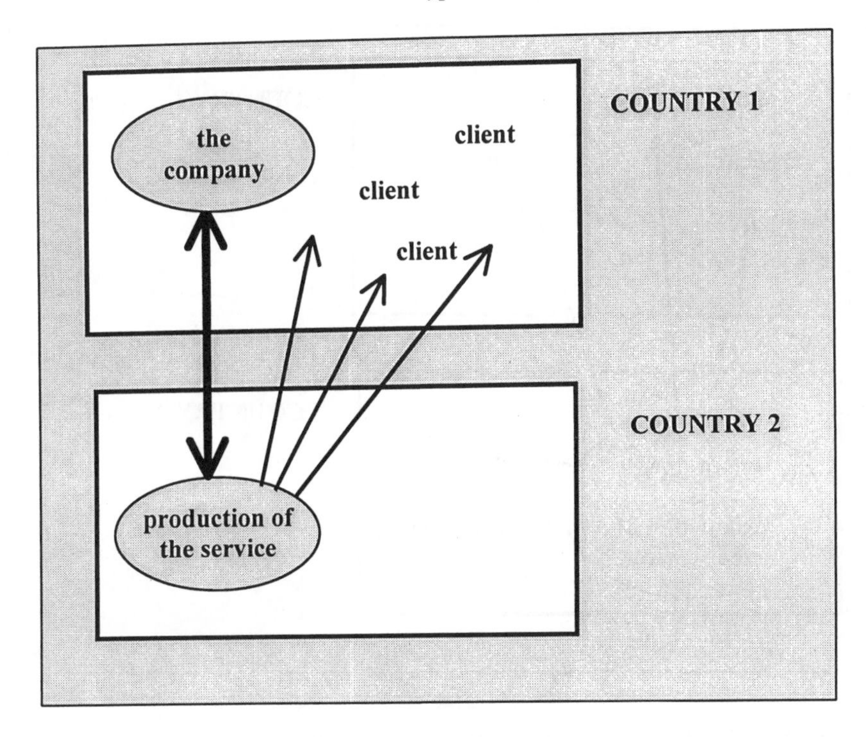

Figure 15.5 Distance work with customers and the company at distance

Source: Jönsson (1999)

Distance Work with the Customers Located in Another Area

Figure 15.6 illustrates a further form of distance work where the customers are located far away. The most typical of these are real estate agents and investment and insurance brokers. These kinds of services use ICT intensively to facilitate this form of working.

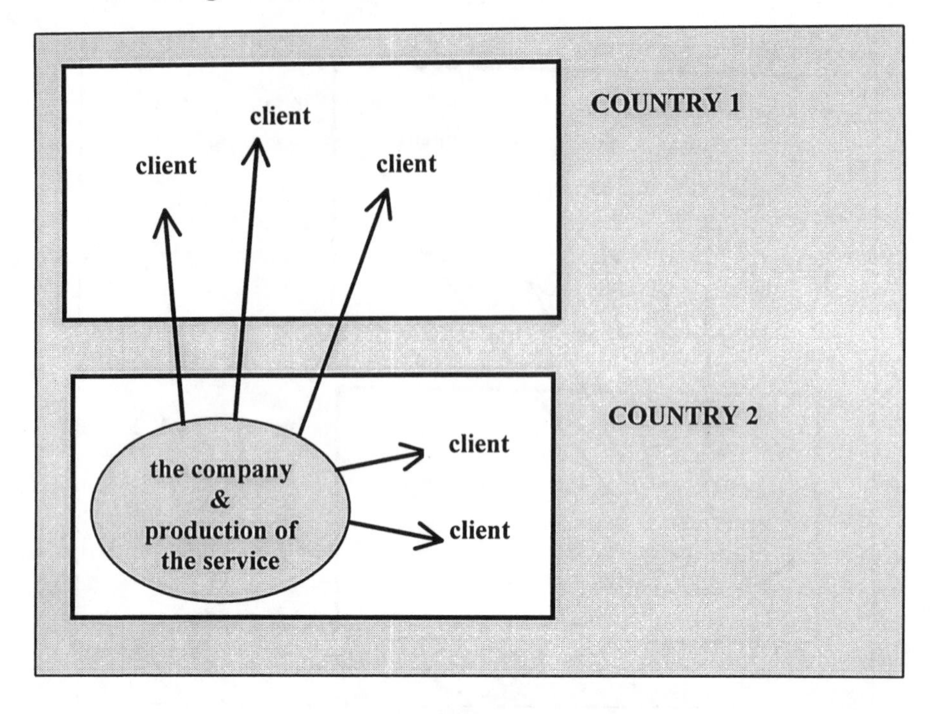

Figure 15.6 Distance work with customers at distance

Source: Jönsson (1999)

Real estate agents: as a popular resort area, Costa del Sol real estate agents service many non-local clients looking for properties for vacation or migration purposes. To operate this service at a distance, the agents make frequent use of advertising in foreign newspapers, intensive internet presence and offer multi-lingual services.

The investment brokers: are heavily represented in European sun-belt locations and in locations with large ex-patriate populations. In this instance, funds are often managed in a third location, such as off-shore tax havens like the Channel Islands or Gibraltar. Again, extensive use is made of telecommunications and face to face contact is often maintained on the basis of short visits to new clients once a month.

Insurance brokers: these service providers are again very dependent on ICT and proximity to a major airport in order to facilitate distance working. In the European sun-belt, many of these services, represent companies located elsewhere and offer facilities to ex-patriates now living at resort locations. While client networks are usually on a national basis, funds and investments are not limited to any one company or place. Clients, similarly, while networked on a national basis, can be residents anywhere in Europe.

Other forms of distance-based services: these include import and export services, business management consultants, journalists and translators. As with the fore-going services, location of service-provider and client are not really important, as long as telecommunications infrastructure and convenient (air) transportation access, are available. In all these cases there is often a national (or linguistic) basis for breaking into the market and building up a client base. After that however, location becomes marginal in both the variation of services and expansion of the network.

Barriers to Distance-Working in the European Sun Belt

One problem of distance working, especially in the European Sun-Belt relates to issues of image in the country of origin. This form of activity is seen as a rather lazy way of living and working, and distance workers are not always taken seriously by customers or competitors. In order to obscure this fact, several of the respondants have their mobile phones connected to the Swedish office numbers. By operating in this 'virtual' kind of space, location can become totally obscured with phone numbers, web sites and email addresses having little bearing on the actual location from which the service is provided.

Other problems related to distance working are social isolation, and dependence on a communications infrastructure at the destination that is inferior to that at the origin location.

Some respondents believe that telephone-connections do not work as well as they should and there have been problems with the Internet-connections. This should become better after the deregulation of the phone system that was started in December 1998, through which, even foreign providers, such as the British Telecom, are allowed to provide services on the Spanish market.

Attractiveness of the Mediterranean Region

There are, obviously, many reasons why it is attractive for people from northern Europe to engage at distance work based in the Mediterranean region. From the foregoing, the obvious question arises as to the competitiveness and economic sustainability of this form of activity. While it may satisfy the particular circumstances of the present service products, the real issue is can the Costa del Sol, the Côte d'Azur and the cities of Barcelona and Milan really compete with Swedish locations when discussing where to locate service companies within hyper-footloose sectors?

The main attractive location factors of the southern Mediterranean region to Swedish and other European 'footloose' service providers, are ease of accessibility and low transportation cost, attractive environment and quality of life, lower cost of living and the search for a different life style. Very few movers plan, a-priori, to engage in distance working. Rather, the decision to move is primate and is followed by a work-style that complements this decision.

Conclusions

The hyper-footloose producer of various kinds of professional business services has come to stay. This is due to the combined effects of outsourcing, downsizing and lean production, together with the impact of ICT and the emergence of new values, life-styles and attitudes among the new generation entering the labour-market. Together these make for a forceful shift in the direction of decentralisation and the spatial dispersal of the production activity, matched with a possible centralization of power and decision-making on a global scale.

In contrast to pure teleworking, that is usually intended to preserve a spatial pattern of settlement and labour commuting, the hyper-footloose business services producer, is extremely mobile. This business proprietor has no particular prejudices towards a certain location, and no sentiment towards a place that does not fulfil a given set of requirements needed for successful professional working and private 'quality of life'. Good accessibility to an international airport can, in most cases, compensate for a peripheral business location in other respects, although it seems that the combination of an agreeable year-around climate, proximity to a wide variety of cultural recreational facilities, and a network of people for business as well as private contacts, make for the optimal combination. This type of new company may be an attractive target for regions, that up to now have been considered to be peripherally located with respect to traditional manufacturing centres.

This study reveals that the number of enterprises representing this new type of business organization is sharply growing. It also shows that there are a number of different types of hyper-footloose business services concentrated along the Mediterranean coastline and in attractive city environments, such as Barcelona and Milan, thus requiring further work on classification and conceptualisation of these activities. The major distinction should be made between those who engage in distance work when their company is located in another area, and those using distance work when their customers are located in another area. Both these groups have their own particular characteristics, and are represented within many different sub-sectors. While no company has completely substituted face-to-face contacts with electronic communication, it does seem that personal communication can be reduced to roughly once per month. Access to new ICT, particularly ISDN, Internet and e-mail contacts, however is likely to increase the intensity of communication between company and client.

References

Alvstam, C.G. and Jönsson, A. (2000), 'On Promoting Regional Economic Growth in Industrial Peripheries: Hyper-Footloose Business Services – The Case of Swedish Teleworkers in Spain', *NETCOM*, vol. 14, no. 1-2, pp. 25-52.

Bailly, A.S. (1995), 'Producer Services Research in Europe', *Professional Geographer,* vol. 47, no. 1, pp. 70-74.

Beyers, W.B. (1991), 'Trends in the Producer Services in the USA: the Last Decade', in P.W. Daniels (ed.), *Services and Metropolitan Development – International Perspectives,* Routledge, London, pp. 146-172.

Bryson, J.R. (1997), 'Business Services Firms, Service Space and the Management of Change', *Entrepreneurship and Regional Development,* vol. 9, pp. 93-111.

Bryson, J.R. and Daniels, P.W. (eds) (1998), 'Service Industries in the Global Economy', in J.R. Bryson and P.W. Daniels (eds), *Service Theories and Service Employment,* vol. 1, Edward Elgar, Cheltenham.

Burt, R.S. (1992), *Structural Holes: the Social Structure of Competition*, Harvard University Press, Cambridge, MA.

Castells, M. (1996), *The Rise of the Network Society,* Blackwell Publishers, Oxford.

Coffey, W.J. (1995), 'Producer Services Research in Canada', *Professional Geographer,* vol. 47, no. 1, pp. 74-81.

Cornish, S.L. (1997), 'Product Innovation and the Spatial Dynamics of Market Intelligence: Does Proximity to Markets Matter?', *Economic Geography,* vol. 73, no. 2, pp. 143-165.

Daniels, P.W. (1985), *Service Industries*, Methuen, London.

Daniels, P.W. (1991), 'Producer Services and the Development of the Space Economy', in P.W. Daniels and F. Moulaert (eds), *The Changing Geography of Advanced Services. Theoretical and Empirical Perspectives*, Belhaven Press, London, pp. 135-150.

Daniels, P.W. (1993), *Service Industries in the World Economy*, Blackwell Publishers, Oxford.

Daniels, P.W. (1995a), 'Producer Services Research in the United Kingdom', *Professional Geographer,* vol. 47, no. 1, pp. 82-87.

Daniels, P.W. (1995b), 'Services in a Shrinking World', *Geography*, vol. 80, no. 347, part 2, pp. 97-110.

Daniels, P., Illeris, S., Bonamy, J. and Philippe, J. (eds), (1993), *The Geography of Services*, Frank Cass and Co, London.

Daniels, P.W. and Moulaert, F. (1991), *The Changing Geography of Advanced Producer Services*, Belhaven Press, London.

Dicken, P. and Thrift, N. (1992), 'The Organization of Production and the Production of Organization: Why Business Enterprises Matter in the Study of Geographical Industrialization', *Transactions of the Institute of British Geographers,* vol. 17, pp. 279-191.

Garvi, M. and Kullenberg, M. (2000), 'ICT, an Enabling Force – Service Location, Work Forms, and the Individual: A Case Study of Scandinavian Teleworkers in the Sun Belt', *NETCOM,* vol.14, no. 1-2, pp. 53-76.

Gertler, M.E. (1995), 'Being There: Proximity, Organization and Culture in the Development and Adoption of Advanced Manufacturing Technologies', *Economic Geography*, vol. 71, no. 1, pp. 1-26.

Graham, S. (1998), 'The End of Geography or the Explosion of Place? Conceptualizing Space, Place and Information Technology', *Progress in Human Geography,* vol. 22, no. 2, pp. 165-185.

Granovetter, M. (1985), 'Economic Action and Social Structure: The Problem of Embeddedness', *American Journal of Sociology*, vol. 91, no. 3, pp. 481-510.

Gummesson, E. (1983), *On Marketing and Management of Professional Services*, Stockholm University Department of Business Administration, Stockholm.

Harrington, J.W. (1995), 'Empirical Research on Producer Service Growth and Regional Development: International Comparisons', *Professional Geographer,* vol. 47, no. 1, pp. 66-69.

Håkansson, H. (1989), *Corporate technical behaviour, Co-operation and networks*, Routledge, London.

Illeris, S. (1989), *Services and Regions in Europe*, Avebury, Aldershot.

Illeris, S. (1994), 'Proximity Between Service Producers and Service Users', *Tijdschrift voor Economische en Sociale Geografie*, vol. 85, pp. 185-196.

Illeris, S. (1996), *The Service Economy: A Geographical Approach*, Wiley, Chichester.

Jönsson, A. (1998), *El Turismo Como Generador para el Establecimiento de Empresas Nuevas en el Sector Servicios – Empresas Suecas en la Costa del Sol*, Master Thesis in International Business (unpublished), Göteborg University, Göteborg.

Jönsson, A. (1999), *Service Industries and the New Information Technology: Swedish Distance Workers in Spain*, Occasional Papers 1999:3, Department of Human and Economic Geography, Göteborg University, Göteborg.

Kitchin, R.M. (1998), 'Towards Geographies of Cyberspace', *Progress in Human Geography*, vol. 22, no. 3, pp. 385-406.

Lindahl, D.P. and Beyers, W.B. (1999), 'The Creation of Competitive Advantage by Producer Service Establishments', *Economic Geography*, vol. 75, no. 1, pp. 1-20.

Lorentzon, S. (1997), 'The Uses of ICT in Swedish Business: Prerequisites for Local and Regional Competitiveness', in H. Bakis and E.M. Roche (eds), *Developments in Telecommunications. Between Global and Local,* Ashgate, Aldershot, pp. 255-281.

Lorentzon, S. (1998), 'The Role of ICT as a Locational Factor in Peripheral Regions – Examples from 'IT-active' Local Authority Areas in Sweden', in H. Bakis and J.M. Segui Pons (eds), 'Geospace and Cyberspace', *NETCOM,* vol. 12, no. 1-2-3, pp. 303-331.

Lorentzon, S. (2000), 'Guest Editor's Introduction', *NETCOM,* vol.14, no. 1-2, pp. 1-4.

Malmberg, A., Sölvell, Ö. and Zander, I. (1996), 'Spatial Clustering, Local Accumulation of Knowledge and Firm Competitiveness', *Geografiska Annaler*, vol. 78 B, no. 2, pp. 85-97.

Markusen, A. (1994), 'Studying Regions by Studying Firms'. *Professional Geographer,* vol. 46, no. 4, pp. 477-490.

Markusen, A. (1996), 'Sticky Places in Slippery Space. A Typology of Industrial Districts', *Economic Geography*, vol. 72, no. 3, pp. 293-313.

McCall, S. (1975), *Quality of Life*, McGill University, Montreal.

Mokhtarian, P.L. and Salomon, I. (1997), *Emerging Travel Patterns: Do Telecommunications Make a Difference?* Paper prepared for the Eighth Meeting of the International Association of Travel Behavior Research, September 1997; Revised November 1997, Austin, Texas.

Nunn, S., Warren, R. and Rubleske, J.B. (1998), 'Software Jobs go Begging: Threatening Technology Boom: Computer Services Employment in US Metropolitan Areas 1982 and 1993', *Professional Geographer*, vol. 50, no. 3, pp. 358-372.

Skåmedal, J. (1999), *Arbete på Distans och Arbetsformens Påverkan på Resor och Resmönster,* (Work at Distance and the Impact of the Mode of Work on Travels and Travel Patterns (Swedish)), Licentiate Thesis No. 752, Linköping Studies in Science and Technology, Linköping.

Törnqvist, G. (1996), *Sverige i Nätverkens Europa – Gränsöverskridandets Former och Villkor*, (Sweden in the European Network: Modes and Preconditions for the Crossing of Boundaries (Swedish)), Liber-Hermods, Malmö.

Törnqvist, G. (1998), *Renässans för* *Regioner*, (Renaissance for Regions (Swedish)), SNS, Stockholm.

Warf, B. (1995), 'Telecommunications and the Changing Geographies of Knowledge Transmission in the Late 20th Century', *Urban Studies*, vol. 32, pp. 361-378.

PART III
POLICY

Introduction to Part III

Some of the multifarious forms in which local growth takes place in practise, have been analysed in Part II. Two implications now arise with respect to public policy for promoting local growth. The first relates to the instruments of policy. The available arsenal invariably scorns heavy-handed government intervention in terms of direct investment in hard infrastructure, loans, grants and guarantees and promotes more indirect policies. These 'third-wave' policies are often aimed at providing 'soft' infrastructure such as consultancy and management inputs, incubator facilities, technology transfer, formation of local networks and linkage mechanisms and export promotion. The second implication relates to the changing focus of policy. Promoting growth is no longer associated with enticing inward investment or 'smoke-stack chasing' in general. Instead more emphasis is placed on promoting indigenous capacity, up-grading local human capital, utilizing under-utilised resources and generating local value added. Policy response has thus moved from an emphasis on a supply-side response (roads, buildings, capital) to that of promoting growth though demand-side activity.

The chapters in this section reflect these changing emphases. Using the Nordic countries as an example, Maskell (*Regional Policies: Promoting Competitiveness in the Wake of Globalisation*) shows how small, regional economies with high labour costs and a low-tech base can gain a competitive edge through suitable regional policy instruments. In his account, much of the answer lies in upgrading local human capital and embellishing local social capital. The policy prescriptions to consolidate these processes point to institutes of higher education as harbingers of a regional growth dynamic. The key question here would be whether the university functions as a regional 'anchor' attracting non-local students, research funds and technology in order to enhance the local economy or simply as a regional 'conduit' facilitating the temporary flow-through of students and funds.

Bar-El's chapter (*Promoting Regional Growth and Convergence in the Long-Term Master Plan for Israel*) takes a very different approach to regional growth policy. He examines regional disparities in Israel and differentiates between the 'old paradigm' in which struggling regions were perceived as a national burden and an emerging 'new paradigm' grounded in endogenous regional growth. The policy instrument for effecting this switch in thinking is the national long-term master-plan for Israel 2020. This strategic approach views less-favoured regions as a potential resource, considers government regional transfers as leverage for promoting regional competitiveness and stresses that public assistance should be invested in physical and human infrastructure rather than in the direct stimulation of private business activity through regional grants and loans. The chapter also provides some empirical estimates of the re-allocation of resources needed in order to set in motion a process of endogenous growth in Israel's weaker regions.

The next two chapters deal with labour market policy. Van der Laan (*Knowledge Economies and Transitional Labour Markets: New Regional Growth Engines*) looks at the change in labour supply and demand resulting from the knowledge economy. On the demand side, this is reflected in the increasing emphasis on communicative and creative abilities. On the supply side, the main feature is emergence of the transitional labour market with individuals moving between work, study, retirement and care, These, rather than gender or age categories, are likely to be the main labour market cleavages in the future. The policy implications of the demands of the knowledge economy on the one hand and the supply of transitional labour on the other are that a new view of labour skills and the policies to nurture them, is needed. This point is taken up further in Leonard's chapter (*Regulating Labour, Transforming Human Capital and Promoting Local Economic Growth: The Case of the UK Skillcentres Initiative*). His case study of the UK Skillcentres Initiative shows how the policy of handing over the market for skills to the private sector, has failed to address the existence of market imperfections in this area. An equity-based policy such as this can not be run on the lines of market efficiency. While promoting local economic growth often involves a measure of 'business first' rhetoric, in the case of education, training and labour skills, privatisation invariably fails to deliver the goods.

Finally, the policy responses of a local economy, grappling with the challenges of globalisation, are analyzed in the chapter by Le Heron and McDermott (*Rethinking Auckland: Local Response to Global Challenges*). They argue that the city of Auckland needs to adopt a series of policies that will improve its positioning in global networks. In examining the public policies needed in a host of sectors (infrastructure, telecommunications, research, leisure and entertainment) in order to jump-start this process, they present the case that market forces and deregulation are incapable of bringing about the required transformation. In this respect, this chapter brings us full circle to the opening paradox of local growth: the revival of interest in local and regional growth dynamics in an ever-globalising world.

16 Regional Policies: Promoting Competitiveness in the Wake of Globalisation

Peter Maskell

Introduction

That the global market-based economic system is an efficient device to generate economic growth seems presently to be beyond any reasonable doubt, but the market-based economic system does not function independently of society's institutions and regulatory frameworks. The development in Eastern Europe and Russia in the 1990s has made many spokesmen of rapid liberalisation relearn the old truth that the market economy can only function if carefully regulated and the interests of the different economic agents carefully balanced. No 'instant capitalism' is sustainable and freer markets depend on a 'strong society' to ensure that the results are beneficial. Adam Smith's much quoted invisible hand of the market economy functions in a satisfactory manner only if firmly fixed on the long arm of the law (Wigren and Persson, 1996).

The crucial role of specific and carefully adjusted market regulations for the economic development of nations has probably never been more obvious than in the case of the Nordic countries. This story will be briefly told in the next section in order to emphasise the changing roles of the nation state, which make similar economic transformations unlikely in the future. The crucial question is therefore how regional and national policy makers of today might enhance competitiveness. The following sections highlight several current approaches, beginning with the slippery concept of social capital and the way it is used to explain economic growth. The section after that briefly examines policies to strengthen the university-industry linkage to promote national and regional economic development. The chapter concludes with a number of reflections on competitiveness.

Moving from Periphery to Centre: the Role of Strong Institutions

The Nordic Countries' transition from a place on the economic periphery of Europe to a place in the most economically developed part of the world was not at all self-evident. Like many other nations, before the First World War, the Nordic countries were linked to the world market through the export of staples: fish from Iceland and Norway; agricultural products from Denmark; iron from Sweden; and forest product from Norway, Sweden and Finland. The absence of any single European currency or any system of fixed exchange rates enabled the countries to adjust their currency to changes in prices and productivity vis-a-vis the main foreign markets. More important were perhaps a series of on the whole successful selective *interventions*. Through these, the various Nordic governments moved on an integration-related knife edge between *too little exposure* to foreign competition (which could lead to national inefficiency in the whole industrial production system and thus necessitating further protection) and *too much exposure* (where superior foreign competitors discourage domestic firms by demonstrating their inability, thus leading to economically destructive 'import of helplessness') (Ropke, 1978).

While countries in South America, with abundant natural resources, a similarly qualified workforce and equally high growth rates of export, degenerated into *peripheries,* together with Ireland and other marginally located European countries like Portugal and Spain, the Nordic countries alone escaped, becoming *mature* capitalist *national* economies in the early 20th century (Menzel, 1980; Senghaas, 1982). A specific Nordic development trajectory can be identified, characterised by autocentric development despite world market integration (Mjoset, 1992; Adelman, 1999).

Senghass and Menzel view egalitarian income distribution and similarly egalitarian distribution of land holdings as the *crucial feature* of the Nordic development path, because it guaranteed that the *income generated* by staple exports was distributed across social strata within the various regions, just as it favoured the growth of manufacturing industries through forward and backward linkage. The Scandinavian societies' general aversion to conspicuous consumption (Veblen 1899) helped to ensure that accumulated capital was directed into productive use. The income distribution also played an important part in the process of consolidating domestic markets, which developed further in the inter-war years.

Other scholars have emphasised the way in which a common and distinct language, a shared cultural heritage, a sense of unity and participation have all reinforced a tendency towards consensus seeking in the Nordic countries. Within the consensus-seeking framework, it can be recognised both implicitly and explicitly, that criticism and conflict play a positive role in identifying arenas of interest where a compromise must be found between stakeholders in business, politics and public life in toto. This 'negotiated economy' (Hernes, 1979; Nielsen and Pedersen, 1988b) is rooted in a special sort of collective learning that takes place when all participants know that their chance of success among larger players critically depends on the degree of domestic, regional and local unity. Dissatisfied partners or neighbours means continuous problems, which will have a negative effect on everyone. The collective learning taking place when living together - yesterday, today, tomorrow and the day after - seems to convey the message that yielding on some point in order to reach a compromise will often give better long-term results than taking full advantage of a temporarily strong bargaining position. The low mobility rate among and within Western Europe thus plays an economically important though often overlooked role when investigating social capital. Few who have witnessed the functioning of the negotiated economy at close quarters will necessarily feel tempted to praise its simplicity or effectiveness. Its merit lies, however, on another level: in the way the process of reaching an agreement or decision simultaneously increases the insight into, and understanding of, the other participants positions, interests and visions. Negotiation does, in this sense, imply learning, which makes the next rounds of negotiation slightly easier and which enables not just the elite but sometimes even the society at large to reach a common perception of present and future challenges and the way the society might proceed. When imported disruptions necessitate rapid restructuring and unlearning, the necessary framework can already be in place.

Taken together, these rather specific political and social factors have been seen as contributing to the placing of the Nordic Countries among the more economical advanced areas of Europe, with levels of GDP at the high end of the global spectrum. The production and reproduction of a specific political, cultural and economic identity has been a decisive factor in the local processing of growth impulses from the world market.

The specific path of this development process has further helped to develop a strong tradition of public intervention at all levels of society in the Nordic countries, adding to an existing belief (almost regardless of political inclination) that economic development does not just happen as the unintended and unanticipated outcome of pure market processes, but that things can be changed to the better as long as political control remains in the hands of the nation state. In fact, we can note that the dissimilar form and unequal distribution of such factors across nations might account for at least some national differences in economic growth rates that remain unexplained by old and new growth theories.

Withering Nations

While belief in government has historically helped the Nordic countries successfully to accomplish political goals related to welfare and economics, clouds are building in the horizon as citizens are increasingly feeling that the government cannot fully deliver the services they request or meet the expectations they might harbour. Governments often feel less and less able to provide solutions that previously were more easily forthcoming.

It might at this stage be worth recalling how the whole idea of the nation state - combining power and policy with a defined territory - is a rather new invention in the history of humankind, even though the first articulated idea of the concept can be traced back to 1312, when Pope Clement V was called in to arbitrate in the conflict between Henry VII and Robert the Wise of Sicily. In his decree, *Pastoralis Cura,* the Pope stated that a ruler (lord, king) had full sovereignty within his territory and that no ruler was subordinate to another ruler. But, as a general principle it did not gain ground until much later and was only legitimised when the Peace of Westphalia in 1648 liquidated the personal supremacy of the Pope and the Holy Roman Empire, and initiated an "era of sovereign absolutist states which recognised no superior authority" (Gross, 1968 p. xii). By the ruling strata's subsequent intentional formation of specific national institutions based on a (more or less) shared language and a partly authentic and partly produced common heritage (Grabher and Stark, 1997), outsiders could be distinguished from insiders, and the latter came to conceive a communality with others whom they did not know and of whose specific identities they were unaware. So powerful were the resulting sentiments that nationalism has become probably the most significant mobilising principle in the last two centuries (Hyman, 1999).

At the end of the millennium the advance of globalisation has, however, changed all this through a number of interrelated processes. First, there has been a *vertical transfer of power upwards* from weak to strong nation states. Increasingly, the latter group sets the agenda and calls the shots on an growing number of issues *outside* their own borders to build systems of global governance that further secure and channel the processes of globalisation to accord with their own economic or political interests and idiosyncrasies. This tendency is especially apparent in economic policies, where the possibility for even medium sized countries to follow an independent course has become seriously restricted. The sovereignty of the small Nordic countries has, however, been under threat for much longer as the economic and political strength of their trading partners and larger neighbours has severely limited their room to manoeuvre.

This upward power transfer has also involved the setting up of a number of super-national bodies, as the unrestrained liberty of states in the 20th century has increasingly been recognised as ultimately self-destructive. A number of efforts have been made to establish a community of states, based on the will of states or at least on the will of the great powers. This effort has been influenced by not only a revival of natural law thinking but also with an eye on the potential economic benefits of the globalisation process. One result has been the setting up of the European Court of Human Rights, the Court of War Crimes and other such broadly recognised super-national bodies. Another has been the international intervention in the wars in the Balkans, especially in Kosova. Also, in the field of commerce, trade agreements (e.g. GATT, WTO, NAFTA) have in reality moved power away from most nation states.

Second, there has been a *vertical transfer of power downwards* from nation states to regions and groups of citizens outside the formal governmental structure. These regions and groups have obtained previously unknown degrees of political autonomy and administrative competence. NGO's and grass roots movements have gained prominence as political forces in their own right. Regions are now increasingly in a position successfully to pursue their own agendas and to articulate visions of development that are divergent from, and sometimes in direct conflict with, national policies. This topic will be developed further in this chapter.

Third, the process of globalisation has highlighted an internal tension that has long lurked beneath the surface of the nation state. In spite of the carefully constructed systems for delegation and accountability by which a certain degree of cohesion between policy decisions and their implementation is maintained in a swelling national bureaucracy, strong countervailing tendencies have always been present, frequently furnishing some sections of the administrative apparatus with considerable autonomy. Ministries and other offices of government develop over the years a distinct style, having their own channels of recruitment, internal training systems, career paths, administrative traditions and policy practices. Such autonomous tendencies have constantly inhibited cross-departmental co-ordination within the state apparatus. What is new, however, is how this inherent tension has been radically amplified as the process of globalisation has necessitated an increase in cross-national co-ordination within administrative sectors. The traditional rivalry between departments on the national arena is affected, with final decisions being influenced by departments' abilities to establish and maintain international networks and alliances. As globalisation progresses former loyalties to the nation state are increasingly made to compete with new loyalties to network partners and external allies in Brussels, Paris, Washington, and a number of other places.

The changes from inward to more outward looking national bureaucracies is of substantial significance for regions that will themselves increasingly have to coordinate national and super-national (EU) policies within their defined territories. If, additionally, regions have the ambition to influence national decision making, they will at least to some extent have to follow the same logic as their national counterparts and develop international connections in order to regain levels of information that are no longer obtainable nationally.

Fourth, an even more important instinctive tension in the national state that has been amplified by the processes of globalisation is associated with the state's sanction and warranty of private property, and the consequent hierarchy developed among its citizens. While the monistic state by definition claims to represent the source of all power relations within its borders, it also guarantees the power structures that do not emerge from the state as such and which the state does not fully control. The process of globalisation highlights this contradiction as new opportunities arise for utilising the ownership of private capital to create and benefit from activities that take place outside the jurisdiction and territory of the nation state. The transnationally operating company is a very obvious and visible manifestation of this process, but the ability of individual citizens and firms to operate freely and to engage in economic transactions outside their country of residence places severe limitations on the policy options available to national governments.

Fifth, in parallel with the vertical shift of power from the level of the nation to super-national or supra-national levels, a *horizontal transfer* of power can be identified as significant portions of power shift *from nation states to markets,* i.e. to non-public bodies whose power and influence depend on the size and economic importance of their global market share. The voices of citizens are increasingly heard in their capacities as consumers and customers (also of public services) within or beyond their previous national affiliation.

Taken together, these five interrelated processes have substantially weakened the nation state. They make it less able to cope with the challenges posed by market and political changes or to meet the opportunities, which the same processes continuously create. Some former state authority has moved upward to stronger powers or downward to regions, some has moved sideways to non-public organisations and some has simply disappeared (Strange, 1991).

The New Economic Role of Social Capital

With the weakening of the national state, another group of players has gained importance: the sub-national regions. However, the relationship is not as it might appear. While regions, to a certain degree, could benefit from the vertical transfer of power from the nation state, no such beneficial spill-over takes place in relation to the horizontal transfer of power to non-public market organisations. These power shifts are in no way restricted to the level of the nation state but can equally well be identified at the level of the region. There is thus no simple one-to-one correspondence between the weakening of the state and the strengthening of the regions. The attention directed to the regional level is not as much the result of changes taking place at the level of the nation, as the result of changes associated with globalisation. These changes increase the need for firms to innovate and to learn when faced with new low-cost competitors, and create the view that regions might be in a better position than nations to help accumulate and protect the valuable 'social capital' that is instrumental in the process. 'Social capital' refers to the normative values and beliefs that citizens share in their everyday dealings (Bourdieu, 1980; Coleman, 1988; Burt, 1992; Putnam, 1993; Fukuyama, 1995; Woolcock, 1998). These habits provide reason and design for all sorts of rules, of which some have substantial implications for economic performance by enabling firms to improve their innovative capability and conduct business transactions without much fuss. The use of the word *capital* implies that we are dealing with an asset. The word *social* tells us that it is an asset attained through membership of a community. Social capital is therefore accumulated within the community through processes of interaction and learning (Greif, 1994) often as the unanticipated consequence of doing something else.

Social capital contributes to economic performance by reducing inter-firm transaction cost (Coase, 1937; Williamson, 1975), i.e. "search and information costs, bargaining and decision costs, policing and enforcement costs" (Dahlman 1979, p. 148). Lower search and information costs improve efficiency in resource allocation. Reduced costs for bargaining and decision making facilitate the coordination of diverse activities between firms, enabling even further division of labour (Richardson, 1953, 1972). Diminishing costs of policing and enforcement free resources to be used in ways that are more productive. This applies both to inter-firm relations and to the relations within firms between management and employees. Furthermore, it has been observed that "[W]ith different knowledge endowments and different preferences (determining individual responses), each individual will be unable to communicate all he knows, and unable to learn everything other agents know or will do" (Eliasson, 1996, p. 15). The more owners, managers, workers and stakeholders in the community share the same beliefs, judgements and values, the easier it is for them to bridge communication gaps resulting from heterogeneous individual knowledge endowments or heterogeneous preferences, or both.

Even if examples of non-collaborative attitudes might be copious among firms in communities generally characterised by a large stock of social capital, the conduct of firms is usually constrained by their knowledge of the unattractive consequences of misbehaving. Information about any kind of misbehaviour will sooner or later be available to most potential partners in the community, who in the future will tend to take their business elsewhere. Worse still, by becoming a local outcast a firm is deprived of the flow of knowledge, including its tacit elements, which can prove very difficult to substitute. The combination of incentives and penalties acts as the crucial component in a transmission mechanism, preparing new generations to accept the inheritance and to concede to its behavioural constraints. This is why communities such as Silicon Valley continue to accumulate social capital and prosper even though:

> ... nobody knows anybody else's mother there. There is no deep history, little in way of complex familial ties and little structured community. It is a world of independent - even isolated - newcomers. (Cohen and Fields, 2000, p. 191)

Firms in communities with a large stock of social capital will always have a competitive advantage to the extent that social capital helps to reduce malfeasance, induces the volunteering of reliable information, causes agreements to be honoured, enables employees to share tacit information, and places negotiators on the same wave-length. This advantage gets even bigger when the process of globalisation deepens the division of labour and thus augments the need for coordination between and among firms (Loasby, 1992; Foss and Loasby, 1998).

As a repercussion of ongoing globalisation, most formerly localised inputs, crucial to the competitiveness of firms, have been *converted into ubiquities*, making those input equally available at more or less the same cost to all firms almost regardless of location. A large domestic market is, for instance, of no advantage when global transport costs are becoming negligible; when the loyalty of customers toward national suppliers is dwindling; and when most trade barriers are being eroded. Likewise, domestic suppliers of the most efficient production machinery are no longer an unquestioned blessing, when identical equipment is available worldwide, and at essentially the same cost. The omnipresence of organisational designs of proven value makes a long industrial track record less valuable (Maskell *et al.*, 1998).

When an input becomes ubiquitous all competing firms are placed on an equal footing. What everyone has cannot constitute a competitive advantage (Dierickx and Cool, 1989; Prahalad and Hamel, 1990). To contribute to the competitiveness of firms, inputs must, therefore, be valuable, rare, and not easy to imitate, replicate, or substitute (Barney, 1991). Social capital is such an input. It is increasingly valuable, as argued above. It is abundant in less than all communities. It cannot be bought or acquired. And, most significantly, it is impossible to imitate, replicate or substitute for three reasons:

- first, the accumulation of social capital is at least in part the unintended and unanticipated outcome of activities performed to achieve another purpose. This gives rise to all sorts of causal ambiguity, the disentangling or unravelling of which might prove impossible (Lippman and Rumelt, 1982);
- second, attempts to catch up with first movers, already in possession of a large stock of social capital, are faced with time-compression diseconomies. Because social capital accumulation always requires time-consuming reiteration and habituation, no short cuts are generally available; and
- third, social capital only has value in the context within which it was accumulated. Even if we imagine that imitation was possible, it would have little prospect of providing a functional and valuable outcome in an unfamiliar setting.

As the process of globalisation has gradually converted most previously important localised inputs into ubiquities, the competitiveness of firms exposed to international competition has become increasingly associated with the only major inputs which remain largely immobile: labour and social capital.

This is why national as well regional authorities must increasingly focus on how valuable social capital is accumulated, reproduced, and protected when globalisation advances. Basing future growth on existing valuable capabilities and institutions is more likely to be successful than trying to start afresh in some unfamiliar field.

When planning for improvement in receiving systems it turns out that universities in particular act as strategic links between worldwide networks and local environments. These links move in two directions. Universities connect a 'cluster' or a whole region with centres of knowledge throughout the world. At the same time, universities mobilise regional competence in ways that can create an environment attractive to domestic and foreign firms alike (Dahlstrand, 1999).

Building New Universities - a Viable Way to Enhance Regional Learning Processes?

The establishment of new universities and other institutions of higher learning has become instrumental to regional policy in all the Nordic countries. Policy makers increasingly view such organisations as indispensable ingredients for creating attractive regional environments (see for example Patchell and Eastham, this volume). However, this might not necessarily be correct for a number of reasons.

Universities and other institutions of higher learning have certain obviously positive consequences and naturally give rise to multiplier effects. In many areas, students and teachers form the region's largest workforce. In the vicinity of the university, retail, culture and other quality services become established. University towns also often become attractive places to live in both a physical and a social sense. Not just large, but also smaller universities and institutions of higher learning can have an important symbolic value and raise the status of a region. Universities and other institutions of higher learning can also attract new firms, frequently small, as well as the research-intensive departments of large companies. Research villages and science parks are examples of such concentrations. However, it would not generally appear to be the university's own research that acts as a magnet but rather the opportunity of recruiting skilled graduates. From a Nordic perspective, it is fairly evident that the new universities largely recruit students from their vicinity. It is therefore probable that regional universities manage to mobilise a supply of skills that would otherwise not have been equipped with higher education.

Regarding the supply of graduates, the picture is not quite so clear. The extent to which newly examined graduates remain in the region or move outside it depends largely on the local labour market. In large dense regions, many graduates decide to stay on after completing their studies whereas in more sparsely populated regions, new graduates tend to move. Particularly in small regions, the most important source of graduate employment is the university itself.

The main question when assessing the role of universities for regional development is, of course, whether the university can create a greater degree of dynamism in a region. Are there substantial synergic effects when research universities and firms group together in a region? The studies that have been carried out thus far have not provided any clear evidence of synergy effects between university research and entrepreneurial success. The contacts between them are few. Despite their close proximity, universities and research institutes live in one world, small companies in another.

However, in the case of the universities of Stanford, Cambridge, and Boston (MIT), as well as the California Institute of Technology, studies show direct links and a substantial transfer of knowledge between research universities and clusters of companies. Granstrand, for example (1999) present an analysis of the firm-university interaction based on a questionnaire survey of Japanese and Swedish multinationals arguing that technological development tends to become increasingly supply-led, making competitive US universities an substantial advantage. There are numerous institutional and individual networks as well. These networks are frequently held together by key individuals who know each other well. The regulatory framework is minimal and the environments are relatively unplanned. The researchers who have studied these environments emphasise the aspects of time and relation. It takes years for synergy effects to appear and it is vital that university research is in tune with the needs of industry. It is also vital that an intimate interaction between researchers and entrepreneurs is maintained (Castells and Hall, 1994; Hall, 1997; Keeble, 1989; Saxenian, 1994; Scott, 1993; Tatsuno, 1986).

The irrefutable success of the interaction between universities and industry in a handful of places throughout the world has to be balanced against the many cases where outcomes have been more questionable, before policy recommendations are made. Nevertheless, the discourse of university-firm interaction embodies a vision that is not very promising for many rural and peripheral localities, even if local firms might experience steadily easier access to knowledge bases in universities through information and communication technology. If it is unscheduled and difficult-to-foresee knowledge spillovers and face-to-face interactions that matter, such a technology-fix is no real substitute for the real thing. The problem, however, is not only getting the funds and other resources to build a university or a school of higher learning in a remote setting but, even more, attracting teachers and researchers of sufficiently high standing to make a difference. Furthermore, and perhaps most importantly, there will often be severe problems in making graduates stay in the region and not move to the larger cities where jobs are many and often better paid, career opportunities better and the scope for pursuing individual social interests much larger. If, however, graduates are old enough when graduating they may have paired and perhaps even settled down, making them considerably less mobile and, therefore, considerably better motivated to build a future in the region. National interests in shortening students' stays at universities collide with the interests of remote regions.

Fortunately for the remotest and less populated regions most threatened by the high-tech scenario, the learning process by which firms and regions achieve competitiveness is not identical to a general shift towards research and development intensive industries (see Maskell *et al.*, 1998). On the contrary, there seem to be strong economic reasons for expecting most regions in the developed part of the world to maintain a pattern of industrial specialisation dominated by research and development *extensive* industries. Learning implies much more than research and development intensity when seen from the perspective of regional and national economic development. Equally important is the 'low-tech' learning and innovation, that takes place when firms in fairly traditional industries are innovative in the ways they handle and develop resource management, logistics, production organisation, marketing, sales, distribution, industrial relations, and so on (Maskell, 1998).

Many studies reveal how foreign direct investments is closely associated with the location of markets and industries that are already well established and which may support or appreciate the capabilities of the investor. Instead of spending large sums when attempting to become a clone of Silicon Valley, regions with a fairly traditional industrial basis could, perhaps, increase the likelihood of success by improving their existing bases while identifying and eliminating local shortcomings. Without taking into account the region's strength or weaknesses, the often substantial efforts put into acquiring high technology industries are bound to lead to disappointing results (de Vet, 1992).

Final Remarks

While the ability of nation states to close interregional disparities might not be seriously affected by the loss of power to markets, to supranational organisations and to regions, their perception of the problem and its possible solutions has certainly changed. Traditional publicly financed redistribution measures are not only supplemented by, but sometimes even succumb to, local growth enhancing policies as regions attempt to develop their own specific capabilities.

Localised capabilities are not simply the sum of every asset a region might possess. Some assets are found in most other regions, and what everyone has will never constitute an advantage. For instance, regions do need infrastructure of a certain quality but, when all competing regions have the same, high quality infrastructure, it constitutes a threshold to development but does not add much to competitiveness. Basic education, standard legal institutions or labour market policies, off-the-shelf transport systems, and carbon copy 'science parks' are all pedestrian in the sense that they will add very little by way of competitiveness. Only by doing something more, or different, than the others can a region aspire to obtain enhanced competitiveness.

The emerging policy challenge is thus to develop or enhance the value of capabilities already in existence, enabling the firms and the citizens of a region to do something to enhance demand that others cannot. The ongoing process of globalisation creates possibilities for deepening the territorial division of labour and for reaping the subsequent benefits of specialisation. The most prominent of these benefits is associated with learning as specialisation leads to the "perception of detailed anomalies that would be passed over in a broader perspective" (Loasby, 2000, p. 6). The resulting acceleration of the growth of knowledge in a region can be furthered by local universities and other organisations of higher learning. More importantly, localised knowledge creation can be translated into competitiveness, i.e. the ability to generate relatively high factor income and factor employment levels on a sustainable basis, while being and remaining exposed to international competition.

The territorial division of labour and the utilisation of localised capabilities has established the foundation for a process of economic growth – including growth in income levels – in high tech and low-tech regions alike. Actually, some of the Italian regions specialising in clothing or household items like furniture now have some of the highest income levels in Europe. Many regions in the Nordic countries have experienced similarly favourable economic development, but some are lagging behind, burdened by institutions of yesterday that hinder the development of new, valuable, localised capabilities. However, much more research is needed before it will be possible to confirm or reject the emerging suspicion that tomorrows main regional policy measure will be targeted towards eliminating such barriers to growth: by assisting in un-learning past behaviour and routines in order to bring a region on to a track of accelerated creation of valuable knowledge through specialisation accompanied by a deepened territorial division of labour.

References

Adelman, I. (1999), 'Society, Politics and Economic Development Thirty Years After', in J. Adams and F. Pigliaru (eds), *Economic Growth and Change. National and Regional Patterns of Convergence and Divergence*, Edward Elgar, Cheltenham, pp. 71-101.

Barney, J.B. (1991), 'Firm Resources and Sustained Competitive Advantage', *Journal of Management*, vol. 17, pp. 99-120.

Bourdieu, P. (1980), 'Le Capital Social: Notes Provisoires', *Actes de la Recherche en Sciences Sociales*, vol. 31 (Janvier), pp. 2-3.

Burt, R.S. (1992), *Structural Holes. The Social Structure of Competition*, Harvard University Press, Cambridge MA.

Castells, M. and Hall, P. (1994), *Technopoles of the World: The Making of 21st Century Industrial Complexes*, Routledge, London.

Coase, R.H. (1937), 'The Nature of the Firm', *Economica*, vol. 4, no. 16, pp. 386-405.

Cohen, S.S. and Fields G. (2000), 'Social Capital and Capital Gains: An Examination of Social Capital in Silicon Valley', in M. Kenney (ed.), *Understanding Silicon Valley: The Anatomy of an Entrepreneurial Region*, Stanford University Press, Stanford, CA, pp. 190-217.

Coleman, J.S. (1988), 'Social Capital in the Creation of Human Capital', *American Journal of Sociology*, vol. 94 (supplement), pp. S95-120.

Dahlman, C.J. (1979), 'The Problem of Externality', *Journal of Law and Economics*, vol. 22, no. 1, pp. 141-162.

de Vet, J.M. (1993), *Globalisation and Local and Regional Competitiveness*, OECD, Paris.

Dierickx, I. and Cool, K. (1989), 'Asset Stock Accumulation and Sustainability of Competitive Advantage', *Management Science*, vol. 12, no. 35, pp. 1504-1513.

Eliasson, G. (1996), *Firm Objectives, Controls and Organization: The Use of Information and the Transfer of Knowledge Within the Firm*, Kluwer Academic Publishers, Dordrecht.

Foss, N.J. and Loasby, B.J. (1998), 'Introduction: Co-operation and Capabilities', in N.J. Foss and B.J. Loasby (eds), *Economic Organization, Capabilities and Co-ordination*, Routledge, London and New York, pp. 1-13.

Fukuyama, F. (1995), *Trust: The Social Virtue and the Creation of Prosperity*, Hamish Hamilton, London.

Grabher, G. and Stark, D. (1997), 'Organizing Diversity; Evolutionary Theory, Network Analysis, and Post Socialism', in G. Grabher and D. Stark (eds), *Restructuring Networks in Post-Socialism*, Oxford University Press, Oxford.

Granstrand, O. (1999), 'Internationalization of Corporate R&D: a Study of Japanese and Swedish Corporations', *Research Policy*, vol. 28, pp. 275-302.

Gross, L. (1968), 'The Peace of Westphalia, 1648-1948', in R.A. Falk and W.F. Hanrieder (eds), *International Law and Organization*, J.B. Lippencott Company, Philadelphia, pp. 45-67.

Hall, P. (1997), 'The University and the City', *Geojournal*, vol. 41, no. 4.

Hernes, G. (1979), *Forhandlingsĝkonomi Og Blandingsadministration*, Oslo.

Hyman, R. (1999) 'Imagined Solidarities: Can Trade Unions Resist Globalization?', in P. Leisink (ed.), *Globalization and Labour Relations*, Edward Elgar, Cheltenham, pp. 94-115.

Keeble, D. (1989), 'High-technology Industry and Regional Development in Britain: the Case of the Cambridge Phenomenon', *Environment and Planning C*, vol. 7, no. 2, pp. 153-172.

Lippman, S.A. and Rumelt, R.P. (1982), 'Uncertain Imitability. An Analysis of Interfirm Differences in Efficiency Under Competition', *Bell Journal of Economics*, vol. 12, pp. 418-438.

Loasby, B.J. (1992), 'Market Co-ordination', in B.J. Caldwell and S. Boehm (eds), *Austrian Economics: Tension and New Directions*, Kluwer Academic Publishers, Boston, pp. 137-156.

Loasby, B.J. (2000), 'Organisations as Interpretative Systems', Stirling University, unpublished paper.

Maskell, P. (1998), 'Successful Low-tech Industries in High-cost Environments: the Case of the Danish Furniture Industry', *European Urban and Regional Studies*, vol. 5, no. 2, pp. 99-118.

Maskell, P., Eskelinen, H., Hannibalsson, I., Malmberg, A. and Vatne, E. (1998), *Competitiveness, Localised Learning and Regional Development. Specialisation and Prosperity in Small Open Economies*, Routledge, London.

Maskell, P. and Tornqvist, G. (1999), *Building a Cross-Border Learning Region. The Emergence of the Northern European Oresund Region*, Copenhagen Business School Press, Copenhagen.

Menzel, U. and Senghaas, D. (1980), 'Autocentric Development Despite International Competence Differentials', *Economics - A Biannual Collection of Recent German Contributions to the Field of Economic Science*, vol. 21.

Mjoset, L. (1992), 'Comparative Typologies of Development Patterns', in L. Mjoset (ed.), *Contributions to the Comparative Study of Development*, Institute for Social Research Report, Oslo.

Nielsen, K. and Pedersen, O.K. (1988), 'The Negotiated Economy. Ideal and History', *Scandinavian Political Studies*, vol. 2, no. 11, pp. 79-101.

Nielsen, K. and Pedersen, O.K. (1991), 'From the Mixed Economy to the Negotiated Economy: The Scandinavian Countries', in R.M. Coughlin (ed.), *Morality, Rationality and Efficiency: New Perspectives on Socio-Economics*, M.E. Sharpe, New York, pp. 145-168.

OECD (1999), *Boosting Innovation: The Cluster Approach, OECD Proceedings*, The Organisation for Economic Co-operation and Development, Paris.

Prahalad, C.K. and Hamel, G. (1990), 'The Core Competence of the Corporation', *Harvard Business Review*, vol. 3, pp. 79-91.

Putnam, R.D. (1993), 'The Prosperous Community: Social Capital and Public Life', *The American Prospect*, vol. 38.

Richardson, G.B. (1953), 'Imperfect Knowledge and Economic Efficiency', *Oxford Economic Papers*, vol. 5, no. 2, pp. 136-156.

Richardson, G.B. (1972), 'The Organisation of Industry', *Economic Journal*, vol. 2, pp. 883-896.

Ropke, J. (1978), 'Der Einfluss des Weltmarkts auf die wirtschaftliche Entwicklung', in H. Giersch *et al.* (eds), *Weltwirtschaftsordnung und Wirtschaftswissenschaft*, Stuttgart, pp. 30-52.

Saxenian, A. (1994), *Regional Advantage: Culture and Competition in Silicon Valley and Route 128*, Harvard University Press, Cambridge MA.

Scott, A.J. (1993), *Technopolis: High-Technology Industry and Regional Development in Southern California*, University of California Press, Berkeley.

Senghaas, D. (1982), *Von Europa Lernen*, Suhrkamp, Frankfurt a.M.

Strange, S. (1991), 'New World Order: Conflict and Cooperation', *Marxism Today*, January, pp. 30-32.

Tatsuno, S.M. (1986), *The Technopolis Strategy: Japan, High Technology, and the Control of the Twenty-First Century*, Prentice-Hall, New York.

Veblen, T. (1899), *The Theory of the Leisure Class*, Dover, New York.

Wigren, A. and Persson, J.T. (1996), 'Many Deregulations have been Less Successful than Expected', in Anon. (ed.), *Swedish Industry and Industrial Policy 1996*, Annual report from the Swedish Board for Industrial and Technical Development (NUTEK), NUTEK, Stockholm.

Williamson, O.E. (1975), *Market and Hierarchies: Analysis and Antitrust Implications: A Study in the Economics of Internal Organization*, The Free Press, New York.

Woolcock, M. (1998), 'Social Capital and Economic Development: Toward a Theoretical Synthesis and Policy Framework', *Theory and Society*, vol. 27, pp. 151-208.

17 Promoting Regional Growth and Convergence in the Long-Term Master Plan for Israel

Raphael Bar-El

Introduction

Economic regional structures in Israel do not reflect any long-term natural free market development. For many decades regional structures have been primarily a result of public policy that was oriented to the accomplishment of non-economic objectives. This policy was consolidated around three main themes. First, was the need following the establishment of the State of Israel in 1945 to create agricultural settlement to establish a Jewish presence in all areas, to disperse population and thereby to secure the frontiers of the state. Second, was the need to achieve a demographic balance in those regions of the country with large non-Jewish populations. Third, was the establishment of new development towns in the northern and southern regions of Israel to absorb the masses of immigrants who arrived in the 1950s and 1960s.

This policy has led to a spatial distribution of population and economic activity that is not always consistent with natural economic patterns (Lipshitz, 1996; Kipnis, 1996; Gradus, Krakover and Razin, 1993). Many examples illustrate this situation. Urban places in more peripheral regions do not play the traditional role of service and industrial centres for their rural hinterlands. Rural settlements have developed their own services and their own non-agricultural activities. Many of them are more established than the new towns, and do not need the support of a local urban centre. In most cases, rural settlements are even providers of non-agricultural employment to the urban population. Entrepreneurship patterns follow different directions and present different characteristics in rural areas and in the development towns of the periphery, again with no real links between them (Bar-El and Felsenstein, 1990; Bar-El, Erickson and Nesher, 1987).

Industrial activity does play a major role in employment supply in the peripheral areas, but this is characterised by high turnover. Many manufacturing plants only survive in the remoter regions as long as they benefit from public subsidies and incentives. The types of industrial activities that are attracted to the periphery are more of a traditional nature. A recent empirical analysis of the Negev region in the South of Israel indicates the existence of a "fragmentation and remoteness [of industrial development] that continues to be one of the major deterrents to the robust growth and sustained development of the Negev" (Lithwick *et al.*, 1997).

This chapter outlines the main features of the old paradigm for promoting regional growth in Israel and its limited success. It describes the long-term master plan for Israel and its proposals for endogenous regional economic growth. This forms the basis for a new paradigm for promoting regional growth whose implications are analysed.

An Old Paradigm and the Related Economic Policy

Prevailing regional policy is shaped by a paradigm that can be formulated in the following terms:

- Peripheral regions cannot economically compete with central regions, because they lack scale economies, agglomeration economies and skills, and because of their remoteness from centres of research, markets and advanced services.
- Public support should be available to ensure the economic survival of the periphery, in order to achieve the non-economic objectives of population dispersal.
- The appropriate policy approach is therefore allocating resources that allow for more support bringing local incomes up to the minimum survival level.

In practical terms, policy measures that have been implemented over the last decades have followed this main line. Amongst these we can notice the use of heavy support to agricultural economic activity, through the public allocation of land (most land in Israel is owned by the State), subsidised prices for land use, capital and water, protection against imports, insurance against agricultural disasters, and so on. Substantial assistance has been made available to support industrial activity through the law of capital incentives providing grants for capital investments and tax breaks. Finally, heavy support has been provided to the population through transfer payments, income tax reductions, subsidies to housing and financial support to municipalities.

This policy has not led to any long-term endogenous growth in the peripheral regions in Israel. Also, it has not resulted in the short-term to a closing of the economic gaps between peripheral and central regions. The reason is the inadequacy of the policy measures that have been adopted to create structural change in the economy of the peripheral regions.

The heavy support to agricultural activity has actually been efficient in stimulating the development of new agricultural technologies. It has encouraged not only growth in agricultural productivity, but also in agri-business activity, and in the export of agricultural equipment. However, regional and even national benefits resulting from this policy have decreased over the years. Agricultural activity now provides about 8% of employment in the Northern region, and 5% of employment in the Southern region, and around 1-2% in central regions. Its relative share in total employment has decreased continuously, and over the last few years there has even been an absolute decrease in employment.

The efficiency of support to agriculture as a measure to build an economic base for the growth of peripheral regions has decreased continuously with the trends of economic and political change in Israel. The process of increasing openness to the global economy, that has characterised national economic policy since 1985, has imposed many difficulties on the country's ability to compete in agricultural production. The peace process and the trade agreements signed by Israel with countries in the Middle East and the rest of the world, make competition much harder. The continuing direct and indirect subsidies to water and land evoke many questions amongst economists about the wisdom of such policies that create distortions in the allocation of production factors, therefore decreasing the potential for national economic growth.

The long-term master plan for Israel expects agriculture's share of employment to fall to about 1.5% nationally, and to less than 3% in the peripheral regions of the North and the South. Agricultural policies are not therefore expected to provide a base for long-term growth in those regions.

Industrial activity has always been a major source of employment in peripheral regions. The share of manufacturing in total employment in the Northern and in the Southern regions has maintained a level of about 25% in the last decades. The share of manufacturing employment nationally has declined from about 24% in the past to about 20% in the last few years.

Consequently, the share of the peripheral regions in total manufacturing employment at the national level has increased from about 26% 20 years ago to 33% in the recent years.

This does not mean that policy measures to stimulate industrial activity through the law of capital incentives have been successful. The growth of manufacturing activity does not reflect any process of endogenous well-rooted economic growth in the peripheral regions. An in-depth analysis of the influence of the capital incentives conducted by Schwartz (1987) found that capital incentives had little impact on industrial development in peripheral regions in Israel (See also Wiewel, Persky and Felsenstein, 1994). Empirical findings clearly indicate that:

- most manufacturing plants that were established in the peripheral regions were attracted by factors such as the relative low cost of land and the supply of unskilled labour and, therefore, subsidies did not play a major role in their attraction;
- those plants attracted by the incentives lasted in the peripheral regions as long as the incentives were provided. Once the incentives ended the firms left the region and were replaced by a new round of firms eligible for incentives; and
- there is a negative aspect to the manufacturing plants that decide to locate in the peripheral regions. The structure of capital incentives is such that it provides more stimulation to relatively weak plants, while the stronger plants tend to prefer central locations. Capital incentives are also provided to plants that locate in the central regions, provided they produce mainly for export. Plants that are generally more competitive, therefore find a higher relative advantage in a central location.

This leads to the conclusion that regional industrial policy as implemented in Israel over the last three decades has not really contributed to generating an endogenous long-term process of regional economic growth.

Transfer payments, in terms of unemployment allocations, social security payments, public allocations to municipalities or income tax deductions, have probably helped to maintain minimal standards of living in the peripheral regions. They may even have helped to increase local demand for non-tradable goods and services. However, they can play only a marginal role in the creation of a process of endogenous regional economic growth.

The consequences of regional policy based on the old paradigm are quite clear: public policy has enabled the economic survival of the peripheral regions but has not created any substantial base for endogenous long-term economic growth.

Unemployment figures for Israel show the existence of unemployment cycles, with a trend of increasing unemployment. The changes in unemployment rates for each district (region) and for the whole nation are shown in Figure 17.1.

We can see that the cycle of unemployment in the Southern district is higher at all times than the national cycle. Moreover, the gap between the levels of unemployment in the Southern district and those of the nation as a whole has a tendency to grow over time: the trend line of the Southern unemployment data has a steeper slope than that of the national unemployment data. Unemployment data for the Northern region are not generally significantly higher than the national average, probably due of the proximity of this region to the labour pool of the Haifa metropolitan region.

Table 17.1 provides basic information about the relative economic welfare of the various regions in Israel for the most recent year for which data are available (1997). This table and all subsequent tables are taken from the economic development projections in the Long-Term Master Plan for Israel.

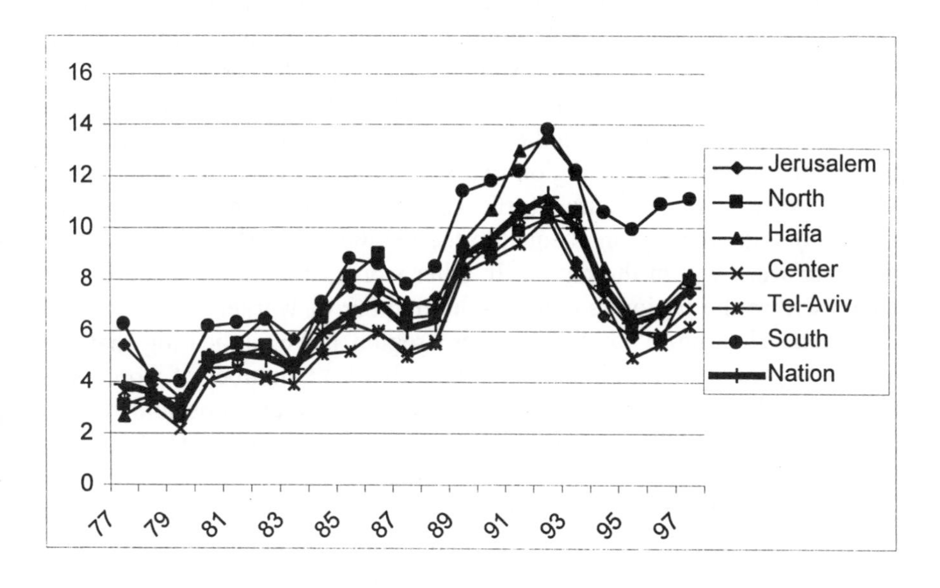

Figure 17.1 Unemployment rates by region, 1977-1997

Table 17.1 Indicators of unemployment, employment sufficiency, income and poverty, by district, 1997 (total=100)

District	Unemployed	Employment regional sufficiency	Income per standard person	Poverty level
Total	100	100	100	100
Jerusalem	97	106	103	113
North	***104***	***86***	***75***	***126***
Haifa	106	104	98	101
Centre	90	84	107	86
Tel Aviv	81	120	115	77
South	***144***	***93***	***83***	***136***

The reported unemployment relates to average data for 1997, and the unemployment rate nationally reached 7.7%. In the same year, the unemployment rate in the Southern region was 44% higher, and reached 11.1%. In comparison to the metropolitan region of Tel Aviv, the unemployment gap is very high.

The Northern region also has an unemployment rate that is higher than the national average, but lower than that of the Southern region. This may be explained by the lower rate of employment sufficiency. Employment sufficiency is defined as the ratio between the number of jobs that exist in a region (the number of workers within the region) and the number of residents of the region that are employed (within the region or outside the region). A figure of 100 shows that the number of residents working outside the region is equal to the number of workers coming from outside to work within the region. In other words, it shows that the supply of jobs within the region is equal to the number of workers that reside in the region. The figure for the Northern region is 86, indicating that the number of jobs in this region is equal to 86% of the number of workers in the region. This is quite a low figure, which is made possible by the proximity of the Northern region to the metropolitan region of Haifa. A low indicator is also present at the Central district, which is explained by the extreme proximity of metropolitan Tel Aviv. This same proximity also facilitates a rather low level of unemployment in the region. The employment sufficiency rate is also lower that 100 in the Southern region, which is located at a greater distance from Israel's metropolitan centres. The lack of job supply in this region, together with the larger distances from other regions, may explain the high rate of employment.

The level of income per 'standard person', as computed by the National Social Security Institute (based on a weighted family average, with different weights for parents, children and elderly people) is lowest in

Israel's two peripheral regions: 75% of the national average in the North and 83% of the national average in the South. The proportion of families below the poverty level, as calculated by the same source, is highest in these peripheral regions.

In summary, we can conclude that Israeli regional policy, based on a decades old paradigm, has not achieved any objective beyond providing peripheral regions with basic resources for survival. Those regions are considered as economic burdens that are needed to achieve non-economic objectives. As a result, the regional policy measures that have been implemented are orientated towards the solution of ad hoc employment and income problems, and not towards the stimulation of long-term, endogenous regional economic growth. This has resulted in an economic base in Israel's peripheral regions that is insufficient to meet the employment needs of the population, with a high level dependence upon employment, high unemployment rates, low income levels and high levels of poverty.

Future Trends and Employment Targets

The Long-Term Master Plan for Israel outlines land use and economic development trends for the period 2000-2020. It is a strategic planning document without statutory force, commissioned by the Israeli government and undertaken by a team of planning practitioners and academics. It is based on a series of long-term projection on all the issues likely to have an impact in a locality: demographic trends, housing, employment, infrastructure, investment and so on. The plan predicts an expected population growth from 5.6 million persons in the 1995 Census to about 8.8 million in 2020, an annual growth rate of 1.8%. The projections for demographic growth in each of the districts show more rapid growth in the Northern and Southern regions. These regions have 31% of the current population, but 48% of the added population until year 2020 is expected to be in those districts. Therefore, the share of the peripheral regions in Israel's total population is expected to grow to 37%.

Employment targets in the peripheral regions are based on the assumptions that unemployment levels will be the same in all districts. The macro-economic component within the long-term plan projects a national level of unemployment of 6% in year 2020, and it is assumed that this level should be equal in all districts. In addition, regional employment sufficiency is expected to reach 100% in the Southern region, which is relatively isolated from the metropolitan centres. It is also expected to increase, without necessarily achieving full sufficiency, in the Northern region. This region is quite close to the metropolitan region of Haifa, and it is assumed that a certain level of employment dependency will continue to exist.

Table 17.2 shows the regional sufficiency indicators for selected years in the past, and the projection to 2020 in the master plan.

Table 17.2 Employment regional sufficiency (ratio of jobs in the district and resident workers): trends and projections

District	1980	1985	1990	1995	2020
Total	1.00	1.00	1.00	1.00	1.00
Jerusalem	0.98	0.99	0.99	1.06	1.08
North	***0.87***	***0.87***	***0.86***	***0.86***	***0.91***
Haifa	1.00	1.03	1.03	1.04	1.05
Centre	0.84	0.81	0.79	0.84	0.93
Tel Aviv	1.05	1.11	1.15	1.20	1.17
South	***0.92***	***0.90***	***0.94***	***0.93***	***1.00***

On the basis of these projections Tables 17.3 and 17.4 show the expected changing proportions of population and employment in the peripheral regions compared to the other regions of Israel.

Table 17.3 Changing distribution of population and employment

District of residence	Population in 1995		Added population from 1995 to 2020		Added employment from 1995 to 2020		Annual % increase	
	Thous.	%	Thous.	%	Thous.	%	Pop.	Empl.
Total	5619	100%	3143	100%	1240	100%	1.8%	2.0%
Jerusalem	663	12%	397	13%	124	10%	1.9%	2.0%
North	***952***	***17%***	***838***	***27%***	***287***	***23%***	***2.6%***	***2.8%***
Haifa	740	13%	360	11%	142	11%	1.6%	1.8%
Centre	1213	22%	587	19%	273	22%	1.6%	1.9%
Tel Aviv	1142	20%	120	4%	57	5%	0.4%	0.5%
South	***770***	***14%***	***670***	***21%***	***269***	***22%***	***2.5%***	***3.0%***
WBG*	139	2%	171	5%	90	7%	3.3%	4.6%

* West Bank and Gaza Strip: Israeli population only

Table 17.4 Distribution of population, labour force and employment (%)

	Population		Labour force		Employed in district	
District	1995	2020	1995	2020	1995	2020
Total	100.0%	100.0%	100.0%	100.0%	100.0%	100.0%
Jerusalem	11.8%	12.1%	10.0%	10.0%	10.8%	10.8%
North	***16.9%***	***20.4%***	***15.0%***	***18.1%***	***13.1%***	***16.6%***
Haifa	13.2%	12.6%	13.1%	12.5%	13.9%	13.1%
Centre	21.6%	20.5%	23.3%	22.8%	19.9%	21.2%
Tel Aviv	20.3%	14.4%	23.5%	16.4%	29.2%	19.1%
South	***13.7%***	***16.4%***	***13.0%***	***16.0%***	***11.8%***	***16.0%***
WBG*	2.5%	3.5%	2.1%	4.1%	1.3%	3.3%

* West Bank and Gaza Strip

The tables point to the fact that almost half of the population added up until year 2020 will be in the peripheral regions of the Northern and Southern districts. This is due to the relatively high natural growth in those districts and to the migration trends to those regions in the last few years. Consequently, their share of national population is expected to grow from 31% to 37%. Additionally, nearly half of the added employment until year 2020 is expected to be created in the peripheral regions, to achieve an objective of equalised unemployment between regions, and regional employment sufficiency. This means an annual increase of 2.8% in the North and of 3.0% in the South in the supply of jobs, as compared with a maximum of 2.0% in other regions. This will practically double the job supply in the peripheral regions.

To meet the objectives of more equitable unemployment levels between regions and a more balanced supply of jobs, a policy needs to be adopted that will lead to the creation of about half of all additional jobs in peripheral regions. Such a goal certainly cannot be achieved by following the present paradigm, and by considering the peripheral regions as marginal to the national process of economic growth. A new approach is, therefore, crucial for the achievement of healthier and sustainable economic growth in those regions.

The New Paradigm for Regional Development

The normative foundations of a new paradigm are that first, a region needs to be considered as a resource, not as a burden and that development of the periphery is needed for national growth. Second, the role of the public sector is to provide the peripheral areas with the necessary means for the development of competitive advantage, and not just to deal with the alleviation of problems regarding the socio-economic welfare of the population. Finally, the public policies needed to achieve these goals relate to the regional allocation of public expenditure on physical and human infrastructure rather than on the direct stimulation of business activities.

Contemporary thinking on regional economic growth considers regions not as a marginal component of the national economy but as economic entities with their own potential to create competitive advantage (O'Donnell, 1997; Porter, 1990). Many studies have dealt with the identification of the various factors that influence the economic development of a region. For example, Wong (1998) confirms, for the British case, the influence of the major traditional factors of land, labour, capital, infrastructure and location. A growing number of researchers find that location in peripheral areas should not necessarily impose a serious problem. McKinnon (1997) reports in the case of Scotland, that a peripheral location need not impose a serious logistical handicap on development, and that "logistical penalties may be more than offset by other locational advantages". Felsenstein and Schwartz (1993) find constraints on the ability of small businesses to develop in the peripheral areas in Israel, but also indicate factors that may overcome these constraints.

The question of the influence of public expenditure has received considerable attention (Aschauer, 1989; Kelejian and Robinson, 1997). The many case studies undertaken, such as Bajo-Rubio and Sosvilla-Rivero (1993), lead mainly to the conclusion that public infrastructure exerts a positive influence on economic development.

This paradigm is actually at the base of most policies adopted by developed countries for the solution of regional inequalities, and is generally formulated in terms of the creation of 'endogenous regional growth' (see for example Report of the Regional Policy Commission, 1996). The approach of attracting large-scale enterprises to peripheral regions as a mean of creating dynamic economic processes through the 'trickle-down effect', has not really proved effective. The new paradigm is coherent with the general approach of a free market economy, where the government does not interfere in the business sector, but concentrates on its role as a major provider of physical and human infrastructure.

The major question related to this approach is to what extent this new allocation of public expenditure is efficient at the national level. In pure economic terms, the question is: is the rate of return on investment in public infrastructure higher in core metropolitan areas or in peripheral areas? In other words, to what extent would the increase in the allocation of public expenditure to peripheral regions improve their competitive potential?

The answer to these questions is beyond the scope of this chapter. The economic return to public expenditure on infrastructure is a widely discussed issue at the national level, but the efficiency of the regional allocation of such expenditures is a much more complex matter. Most research finds a positive contribution of public expenditure to economic growth at the national level. The influence of infrastructure on productivity has been extensively analysed (Aschauer, 1989; Kelejian and Robinson, 1997). For the Israeli case, Bregman and Merom (1993) found an even higher rate of return for investments in infrastructure than for private business investment.

It is important, however, to indicate that most of the literature on the convergence-divergence question points to the existence of some convergence between regions, alluding to the possibility of increasing economic effectiveness in the periphery in relation to the centre. The increase in the regional competitive advantage of regions may be attributed to growing gaps in land prices between peripheral regions and the increasingly crowding metropolitan centres, decreasing returns to agglomeration and the decreasing cost of distance with improvements in transport and communications infrastructure. Recent literature on this subject is abundant, including many case study analyses. For example, Jayet, Puig and Thisse (1996), Barro and Sala-i-Martin (1991), Armstrong (1994) and Fagerberg and Vespagen (1996) identify the existence of a process of regional convergence in Europe, although they find a trend of decline in the speed of that convergence. When such convergence is not identified or is evaluated as too slow, it is explained in many cases by the existence of distorting factors. Fagerberg, Verspagen and Caniels (1997), for example, investigate a number of factors that might have acted against the convergence process in Europe, such as differences across regions in the diffusion of technology, economic regional structures, and unemployment.

It is also important to clearly state that the argument of 'demand pressures' for infrastructure should be rejected. This argument, especially popular amongst treasury officials in Israel where the allocation of public infrastructure is highly centralist, maintains that the existence of a stronger demand for infrastructure in central regions indicates a potential economic contribution to peripheral areas through 'filter-down' process. Since there is no free market for infrastructure (the users do not pay directly for their use), a high level of demand may be the result of various distorting factors. The allocation of public expenditure between regions should, therefore, be determined on a normative basis, as a function of the needs of economic growth in each region.

An Evaluation of Necessary Changes in the Allocation of Public Expenditure

Physical Infrastructure

Physical infrastructure is defined as transportation infrastructure (roads, railways, airports, seaports), water, sewerage and communications infrastructure. The stock value of infrastructure in Israel in 1997 was estimated at 131 billion New Israeli Shekels (NIS), on the basis of data from the Central Bureau of Statistics and the Bank of Israel. The long-term economic plan for Israel estimates a required annual growth of 6.8% in the stock of infrastructure capital to enable an optimal average annual growth of 4.3% in GDP until year 2020.

The normative allocation of public expenditure for physical infrastructure by regions has to be based on a criterion variable. The critical question is which variable should be used to evaluate the 'infrastructure intensity' in any given region. Indicators that are generally used relate to the relation between infrastructure stocks and production factors stocks: infrastructure per worker, or per unit of business capital. Another indicator may be the density of infrastructure, as measured by the relation between the stock of infrastructure and the regional area. It is clear that none of the criteria – labour force or area or other variables – provides a completely reliable and valid representation of infrastructure intensity. For example, the construction of a road linking two major urban centres located in two separate regions could be measured in terms of the relative part of the road within the geographical borders of each region. An arbitrary modification of the border, for example moving it closer to the urban centre in region A and farther from the urban centre of region B, would decrease the measure of infrastructure intensity in relation to regional labour force in region A and increase it in region B, without really having any real effect on the potential for economic growth in each region. In the Israeli case, the Central region has a quite high population density, while the adjacent Southern region is characterised by its vast area and low population density. Any new investment in infrastructure such as the proposed rail link connecting Beer-Sheva, the main urban centre of the Southern region, and the Central region, would mainly be measured as an addition to the infrastructure stock of the Southern region (since most of it would be located there because of the greater area), while such a new investment contributes to the economic development of both regions in proportions that are independent of the location of the regional border.

Elaborating a valid measure of regional infrastructure intensity is beyond the scope of this chapter, but the constraints of the existing measures should be kept in mind. Furthermore, no regional statistics on infrastructure stock exist in Israel. Therefore, the approach taken in this study is to focus mainly in the marginal changes required in infrastructure in each region, in terms of percentage growth, in order to realise the potential for economic growth.

For the nation as a whole, the macro economic component for the Long-Term Master Plan has estimated, as mentioned earlier, a required annual growth of 6.8% in infrastructure stock. This figure is based on a modified national 'Cobb-Douglas' production function where gross domestic product is evaluated as a function of labour, business capital, infrastructure capital and human capital. The coefficients of this function have been evaluated on the basis of the empirical study by Bregman and Merom (1993).

Available regional data relate to employment. There is no available data on the regional distribution of output, capital stocks or infrastructure stocks. The long-term economic plan makes, therefore, the simplifying assumption that the regional distribution of additional stocks of infrastructure in the future should be determined by the relative additional employment in the business sector in each region. Such an assumption implies that no major changes are expected in relative gaps in capital intensity between regions. It also implies that there is no prevailing distortion in the present regional distribution of infrastructure. The results are presented in Table 17.5.

From this table we can see that, at the national level, the stock of infrastructure is expected to grow at an annual average of 6.8%, to enable the realisation of an annual GDP growth rate of 4.3%, with an expected annual growth of 2.3% of employment in the business sector. This requires national growth in 'infrastructure intensity', i.e. a growing rate of infrastructure per worker in the business sector and a growing rate of infrastructure per unit of business capital stock. Table 17.5 also shows that the lowest required growth in infrastructure stock is in the district of Tel Aviv: an annual growth of 4.7% (69% of the national average). This is still much higher than the expected employment growth of 0.5% in the business sector in Tel Aviv. The gap is explained by the required growth in infrastructure intensity. Finally the table highlights the fact that the highest rates of increase in infrastructure stock are required in the districts of the South (8.3%) and of the North (7.8%).

Table 17.5 Required annual growth of stock of physical infrastructure by region, 1997-2020

District	Total	Jerusalem	North	Haifa	Centre	Tel Aviv	South
% annual growth of employment in the business sector	2.3%	2.5%	3.3%	2.0%	2.7%	0.5%	3.7%
% annual growth in stock of infrastructure	6.8%	6.9%	7.8%	6.4%	7.2%	4.7%	8.3%
indicator of annual growth in stock of infrastructure (total=100)	100	102	116	94	106	69	123

Evaluating the findings means recognising the assumptions that underlie them. It should be noted that no assumption is made in relation to the definition of infrastructure intensity or in relation to the 'right' measure of infrastructure intensity in each region. The findings are based on the assumption of the existence of present equilibrium. The methodology utilised recognises the existence of differences in various measures of infrastructure intensity between regions, and evaluates the required distribution of additional infrastructure in order to maintain the given equilibrium.

However, this assumption of no change in the relative capital intensity of economic activities across regions is not realistic. Peripheral regions experience higher levels of unemployment and, therefore, probably higher levels of labour intensity in their economic activities. Further economic development, as projected in these regions by the long-term plan, is expected to lead to higher capital intensity in the new economic activities. Given the expected employment growth, this means higher levels of output, and greater needs for infrastructure.

Furthermore, since no prevailing distortion parameter was used, the relative results for the peripheral regions should probably be considered as lower limits. An annual increase of 8.3% of infrastructure stock in the South (23% higher than the national average) and of 7.8% in the North (16% higher than the national average) are required to meet the needs of additional business activity. This is based on the misguided assumption that the infrastructure intensity today is appropriate in all regions. However, if the argument stated above about the existence of a distortion of regional infrastructure allocation is right, the figures should be higher in the peripheral regions, in order to compensate for such distortions.

The final conclusion, therefore, is that the results reported in Table 17.5 should be considered as lower limits of infrastructure growth required in the peripheral regions. Growth of infrastructure stock in the Northern district and in the Southern district respectively should be more than 16% and 23% higher than that of the national level, in order to accommodate, not just the expected growth of labour force in those regions, but also the expected growth of capital intensity, and the decline of prevailing distortions.

Human Capital

The measurement of human capital and the evaluation of its impact on economic growth has attracted much attention in the literature (Becker, 1964; Welch, 1970; Nelson and Phelps, 1966; Benhabib and Spiegel, 1992). The main variables used to measure human capital are the average number of years of schooling of the labour force, the level of enrolment in high school or in colleges, the percentages of the labour force with a high school degree or with an academic degree, the distribution of professional skills and so on. The long-term plan for Israel uses a composite measure based on a combination of various variables. Fortunately, such data exist at a district level, enabling regional comparisons, which could not be done for physical infrastructure. This measure is presented in Table 17.6 as the relative educational level in 1995.

This measure shows a national level assumed at 100, and relative figures of 77 in the Northern region and of 88 in the Southern region. In other words, the labour force is on the average 23% less well educated in the North and 12% less well educated in the South compared with the national average. In all other districts, the figures are higher than the national average.

At the national level, the macro economic plan estimates a required growth of average education of 13% in the country as a whole. In terms of average schooling years, this means an increase from the prevailing average of 12 years in 1995, to about 13.6 years in 2020. The question now concerns the necessary increase in the educational level of labour force at a regional level.

Table 17.6 Required annual growth in education budgets by districts, 1995-2020

District	Total	Jerusalem	North	Haifa	Centre	Tel Aviv	South
Relative educational level in 1995 (total=100)	100	114	77	106	104	105	88
Gap diminution coefficient		50%	50%	50%	50%	50%	50%
Annual growth of average educational level	13%	6%	32%	10%	11%	10%	22%
Relative educational level in 2020 (total 1995=100)	113	120	102	116	115	116	107
Additional annual average budget (million NIS)	4553	227	1435	450	870	672	898
Distribution of additional budget (%)	100%	5%	32%	10%	19%	15%	20%

The implementation of the suggested policy for regional growth requires a contraction of the regional gap in human capital. Government is committed to the provision of education to the population by the allocation of full subsidies to the basic levels of education and partial subsidies to higher education and to professional training. It is assumed that the gaps between regions cannot be fully eliminated. The long-term regional plan suggests a decrease of 50% in the inter-regional education gap by means of a change in the relative allocation of public funds for education to the regions.

Closing the education gaps across regions requires, as can be seen in Table 17.6, a growth of 32% in the average educational level of the labour force in the Northern region, and of 22% in the Southern region, in the 25 years from 1995 to 2020 (as compared with the national average of 13%).

The public budget that is required in order to achieve the educational growth in the nation as a whole is calculated in the macro economic plan on the basis of average costs of schooling, and is presented in Table 17.6. Our interest, however, is in the distribution of this budget between regions. The results as presented in the table indicate that 32% of the additional public expenditure needed to increase the average level of education nationally should be allocated to the Northern region, and 20% should be allocated to the Southern region. This would achieve a 50% contraction in the educational gap between regions. In total, more than half of the additional expenditures should be allocated to the peripheral regions.

Again, these figures may be considered as a lower limit. The calculations are based on the assumption of an equal cost of education in all regions. If costs in the periphery are higher, a higher relative share of public expenditure may have to be allocated.

Conclusions

This chapter has considered the prevailing regional policy in Israel and has analysed the results of this policy to date. An alternative approach for promoting long-term growth has been presented and the practical measures to be taken in terms of the regional allocation of public expenditure have been outlined.

The old paradigm at the base of Israeli regional policy in past decades considers the periphery as an economic burden that is needed in order to achieve non-economic national targets. The derived policies of transfer payments and of the artificial attraction of industrial activities has not led to any satisfactory results. Most indicators show growing gaps between the peripheral regions and the centre, and no process of regional endogenous economic growth has been launched.

Stimulating endogenous regional economic growth, an approach gaining grounds in Europe, seems more appropriate to the Israeli situation, especially if we consider expected long-term demographic and economic changes.

The analysis of the Israeli case shows that the demographic forecasts in the long-term master plan for the regional distribution of the population and labour force highlights the urgent need to adopt a new policy. Until 2020, the peripheral regions of the country are expected to absorb about half of the added total population (mainly as a result of high levels of natural growth). In economic terms, about half of the additional jobs to be created during this period have to be located in the peripheral regions of the North and the South if we want to achieve a goal of reasonable unemployment levels in the future.

The key to promoting growth is the appropriate regional allocation of public expenditure on physical infrastructure and on education. The quantitative results show that the growth of infrastructure in the peripheral regions should be at least 16% to 23% higher than the national average in order to provide those regions with the minimal tools for endogenous economic growth. A higher allocation is needed in order to compensate for prevailing regional distortions.

The evaluation of the necessary human capital investment shows that more than half of the additional budget for education and professional training should be allocated to the peripheral regions, although the share of these regions in total employment reaches only one third. This allocation of public expenditure for the formation of human capital is required to halve the educational gap between the periphery and the other regions. Public policy for promoting key term growth cannot ignore these critical areas.

The general lesson to be drawn from this study is that policies to foster regional economic growth need to be formulated within an appropriate ideological paradigm that allows national goals to be effectively expressed.

References

Armstrong, H. (1994), *Convergence versus Divergence in the European Union Regional Growth Process 1950-1990*, Working Papers EC19/94, Lancaster University, Department of Economics.

Aschauer, D. (1989), 'Is Public Expenditure Productive?', *Journal of Monetary Economics*, vol. 23, pp. 177-200.

Bajo-Rubio, O. and Sosvilla-Rivero, S. (1993), 'Does Public Capital Affect Private Sector Performance? An Analysis of the Spanish Case, 1964-88', *Economic Modelling*, vol. 10, no. 3, pp. 179-185.

Bar-El, R., Erickson, G. and Nesher, A. (1987), 'Rural Industrialization in Israel: Concluding Considerations', in R. Bar-El (ed.), *Rural Industrialization in Israel*, Westview Press, pp. 169-189.

Bar-El, R. and Felsenstein, D. (1990), 'Entrepreneurship and Rural Industrialization: Comparing Urban and Rural Patterns of Locational Choice in Israel', *World Development*, vol. 18, no. 2, pp. 257-267.

Barro, R.J. and Sala-i-Martin, X. (1991), 'Convergence Across States and Regions', *Brookings Papers on Economic Activity*, vol. 1, pp. 107-182.

Becker, G. (1964), *Human Capital*, Columbia University Press, New-York.

Benhabib, J. and Spiegel, M. (1992), 'The Role of Human Capital in Economic Development', *Economic Research Report*, 92-46, C.V. Starr Center for Applied Economics, New York University.

Bregman, A. and Merom, A. (1993), *Growth Factors in the Business Sector in Israel, (1958 to 1988)*, Discussion Papers Series, 93.02 Bank of Israel (Hebrew).

Fagerberg, J. and Verspagen, B. (1996), 'Heading for Divergence? Regional Growth in Europe Reconsidered', *Journal of Common Market Studies*, vol. 34, pp. 431-448.

Fagerberg, J., Verspagen, B. and Caniels, M. (1997), 'Technology, Growth and Unemployment Across European Regions', *Regional Studies*, vol. 31, no. 5, pp. 457-466.

Felsenstein, D. and Schwartz, D. (1993), 'Constraints to Small Business Development Across the Life Cycle: Some Evidence From Peripheral Areas in Israel', *Entrepreneurship and Regional Development*, vol. 5, pp. 227-245.

Gradus, Y., Krakover, S. and Razin, E. (1993), *The Industrial Geography of Israel*, Routledge, London.

Jayet, H., Puig, J.P. and Thisse, J.F. (1996), 'Economic Issues in Regional and Urban Organization', *Revue d'Economie Politique*, vol. 106, no. 1, pp. 127-158.

Kelejian, H.H. and Robinson, D.P. (1997), 'Infrastructure Productivity Estimation and its Underlying Econometric Specifications: a Sensitivity Analysis', *Papers in Regional Science*, vol. 76, no. 1, pp. 115-131.

Kipnis, B.A. (1996), 'From Dispersal Strategies to Concentration: Alternating Spatial Strategies in Israel', in Y. Gradus and G. Lipshitz (eds), *The Mosaic of Israeli Geography*, Ben-Gurion University, pp. 29-36.

Lipshitz, G. (1996), 'Core vs. Periphery in Israel Over Time: Inequality, Internal Migration and Immigration', in Y. Gradus and G. Lipshitz (eds), *The Mosaic of Israeli Geography*, Ben-Gurion University, pp. 13-28.

Lithwick, H., Gradus, Y., Razin, E. and Yiftachel, O. (1997), 'Industry in the Negev: Policy, Profile and Prospects', *Negev Center for Regional Development*, Ben-Gurion University.

McKinnon, A. (1997), 'Logistics, Peripherality and Manufacturing Competitiveness', in B. Fynes and S. Ennis (eds), *Competing From the Periphery, Core Issues in International Business*, Oak Tree Press, Dublin, pp. 335-369.

Nelson, R.R. and Phelps, E.S. (1966), 'Investment in Humans, Technological Diffusion and Economic Growth', *American Economic Review*, vol. 56, no. 2, pp. 69-82.

O'Donnell, R. (1997), 'The Competitive Advantage of Peripheral Regions: Conceptual Issues and Research Approaches', in B. Fynes and S. Ennis (eds), *Competing From the Periphery, Core Issues in International Business*, Oak Tree Press, Dublin, pp. 47-82.

Porter, M.E. (1990), *The Competitive Advantage of Nations*, The Free Press.

Report of the Regional Policy Commission (1996), *Renewing the Regions, Strategies for Regional Economic Development*, Sheffield Hallam University.

Schwartz, D. (1987), 'The Government's Role in Private Investment', in R. Bar-El, A. Ben David-Val and G.J. Karaska (eds), *Patterns of Change in Developing Rural Regions*, Westview Press, London.

Welch, F. (1970), 'Education in Production', *Journal of Political Economy*, vol. 78, pp. 35-59.

Wiewel, W., Persky, J. and Felsenstein, D. (1994), 'Are Subsidies Worth It?', *Economic Development Commentary*, vol. 18, no. 3, pp. 17-22.

Wong, C. (1998), 'Determining Factors for Local Economic Development: The Perception of Practitioners in the North West and Eastern Regions of the UK', *Regional Studies*, vol. 32, no. 8, pp. 707-720.

18 Knowledge Economies and Transitional Labour Markets: New Regional Growth Engines

Lambert van der Laan

Introduction

Societal change at a supra-national scale leads to opportunities, threats and challenges for individuals, households, corporations, trade unions and governments. Evolving production and technological structures and socio-institutional developments cause these great changes. Some attribute these changes to the processes of the capitalist economy, others to business cycles, technological innovations or 'system-shocks'. Whatever their cause, these fundamental processes result in a reorganisation of space. However, this also depends greatly on the institutional framework: the manner in which the economy and the labour market are organised. The differences in the institutional framework particularly, are likely to generate spatial heterogeneity in the labour market, in the future.

This chapter highlights first some of the major changes in the socio-economic setting of regions and economic actors. Central to this are fundamental changes on the demand and supply side of the labour market, which clearly reflect the macro-changes in the economy and society of Europe. Particularly the information economy has led to a major change in European labour demand. On the supply side, a heterogeneisation of the labour force has developed. What is important is that these demand and supply changes evolve into still further forms. These can be encapsulated in the concepts of the knowledge economy and the transitional labour market (Van der Laan, 1999). In this framework, new resources and new concepts of work are created. The next two paragraphs discuss the emerging knowledge economy and the transitional labour market. The chapter then proceeds to analyse how the new resources of the knowledge economy and transitional labour can be matched. In this the social nature of knowledge and the role of organisations and the state are discussed. Finally, the implications for promoting local growth and regional policies are analysed.

Labour Demand: from the Information to the Knowledge Economy

Inter- and intra-sectoral shifts have resulted in the information economy boom based on the generation and handling of information. While countries in North-Western Europe have already moved towards the information economy, some Southern and Eastern European countries are still dominated by manufacturing. Countries like Romania or Albania are even largely oriented towards agriculture. These differences in economic structure point to possible future employment changes and problems. The lagging countries are more prone to economic restructuring and are candidates for large losses of jobs in specific sectors. This is aggravated when their international competitiveness is less than the countries with which they increasingly trade. The incorporation of East European countries into the European Union (EU) is a case in point.

Economic, Social and Cultural Capital

Demand changes linked to the information economy do not fully reflect changes in the production system. Although the generation and handling of information increases in importance, economic growth becomes less based on this demand and more on communicative and creative abilities. Particularly knowledge intensive functions will grow (Woodall, 1996). Concurrently the meaning of knowledge itself also changes (Grossman and Helpman, 1991). To understand this, it is useful to distinguish between three kinds of knowledge (Klamer, van der Laan and Prij, 1997). Traditionally knowledge is reflected in *economic capital*. This comprises factual knowledge and how to use it. Increased knowledge of this kind, embodied in technological development, generates greater competitiveness and economic growth. R&D in particular is an investment for generating economic capital. Spatially, most economic capital is found in the denser urbanised regions of Europe.

In addition, *social capital*, the manner how to deal with people, groups and organisations, becomes increasingly important. Communicative capabilities in particular, are crucial. Therefore social capital is sometimes also called 'relationship capital' because it reflects the value of the relationships with suppliers, allies, and customers. Social capital is not solely responsible for economic growth. The point is that it becomes more important. This decreases the relative value of 'high-tech' oriented economic capital. In social capital, 'trust' and consensus play a major role (Putnam, 1993; Fukuyama, 1995; Peyrefitte, 1995). There is a positive relationship between trust and economic growth. Regions and countries with higher levels of trust and reciprocal institutional structures have shown higher growth. Significant positive relationships are indicated by an analysis of the 'World Values Survey' (Bomhoff, Den Hartog and Lageweg; 1996). According to these authors, trust is relatively high in EC countries such as Ireland, Denmark and Norway. Most Southern European countries have, however, a lower score.

The third kind of knowledge is embodied in *cultural capital*. This represents the ability to give meaning to information and motivation to behaviour. This also increasingly determines economic success. Not the amount of information as such gains importance, but particular the capacity to generate meaningful ideas out of this plenitude of information (McCloskey and Klamer, 1995; Nonaka, Umemoto and Senoo, 1996). Although costs of information become increasingly cheaper, average quality of this information decreases: the information paradox. Skills related to cultural capital are necessary to select out of the mountains of information. Not the reproduction, but the selection and combination of the information becomes important as a defence against the overkill of information. Within companies cultural capital is sometimes indicated as knowledge assets (Sveiby, 1997). These transform knowledge materials, like data and random know-how, into value-creating qualities. Cultural capital is particularly important for jobs in which convincing other people is central, for example consultants, managers and specialists. Because the awareness of the importance of cultural capital is just emerging, no substantial spatial research results are available.

Knowledge Jobs

The information economy together with the emergence of new forms and resources of knowledge leads to the knowledge economy. In this economy, human capital becomes even more crucial. It is reflected in the emergence of knowledge labourers: software developers, advertisers, consultants, researchers, etc. (McCloskey and Klamer, 1995). Not only is human capital the most significant resource of the firm, firms also may risk the loss of markets of output through its reduction. Because of the social involvement of the knowledge worker with clients, personnel who leave firms may take the clients of the company with them. R&D in the traditional sense is not enough for growth. Technology becomes more 'soft' and increasingly depends on the co-operation between people, i.e. social and cultural capital. Next to knowledge related to products, processes and logistics, market and organisational knowledge determines the welfare of firms, regions and countries. The development of soft technologies calls for the networking within and between companies, trade unions and the government. Technology is generated, diffused and applied with the help of these networks. The knowledge economy implies that non-material investments become more important. Examples are investments in software, marketing, designing, customer dedication, and quality of relationships. Increasingly, corporations, regions and countries compete for non-material investments. Although this side of economic development is sometimes hard to quantify, Minne, Herzberg and Reijnders (1996) estimate for Europe, that the spending on non-material investments is highest in Sweden and Norway. In these countries, more than 10 percent of the GNP is directed at this investment. Germany and the United Kingdom, on the other hand, have amongst the lowest rates of this firm of investment.

In the knowledge economy the content of jobs changes too. Workers do not produce products in a traditional sense, but increasingly participate in knowledge processes with a strong emphasis on conversation and interaction. Codified knowledge becomes relatively less important. Tacit knowledge, in which unwritten rules and uses of the business environment are embodied, becomes important, as does the embeddedness in social and regional networks. Social and cultural capabilities play a major role. It is, however, difficult to estimate the exact production of knowledge workers. Moreover, because this is a process, boundaries between work and non-work become unclear. The job is never finished; there is no nine to five. Also the usual rewards to labour work differently. Of course financial rewards remain important, but in addition the appraisal of peers and customers, the feeling of doing something right and working in a stimulating environment are important. Thus, the role of the price mechanism diminishes. Value depends increasingly on the non-tradable recognition of others. Wage determination is difficult because comparison is often not possible (Carter, 1996). If there is a price, it depends also on subjective estimates in which status and image building are important.

The crucial point of all these changes is that comparative advantage to a large degree is based on networks and knowledge sharing reflecting, alongside economic, also social and cultural capital. Social embeddedness and the giving of meaning to information become crucial for economic progress. Tacit knowledge becomes a main factor and importance of the embeddedness of labour in regional networks and non-material investments increases. High-tech as such certainly does not lead automatically to economic success. Because such a technology-push perspective can have backwash effects on other possible activities, it may even be detrimental to economic growth and competitive strength. Moreover, while 'high-tech' is not the only successful strategy, neither is 'higher education'. There is certainly a bias that equates more knowledge with being highly educated. A popular conception is that people with only basic formal education do not contribute to the knowledge of an organisation or region. In indices of knowledge, these labourers often get a zero weight while academics get a five-point weight. This bias leads to a denial of the potential role of the less educated in economic growth based on knowledge. Economic growth based on traditional productivity growth, educational attainment and 'high-tech', does not assure enduring competitive advantage which lies in locally based knowledge, including relationships and motivation. Regions can become highly productive, irrespective of the educational and sectoral structure (Porter, 1998). New kinds of knowledge mean new chances for all industries. This does not refer to ICT-intensity or high-level educational input, but to all activities with an optimal balance between economic, social and cultural capital. Research on the actual comparative manufacturing advantages of some European countries shows that traditional clusters with knowledge sharing are still very successful activities (Hinloopen and Van Marrewijk; 1999). The main drivers of economic growth are therefore not necessarily embodied in 'high-tech' or 'higher education' but in the way in which high and low tech and high and low education are mixed. This creates new opportunities for economic growth.

Changing Labour Supply: Emerging Life-Styles and the Transitional Labour Market

While on the demand side the knowledge economy arises, labour supply is moulded by the emergence of the transitional labour market. This suggests new resources, investments and concepts of work and renumeration. Major changes in labour supply have emerged: for example, in Europe, aging, changes in participation, feminisation, individualisation and the effect of migration are some of the changes. These changes are expected to continue into the future accompanied by a serious qualitative change: the *transitional labour market (TLM)*. Over the life cycle of individuals there will be repeating transitions between activities like working, learning, retirement or care. The TLM implies a departure from standard labour market distinctions between gender and age categories, participation or non-participation.

Life-Style

The emergence of the TLM has its roots in changing life-styles concurring with different views on the role of work (Van der Laan and Versantvoort, 1998). Traditionally labour supply is determined by factors like age, gender, ethnicity or education in an almost deterministic way. Until a few decades ago, most people indeed followed a similar life course. Children were educated, and went out to work afterwards. They married as young adults and left their parent's home. A woman often continued to work for a few years until the birth of her first child. Then she gave up her job in order to devote several years to the household and child-raising duties over time. However, the pressure to behave uniformly decreased and the wishes of individuals to make their own choices increased along with the possibilities to materialise these choices. The possibilities of the state and the employer to influence uniform labour market behaviour decreased. Multiformity became standard.

The concept of life-style can be used to structure the increased possibilities of choice and their effects on the size and structure of the labour supply. Life-style is the orientation of an individual to structure his or her life and is the value someone attaches to the different aspects of life (Bootsma, 1995). Each life-style has a specific orientation and results in specific careers, way of living, number of children, specific type of house, location, and job. The life-style depends on the individual's character and background (micro basis) and the (institutional) environment in which he or she lives and works (macro basis). Increasingly it is also contingent on the way individuals choose to mix these micro and macro variables. So, individuals with, for example, the same income level, gender, ethnicity and living in the same city behave differently because they value these variables differently. It is the valuation of the micro- and macro elements in *relation* to each other, which results in different life-styles. Because of this, people also exhibit different labour market behaviour. Variations in life-style therefore explain labour supply variations between individuals. The life-style concept broadens the traditional analytical framework of labour supply by stressing the possibilities individuals have to choose a career in pursuing specific goals. This is in line with the broadening of economic factors as pioneered in the human capital approach (Mincer, 1974) and the social-environment approach (Kazamaki-Ottersten, 1998). The life-style concept has been applied in sociology and consumer research in marketing, and more recently also in spatial and economic sciences. However, it is mainly applied as a concept without an institutional framework. Because life-style depends also on institutions, the challenge is to incorporate institutional restrictions and possibilities.

The Transitional Labour Market

The TLM is the macro-economic result of different life styles on the labour market. It includes changes like more flexible labour relations, more time for caring of children or learning and more time for recreational activities. All these changes indicate the obsolescence of mechanistic concepts of labour market behaviour. Activities like working, learning, retiring and care, should be evaluated in relation to these broader societal changes. This gives opportunities for individuals and society to counteract deterministic views of work. New models should include manners in which activity patterns of various labour market categories can change. The transitional labour market (TLM) is such a model (see Figure 18.1).

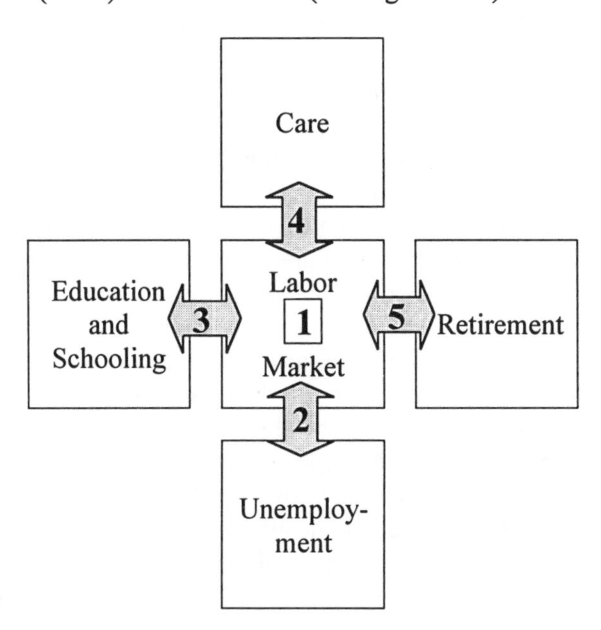

Figure 18.1 The transitional labour market

Source: Schmid (1995)

Figure 18.1 distinguishes between five transitions (Schmid, 1995). Each transition can be temporary and repetitious. Transitions can happen during a working week, month or year, but can also happen during different phases of the life cycle. Possibilities of transitions exist between full-time and part-time work; employment and unemployment; work and learning;

work and care taking and, last, between work and retirement. The TLM presents an emancipating model stressing the individual choice of people freed from deterministic behavioural obligations. People increasingly want to combine various activities during every life stage. Not only work, but also learning and periods of (temporary) retirement. Sabbatical and parental leaves are examples. Older people do not want to totally quit work but apply for gradual forms of retirement. Employees desire periods of work and non-work, combinations of work and schooling (life-long learning), combinations of work and recreation, and of work and taking care of children or elderly.

Indeed, this is the workers' complement to the flexible market, which up till now was dominated by an employer's perspective. Flexibility certainly has an ideological basis and reflects the distribution of power in economic relations (Lagos, 1994). After the fall of the regulation model in industrial relations in the 1970's, flexibility and deregulation became the major cores of economic and labour market policies (Van der Laan and Ruesga, 1998). Labour market institutions, social policies, specific legislation, collective agreements, centralised bargaining, union presence all came under attack. It was claimed that this would generate a reduction of high unemployment rates and an improvement of the functioning of the system of production. Paradoxically, these proposals would eliminate those prerogatives workers had acquired during the eras of regulation and which were supposed to protect their welfare. The imperative of technological development urged for greater *flexibility* in allocating factors of production. This would create higher mobility thereby easing the labour market adjustment to quickly changing global circumstances. Theoretical and ideological support was provided by new forms of the classical-marginalist paradigm, in its neo-classical and monetarist versions (Hall, 1992).

While these measures reflect employer's interests in particular, the other side of the ideological basis of flexibility was hardly heard. This side stresses the possibilities of workers to rearrange industrial relations towards their goals. One of these is the transitional market in which flexible workers can arrange various activities according to their needs. Industrial unrest in Denmark in 1998 illustrates that the attainment of these goals is truly about the distribution of power in industrial relations. Strikes took place, not due to a demand for higher wages, but due to the wish to have more time for care included in collective agreements. As a result, the female participation rate of the 25-35 year age category actually went down in the period 1994-1996 (De Jong, 2000). Organisations are not automatically willing to allow for work practices implied by the TLM. They are quite rigid. However, there are a number of economic reasons why these arrangements should become more flexible. The emerging knowledge economy presupposes this change. The next section discusses these reasons and focuses on how to match transitional labour with the knowledge economy.

Matching the Knowledge Economy and the Transitional Labour Market

This chapter so far has dealt with some of the main changes in the European labour market, which not only frame the position of countries and regions, but also determine their future. On the *demand side* the most important change is from the information to the knowledge economy in which the stress on traditional knowledge incorporates social and cultural capital. On the *supply side* there is a shift from standard models of supply behaviour towards a transitional model in which repetitive transitions between activities like working, learning, retirement and care become crucial. How are these demand and supply changes matched? An important starting point for this is a reflection on the social nature of knowledge.

The Social Nature of Knowledge

If knowledge is understood as a social process, the road to matching demand and supply is to be found in the way in which knowledge-oriented organisations and workers deal with their resources. Although knowledge is also a functional resource, its focal role is in creating optimal conditions for elaborating social and cultural capital. This means the creation of communicative capabilities related to the selection and interpretation of information. Not only the development of codified knowledge is important, but also the organisational relationships of employees as bearers of knowledge. Therefore the matching of demand and supply changes needs new ideas concerning 'human resource management', and 'knowledge management'. Changes in labour demand and supply imply that organisations need different employees and that employees need different organisations. New contracts and new forms of human resource management are needed. The economic success of organisations depends too a large degree on this 'proven connection between people and profit' (Shuster, 1986; Parker, Mullarkey and Jackson, 1994; Leyten, 1998). In the knowledge economy, the stress is more than ever on people. The competitive advantages of organisations lie in their effective use of human resources (Pfeffer, 1994; Miles *et al.*, 1997). Knowledge management is one of the key-factors in this. A recent survey finds that European companies now spend 3.3% of the revenues on knowledge-management.

This will rise to 5.5% in three years time. This is more than EU-companies spend on R&D (Stewart, 1998).

In contrast to the emphasis on the crucial role of labour, less attention is paid to its supply. Although the social nature of knowledge creation is emphasised, the social nature of knowledge itself is not taken seriously (Alvesson, 1998). Although one can point to the power-related structure of knowledge as such, it is important to analyse how a social perspective on knowledge and work bridges the knowledge economy and generates change in labour supply. As discussed, this applies both to high and low educated workers. The role of institutional arrangements is crucial to understanding this bridging function.

The Role of Organisations

In relation to the changes in demand and supply and in particular to the fuzzy determination of productivity and wages and the importance of the valuation of peers, a most important goal of management in the knowledge industries is the commitment and loyalty of their personnel (Alvesson, 1995). This is also important as a condition for the growth and exchange of tacit knowledge. Social effectiveness of organisations is crucial here and is reflected in personnel dedicated organisations. In relation to this Hall and Moss (1998) make a distinction between old and new types of contracts. The old type is based on long term, relational contracts. In the new type (which they call the 'protean career'), there are changing relationships, stressing employee and job development. The career is steered by the employee and there is a lifelong sequence of experiences, capabilities, learning and identity-changes. Development is based on the challenges of the job. The implementation of the new type of management, which fits in quite clearly with the discussed changes of demand and supply, does not come overnight. Some organisations are still quite traditional and do not know how to adapt. They are 'lost in the trees'. Some see the changes: they 'see the forest', while others based on continuous learning adapt to the new circumstances. They are 'comfortable in the woods'.

In the new type of personnel management system, commitment and loyalty of the organisation and employee are important. In this *identity* plays a crucial role. Identities are shaped by social interaction, including material practices. Identity is important for both employees and managers in knowledge organisations (Alvesson, 1998; Leyten, 1998). For employees, identity represents a mechanism for safeguarding competence and self-esteem. For managers in knowledge organisations, identity is an important regulating mechanism. Such mechanisms can be based on a broad range of means such as the development of a corporate identity, and by cultural control and normalisation.

Whatever system is used, the control mechanisms depend not only on efforts of the organisations such, but also whether and how the employee adapts. This implies that the character of labour supply is a crucial determinant for shaping the control mechanisms and by this, the control of knowledge and competitiveness. The social effectiveness based on commitment and loyalty increases if organisations adapt to the transitional labour market. Measures involved are, for example, possibilities for sabbatical and parental leaves, part-time working, on-going educational programmes and gradual and flexible retirement schemes. Moreover adaptations in pension systems and other financial regulations are needed. The key to success is to adapt management structures and work organisation in a way that balances the importance of new factors of production in the knowledge economy with security and flexibility for workers (see also EC, 1998). Although this flexibility will incur higher transaction costs, the overall result is a larger commitment of workers to organisations and through this, stronger competitiveness.

The Role of the State

In understanding the role of the state in the new labour market, one has to look at the ways governments react to the change towards the knowledge economy and transitional labour markets. More specifically, how far do European labour market policies incorporate the growing importance of social and cultural capital, changes in life-styles and transitional labour? Although there is a greater awareness of social and cultural capital, in general *demand-side policies* throughout Europe are still quite wholly focused at R&D reflecting economic capital. This leads to a strong preference for the role of universities and research institutes for economic growth. This perspective mirrors technology-push based starting points in which "we have marvellous technological solutions but do not know for which problem" (Jacobs, 1999, p. 363). However, high-level technology and research cannot be translated directly into market strength. The latter implies attention to the problems and needs of (possible) users. This leads more effectively to real innovative products and services that offer solutions to economic and social challenges. A bias for high-tech solutions and research is clearly besides the crucial point of knowledge building that stresses its multi-dimensional character.

On the *supply side* more attention is paid to policy variables related to life styles and transitional markets. Individualisation, feminisation, childcare provisions and retirement schemes are examples of these.

However, this policy attention reduces to efforts that are directed at the attainment of work (Addison and Siebert, 1997; Van der Laan and Ruesga, 1998). Although attention is paid to combining different activities, the starting point is the function these other activities have in relation to work. Childcare provisions or retirement schedules are, for example, appreciated for their role in making active participation on the labour market possible. The starting point is not the combination of activities with each having its own intrinsic value according to specific life-styles. Transitional starting-points as such have no function. This does not mean that within Europe all things are equal, or that the outlook for transitional based policies is the same. In reference to this, a typology of the European welfare states is useful (Rubery and Smith, 1999). This sets out the main variations among European countries according to key features like the basis for welfare rights and pensions, taxation systems, childcare provisions and parental leave. Four basic models are distinguished: a Nordic, a Continental, a Liberal and a Mediterranean model. Without discussing all these types, it appears that at one end of the spectrum, the individual perspective is particular important (the Nordic policy type). Care provisions are well established and public services are wide spread. At the other end of the spectrum, the Mediterranean type has little benefits for caring and has poorly developed public services. If transitional labour markets become a common future phenomenon in Europe, then the Nordic type is better equipped to meet this development. However, this model also falls short in the sense that all arrangements focus on accommodating the workplace. There is not, as yet, room for institutional arrangements based on life-styles based on a mix of various activities.

Implications for Regional Policies

What do the general changes in the labour market mean for the spatial economic policies in the regions of Europe and what is the role the region of private and public actors? Three main perspectives can be distinguished in relation to this (see also Lagendijk, 1999). The *agglomeration view* stresses that economic growth is fundamentally the result of agglomeration economics. Logistic and social proximity of institutional actors result in economic advantages (Malmberg, Sölvell and Zander, 1996). For economic growth and competitiveness spatial clustering is imperative. The *contingent view* focuses on the role of regions in specific stages of industrial development (Storper, 1997). The intersection between industrial and regional routes governs the economic development of regions. The *indifference view* is more modest in the sense that the link between regional development and enterprises is more arbitrary. Growth is neither the result of 'industrial spaces' nor 'agglomeration'. This view accepts that the strategies of corporations are quite indifferent to spatial circumstances and that regional growth is to a large degree a matter of coincidence (Curran and Blackburn, 1994).

The three perspectives imply different roles for regional factors of production, including the basic factors governing the knowledge economy (economic, social and cultural capital) and the transitional labour market. From the agglomeration view, the basic question is whether the proximity of these factors adds to the external economies of organisations. The importance of spatial clustering is also recognised in the contingent view, but depends on the specific stage of industrial development. In the indifference view, the new production factors play an important role, but do not necessarily result in a specific regional pattern. From all perspectives, the policy problem is how to 'lock-in' economic activities. For the agglomeration view this means a deepening of proximity sensitive aspects of the new factors of production. For the contingent view this means the regional anchoring of potentially mobile activities. The presence of a particular economic activity should lead to the emergence of unique assets by the creation of hard and soft infrastructures. These may anchor the activities in the region even if the industry develops globally. The regional anchoring may also be widened to the next stage of the production process. In the indifference view the focus is on making specific spaces important. It should matter where you are. In accepting that firms as such are spatially indifferent to location, it forms the basis for policy. By embedding active growth efforts in regional knowledge networks and in regional transitional labour markets, the regional labour market itself becomes an important competitive asset. Strategic regional co-operation makes it possible to induce economic advantages.

Whatever choice is made, a common feature of all is that regional efforts should not only be directed at traditional 'know how', but also at organisational and institutional knowledge in which social and cultural capital play a major role. This should not only be limited to traditional 'high-tech' activities, but at other sectors too. A different view on labour supply and particular on skills is also needed. Knowledge is still mainly seen in a material and technological way as a functional approach to problem-solving and as a building block for competence that can be stored in organisations or regions. However, lower qualified labour can have a crucial role in knowledge creation. In addition, the way governmental policies deal with social and cultural capital and with the needs of the transitional labour market, are becoming increasingly important. These are the new engines of regional growth.

References

Addison, J.T. and Siebert, W.S. (1997), *Labour Markets in Europe: Issues of Harmonization and Regulation*, The Dryden Press, London.

Alvesson, M. (1995), *Management of Knowledge-Intensive Companies*, De Gruyter, Berlin/New York.

Alvesson, M. (1998), 'Knowledge Work: Ambiguity, Image and Identity', Paper presented at the WESWA annual congress, Rotterdam (mimeo).

Bomhoff, E.J., den Hartog, R. and Lageweg, I. (1996), *Instituties, Waarden, Normen en Groei* (mimeo), Nyfer, Breukelen.

Bootsma, H. (1995), 'The Influence of a Work-oriented Life Style on Residential Location Choice of Couples', *Netherlands Journal of Housing and the Built Environment*, vol. 10, no. 1, pp. 45-63.

Carter, A.P. (1996), 'Measuring the Performance of a Knowledge-Based Economy', *Employment and Growth in the Knowledge-based Economy*, Paris, OECD, pp. 61-68.

Curran, J. and Blackburn, R. (1994), *Small Firms and Local Economic Networks, The Death of the Local Economy?*, Paul Chapman Publishing, London.

EC (1998), 'Job Opportunities in the Information Society', Report to the European Council, European Communities, Luxembourg.

Fukuyama, (1995), *Trust, the Social Virtues and the Creation of Prosperity*, Free Press, New York.

Grossman, G.M. and Helpman, E. (1991), *Innovation and Growth in the Global Economy*, MIT Press, Boston.

Hall, D.T. and Moss, J.E. (1998), 'The New Protean Career Contract: Helping Organizations and Employees Adapt', *Organizational Dynamics*, vol. 26, pp. 22-37.

Hall, P. (1992) 'The Movement from Keynesianism to Monetarism', in S. Steinmo *et al.* (eds), *Structuring Policies: Historical Institutionalism in Comparative Analysis*, University of Cambridge Press, Cambridge.

Hinloopen, J. and van Marrewijk, C. (1999), 'On The Empirical Distribution of the Balassa Index', *Research Memorandum 9902*, Ministry of Economic Affairs, Hague.

Jacobs, D. (1999), 'Vlaamse Valleien', *Economische Statistische Berichten*, vol. 14, no. 5, p. 363.

Jong, A. de (2000), *Ontwikkelingen van het Arbeidsaanbod in de Europese Unie*, Bevolking en Gezin (forthcoming).

KazamakiOttersten, E. (1998), 'Regional Return to Education in Sweden', in L. van der Laan and S.M. Ruesga (eds), *Institutions and Regional Labour Markets in Europe*, Ashgate, Aldershot.

Klamer, A., Laan, L. van der and Prij, J. (1997), *De Illusie van de Volledige Werkgelegenheid*, Van Gorcum, Assen.

Laan, L. van der (1999), 'Labour Markets in Europe at the Age of New Century: Knowledge Economy and Transitional Labour', *Tijdschrift voor Economische en Sociale Geografie*, vol. 90, no. 4, pp. 428-432.

Laan, L. van der and Ruesga, S.M. (1998), *Institutions and Regional Labour Markets in Europe*, Ashgate, Aldershot.

Laan, L. van der and Versantvoort, M. (1998), 'Analyse en Voorspelling van het Regionale Arbeidsaanbod in een Leefstijl-perspectief', in J. van Dijk and F. Boekema (red.), *Innovatie in bedrijf en regio*, Van Gorcum, Assen, pp. 143-163.

Lagendijk, A. (1999), 'Regional Anchoring and Modernisation Strategies in Non-core Regions: Evidence from the UK and Germany', *European Planning Studies*, vol. 7, (forthcoming).

Lagos, B. (1994), 'Labor Flexibility: an Open Debate', *International Labor Review*, vol. 132, no. 3, pp. 89-101.

Leyten, T. (1998), 'Stimulerend personeelsmanagement', *Tijdschrift voor HRM*, vol. 3, pp. 23-36.

Malmberg, A., Sölvell, O. and Zander, J. (1996), 'Spatial Clustering, Local Accumulation of Knowledge and Firm Competitiveness', *Geografiska Annaler B*, vol. 78, pp. 85-97.

McCloskey, D.N. and Klamer, A. (1995), 'One Quarter of GDP is Persuasion', *American Economic Review*, vol. 85, no. 2, pp. 191-195.

Miles, R. *et al.* (1997), 'Organizing in the Knowledge Age: Anticipating the Cellular Form', *Academy of Management Executive*, vol. 11, no. 4, pp. 7-19.

Mincer, J. (1974), *Schooling, Experience and Earning*, Columbia University Press, New York, for the National Bureau of Economic Research.

Minne, B., Herzberg, V.P. and Reijnders, J. (1996), 'Immateriële Investeringen', *Economisch Statistische Berichten*, vol. 31, no. 1, pp. 106-107.

Nonaka, I., Umemoto, K. and Senoo, D. (1996), 'From Information Processing to Knowledge Creation: a Paradigm Shift in Business Management', *Technology in Society*, vol. 18, no. 2, pp. 203-218.

Parker, K.P., Mullarkey, S. and Jackson, P.R. (1994), 'Dimensions of Performance Effectiveness in High-involvement Work Organizations', *Human Resource Management Journal*, vol. 4, pp. 1-21.

Peyrefitte, A. (1995), *La société de Confiance. Essai sur les Origines et la Nature du Développement*, editions Odile Jacob, Paris.

Pfeffer, J. (1994), *Competitive Advantage through People*, Harvard Business Press, Boston.

Porter, M. (1998), 'Clusters and the New Economics of Competition', *Harvard Business Review*, Nov./Dec., pp. 77-90.

Putnam, R.D. (1993), *Making Democracy Work, Civic Traditions in Modern Italy*, Princeton University Press, Princeton, New Jersey.

Rubery, J. and Smith, M. (1999), 'The Future European Labour Supply', Employment and European Social Fund; Employment and Social Affairs, European Communities, Luxembourg.

Schmid, G. (1995), 'A New Approach to Labour Market Policy: a Contribution to the Current Debate on Efficient Employment Policies', *Economic and Industrial Democracy*, vol. 16, pp. 429-456.

Shuster, F.E. (1986), *The Shuster Report: the Proven Connection between People and Profit*, New York.
Stewart, A. (1998), 'The Next Big Idea. Special Report: Visions of Europe', *Fortune*, vol. 138, no. 12, pp. 81-83.
Storper, M. (1997), *The Regional World*, the Guildford Press, New York.
Sveiby, E. (1997), *The New Organizational Wealth: Managing and Measuring Knowledge-based Assets,* Berret-Koehler, San Francisco.
Woodall, P. (1996), 'The Hitchhiker's Guide to Cybernomics', *The Economist*, September 28.

19 Regulating Labour, Transforming Human Capital and Promoting Local Economic Growth: The Case of the UK Skillcentres Initiative

Simon Leonard

Introduction

Since the 1980s, Britain's developing neo-liberal state has experimented with skills training initiatives which have sought to transform human capital in order to meet the perceived needs of local employers and thereby promote local economic growth. These policy experiments, however, created 'new' institutions of labour market regulation and governance at the level of the local labour market, which were flawed and unable to effectively support local economic development. At the heart of this problem was that, whilst the redefinition of the role of the state during this period was characterised by policy devolution, deregulation, marketisation and privatisation, policies were driven by an increasingly controlling and centralised national government. Within the broader context of the structural crises engendered by industrial restructuring and change and the enterprise culture policy objectives of promoting local growth, the new institutional forms of local labour market regulation and governance were unable to exploit this devolved local responsibility. This was because national government delivered local labour market 'solutions' which were ironically inappropriate to the needs of local business.

This chapter examines the 'skillcentre' programme initiative of the 1980s and 1990s which embraced policies of marketisation and privatisation. It details the intention, preparation and actuality of national government handing the regulatory responsibility for skills training back to the market as a means of promoting local economic development and growth. In so doing, government failed to recognise the historic absence of any real market for skills training within local labour markets suffering the consequences and costs of industrial change. The outcomes of these misdirected policy initiatives are detailed both in terms of the social and economic consequences of the changing nature of skills training provision, as well as the geographical consequences of marketisation and privatisation.

In detailing these outcomes, a number of related shifts in policy objectives during this period are identified. These may be seen as fundamental to any understanding of the failure of these policy programmes to achieve their stated economic objectives of promoting local economic growth. First, these policies represent a major shift within government from the development of a state-provided and directed manpower strategy to a market and business-led strategy. As a consequence, this also represents a shift in the balance between social welfare and economic objectives, dramatically towards the interests and needs of business. These policy changes, therefore, are as good an example as any of the shift from a 'dependency' culture to an explicit and dominating 'enterprise' culture. The movement away from tri-partite corporatism to a neo-liberal state, also has significant geographical consequences. The geographical shift engendered by these policy changes illustrates the move away from national and regional policy aimed at meeting both social and economic objectives to policy supposedly more responsive to perceived and actual local needs aimed at supporting business and promoting local economic growth. This chapter, details these changes in policy and geography by charting the marketisation process, as preparation for privatisation, as well as the outcomes of the privatisation process itself, and the consequent failure of this policy to support business and enterprise within local economies.

Market-Centred Policy as a Means of Promoting Local Economic Growth

The central narrative of this chapter is that government in Britain has, since the 1980s, sought to support and promote local economic growth through deregulation, marketisation and ultimately privatisation. In so doing, regulatory power has been replaced by market forces. Within this situation, the supply, both nationally and locally, of skills training becomes an issue of labour market demand, with the skills training providers emerging where and when needed to offer competitively priced and technologically appropriate training services to local employers. The outcome should be locally specific skills training infrastructures, geared to an efficient response to fluctuations in local labour market demand. Where the training providers fail to respond, within this competitive market, new suppliers will emerge to deliver the 'goods'.

This pristine neo-classical model appeared to be very attractive to government in the neo-liberal Britain of the 1980s and 1990s. This thesis, however, had at least one central flaw, which was that government intervention into the area of adult skills training provision had in many instances come about due to the very failure of that same market. Employers, historically, have been unwilling to meet the costs of skills training provision, particularly during periods of economic recession and industrial restructuring and especially as a counter-cyclical measure to prevent skill shortage 'bottlenecks' at the onset of economic and business growth. Government in Britain in the 20th Century, as elsewhere, repeatedly intervened through a range of policy programmes and initiatives to support or directly create the skills 'pool' required by local business to remain internationally competitive. The skillcentre programme in this chapter represented just one initiative which had existed in Britain, in one form or another, for over 75 years (Leonard, 1999b).

This repeated failure of British industry to provide adequate and appropriate skills within the labour force serves to undermine not only the neo-liberal policy response, but also more fundamentally, the neo-classical description of labour market exchanges. The critique of these exchanges and the recognition of the social constitution of labour markets has been well developed elsewhere (Jones, 1996; Peck, 1996). Central to this critique is the view that contrary to the ideal, real-world labour markets do not act like commodity markets. The labour market is a social and not an exclusively economic institution. Employers do not control or dictate the supply of skills to the labour market, nor are those skills created purely in response to labour market demand, for those same skills are inseparable from the motivations, experiences and purposes of individuals.

Equally individual employers, through their labour market actions do not necessarily act in a manner which supports the operation of that same market. Their actions are much more likely to reflect their immediate needs for skills acquisition rather than broader issues of skill formation within the labour force and labour market. Therefore to construct the labour market as a commodity market is to deny the social nature of human labour and productive activity. From this perspective, the inadequacies of the neo-classical approach to understanding labour market mechanisms points to the likely failings of the neo-liberal policy prescriptions. If the labour market is not a self-regulating equilibrium mechanism focused around labour supply and demand, then why should market forces, unfettered by state intervention, supply the skills necessary to promote local industrial growth. Put even more simply, why should deregulation, marketisation and privatisation by government, coerce a market into existence, especially in areas where that same market has (repeatedly) failed, which was why government was there in the first place.

In his assessment of the roles of the state in the economy, Block (1994) confronts this limited view of the state and the economy as analytically separable entities. Distinguishing between the 'old' and the 'new' paradigms, Block calls into question the ways of defining the state's role in the economy that have been inherited from 19th Century social theorists. Under the 'old' paradigm, state intervention attempts to stabilise the economy by offsetting the impact of the business cycle, the state 'overcoming' the distributive consequences of market processes. The new paradigm, in contrast, rejects the idea of state intervention in the economy. It insists instead, "that state action *always* plays a major role in constituting economies, so that it is not useful to posit states as lying outside of economic activity" (Block, 1994, p. 696). From this perspective, economic activity, and consequently the promotion of local economic growth, will always involve some combination of state action and markets. Given that economies and states are 'profoundly interdependent', the alternative to the old paradigm is to "focus on the specific ways in which states and economies intersect and to begin to examine the variations in these intersections across time and space" (Block, 1994, p. 698). From this position, understanding and promoting local economic growth does not come from neo-classical analyses or neo-liberal policy formulation which seeks to restrict the role of the state and 'unleash' market forces.

In terms of labour market analysis, therefore, an alternative to the 'price mechanism' is required which recognises the social constitution of labour markets and thereby recognises the role and activities of government not as 'spoiler activities' restricting the free market, but as an integral and fundamental component of labour market process. Peck (1989, 1996) has repeatedly argued the case that the labour market should be seen as the intersection of key generative, causal and relatively autonomous labour market processes, centred upon production, social reproduction and the regulatory activities of the state. From this perspective, contemporary local labour market outcomes are also both historically and geographically-derived, making local 'specificities' central to any understanding of labour market process.

Much as the critique of the limitations of the neo-classical interpretation of labour market process raises serious doubts over neo-liberal policy prescriptions which seek to exclude government, so this alternative perspective indicates that labour market analysis which focuses upon the 'local' must recognise government as a key and relatively autonomous variable in influencing labour market outcomes. Labour market policy which acknowledges the fundamentals of this complex set of labour market processes, operating over time and at a variety of spatial scales, has at least the potential for formulating regulatory 'experiments' which are attuned to the complexity of processes underpinning economic change and development.

The skillcentre experience detailed below illustrates how government is central to determining labour market outcomes. Even within a neo-liberal policy context of deregulation, marketisation and the empowerment of local business and industry, the stronger, centralised national government did act as a 'market-spoiler' for the new, privatised agency of labour market skill formation. However, this perspective is also seen to be a limited view as the unfettered free-market forces, which had necessitated government intervention in the first place, served to undermine and eventually close that same agency. Market-centred policy as a means of promoting local economic growth, in the case of the skillcentre initiative, represented policy experimentation which was most vulnerable to market collapse and destined to early failure.

Marketisation of State Labour Training Programmes: UK Skillcentres

The national network of skillcentres, providing primarily adult skills training in the traditional craft trades, had formed an important part of an earlier Labour government's (1974-79) plans for a comprehensive national manpower planning policy (Ainley and Corney, 1990; Stringer and Richardson, 1982). With a change of government at the end of the 1970s, the Manpower Services Commission (MSC) announced a "skillcentre rationalisation plan" at the end of January 1980 (Department of Employment, 1980a, p. 108). The MSC planned to close a total of twenty skillcentres or annexes from its existing complement of 69 skillcentres and 32 annexes, claiming the remaining "skillcentres would be better sited for meeting local labour market needs" (MSC, 1980, p. 26). The proposed closures revealed a slight shift of resources away from those regions receiving greatest assistance from regional policy. This shift was explained by the MSC as a move "to locate the skillcentre network where industry can make most use of it", but overall, "provision will remain at its greatest in areas of highest unemployment" (Department of Employment, 1980b, p. 463). Changes in the network were linked to both the needs of industry and the problems being faced in particular local labour markets from the detrimental effects associated with industrial restructuring. The skillcentre network was then still being required to meet both economic and social objectives although the needs of industry, as they related to local labour markets, were to be prioritised.

This was the basis for a shift away from the regional focus, which had been prominent under Labour, and towards the local labour market, regarded as a more flexible and responsive geographical base upon which to structure a local delivery system appropriate to the needs of industry. In their report of 1979-80, the Public Accounts Committee (1980) had considered the performance of the MSC's skillcentres in providing adult training courses and were advised by the MSC that "a long-term shift of the balance of skillcentre provision towards occupations and geographical areas offering the best employment prospects was planned" (National Audit Office, 1987, p. 8).

By 1982, however, despite the rationalisation plan, the skillcentre programme was earmarked for further review, precipitated by the relative failure of the skillcentres in terms of trainee 'placement'. By the end of 1981, just under a quarter of the trainees completing their course were placed 'in trade'. The Commission consequently reviewed its network of skillcentres in 1982 (MSC, 1982), recommending that skillcentres "should aim for the development of a more flexible and responsive provision" and "a rapid and sustained improvement in the value for money offered by skillcentre training" (MSC, 1983, p. 24). In November 1982, the Commission agreed to set up the Skillcentre Training Agency (STA) to operate on a cost-recovery basis as a separate arm of the MSC. The STA was set up at the start of April 1983 as a separate management unit, outside of the pre-existing Training Division of the MSC. The STA, as a response to the government's Financial Management Initiative, delegated responsibility and accountability to skillcentre managers with the major objective of cost recovery (MSC, 1983).

By January 1984, when the Commission approved the first Business Plan for the STA, the emphasis upon cost recovery was central. The STA was required to "recover its operating costs in full from trading income for 1986-87 onwards" and "that it should seek vigorously and aggressively to adapt, modernise and diversify its training offerings with a view to reducing the shortfall in income as swiftly as possible" (MSC, 1984; Employment Committee, House of Commons, 1985). The government, via the MSC, was only willing to sustain the skillcentre network if immediate action was taken by the STA to recover its costs directly from employers and employer organisations in order to trade in profit within a relatively short period of time. Given that the Training and Enterprise Councils (TECs), as employer-led local institutions of labour regulation, were only introduced in 1988 (Department of Employment, 1988), this development was an early example of the active neo-liberal labour market policy of *marketisation.* This policy sought to create a quasi-market for state-funded skills training, shifting responsibility for training to local employers and paving the way for the eventual privatisation of the skillcentres and the overall delivery system of labour market regulation (Leonard, 1999a, 2000).

The market-centred character of the STA initiative, with management and responsibility devolved to the local level, was also an explicit rejection of the corporatist tri-partist approach, which had dominated training policy in Britain since the 1964 Industrial Training Act. The Chief Executive of the STA in 1984 described the new organisation as a "competitive modern training facility with nationwide coverage able to supply *what the market (whether the Commission or otherwise) wants*, at a time and place that it seeks at a price that it will pay" (MSC, 1984, p. 3). The STA, was now operating outside of the effective control and direction of the MSC, outside of the influence of trade unions and local authorities, and potentially outside of the influence of government. The skillcentre reforms were consequently based upon these beliefs, with the objective of assuring that "[the] new Skillcentre Training Agency established by the Commission [would] ensure that skillcentres adopt a commercial approach in identifying and supplying the training that the Commission and employers want" (Department of Employment, 1984a, pp. 10-12).

The "proposals for changes in skillcentres" (Employment Committee, 1985), presented by the STA in December 1984, were intended to meet the cost-recovery objectives imposed by the MSC. Amongst a set of STA objectives for its future development, was the call for "a streamlined but nationwide network of fewer but more intensively used skillcentres" (MSC, 1984). Each of the proposals was indicative of the complex political context within which the STA was operating. The repeated reference to the need for a 'national' network was based upon the government's historic and persistent need for a set of national social welfare policies which would be seen to be confronting unemployment across the whole of the country. A pure business or market-orientated plan, which the STA was moving towards, would have most likely relinquished this position and focused upon economic viability and not geographical coverage.

In this context, however, this was not possible as the Training Division of the MSC, with its national network of Area Offices and Area Manpower Boards, was still the greatest purchaser of STA training.

Second, the desire to reduce fixed investment and develop a flexible mobile instructor force (Department of Employment, 1984b), illustrated the market-led shift away from welfare-based objectives. If the STA could "survive without premises" then it was aiming its services directly at employers in their factories and employed workers, and away from the unemployed and the 'traditional' off-the-job training provided within the skillcentres. The revised Business Plan was based on a skillcentre by skillcentre assessment which revealed that the provision in 1984 was generally "unrelated to rapidly changing labour market needs today and even more so the needs of tomorrow" (MSC, 1984, p. 9). Industrial and technological change had increasingly rendered the craft skill training in skillcentres redundant, to the extent that they were now regarded by the state as an inflexible means of delivering industrial training.

The existing skillcentre network, the STA argued, had grown and developed in a manner which was not consistent with the needs of industry, the product of a response by government to various pressures and crises over a considerable period. Overall, the restructuring was in line with the MSC's rationalisation plan of 1980, involving a further shift away from most of the development regions, and a recognition of some of the growth potential, for example, along the line of the 'M4 corridor' covering parts of the South-West and South-East regions. Other regions and cities, many with severe unemployment problems, lost skillcentre provision. In Wales training capacity was to be delivered from just four skillcentres, with three of those in the south-east. The Civil Service Union (CSU) in their evidence to the House of Commons Employment Committee, noted that the "ad hoc piecemeal planning of the past" was set to continue, and believed the closure proposals would "result in the complete withdrawal by the STA from major population centres". The CSU felt that it was inconceivable that the STA should "simply abdicate from a city the size of Liverpool" (Employment Committee, 1985, pp. 193-194). This level of withdrawal from major metropolitan areas was inconsistent with the claim that a credible national network was being retained.

The STA was, therefore, planning to withdraw from a number of areas where major industrial change was taking place. One prominent example at the time was in relation to the coal mining industry where closures and voluntary redundancy schemes left a large number of workers unemployed and seeking retraining opportunities. However, these areas were dominated by these extractive industries and alternative employment, particularly for adult male workers, was difficult to obtain locally regardless of retraining. Given the decline in the demand for traditional craft skills, the general decline in manufacturing industry, and the specifics of the decline in the coal mining industry in these areas it would have been expected that a public-funded skills training network, historically and geographically committed to a social welfare role of supporting disadvantaged workers in 'problem' regions, would have remained prominent in those same regions.

The new STA, however, apart from an ill-defined commitment to maintain a 'credible national network' would have little prospect of either placing trainees in work following the completion of their training in these depressed regions; and little prospect of selling their services to other clients, outside of the Training Division of the MSC. The local labour market specifics, relating to the local intersection of labour demand and labour supply in these areas meant that in terms of the criteria against which the STA was increasingly being judged (namely cost-recovery, placement and support for local business in order to promote local economic growth), the disadvantaged or depressed regions offered little immediate or long-term prospects of productivity. Under the proposals for change, skillcentres were scheduled for closure in coalfield areas including, Northumberland, South Yorkshire, Nottingham, Lancashire, Kent and South Wales (Employment Committee, 1985).

A number of conflicts in policy and place, therefore, became apparent in the distribution of skillcentres. The STA, whilst attempting to respond to market needs, was being constrained by the fact that its largest customer, the MSC's Training Division, was still purchasing training places on the basis of its social welfare role. The STA was involved in a policy and spatial conflict between attempting to fulfil "social welfare objectives" and that of "following the market". In addition, the STA was torn between maintaining a 'national network' and responding directly to "expressed local needs". This was to be a continuing conflict which continued to influence the STA throughout the 1980s (MSC, 1987a) and into the 'privatised' 1990s.

The Employment Committee of the House of Commons commissioned an independent background paper on the proposed skillcentre closures. The report concluded that the decision to close 29 skillcentres was soundly based, given the STA's financial target, its pricing policy and the purchasing policy of the Training Division of the MSC. But it also stated that "if these policies [were] retained, the future of virtually the whole skillcentre network [was] uncertain" (Likierman, 1984). If the social objectives of the STA provision were to be abandoned, then the skillcentre system would be more market-related, and training would then be carried out where it was most cost-effective, meeting local needs, as defined by employers. Towards this end, the restructuring programme of 1984-85 had by April 1986 reduced the skillcentre network from 101 skillcentres to 60 operational centres (MSC, 1986).

The responsibility for this element of state-funded labour market regulation was increasingly, through the STA skillcentre network, being shifted to employers. Consequently labour market regulation within the national economy was reconfigured to reflect the skill shortages and labour market needs of employers working within the specificities of their particular local labour market situation. Previous institutional frameworks had centred upon the national economy, industrial sectors and the firm. The new employer-led and employer-directed institutional framework for labour market regulation was 'located' at the level of the local labour market. The skillcentre network, therefore represented an early example of the new 'localism' which underpinned labour market regulation and governance in Britain in the late 1980s and early 1990s. The privatisation of the network, completed in May 1990 (National Audit Office, 1991) was arguably the logical outcome of the development of the government's active neo-liberal market-centred labour market policy over the preceding ten years.

Promoting and Empowering Business Through 'Localism' and Privatisation

The Skillcentre Training Agency became the Skills Training Agency during the trading year 1985-86. Between then and 1988-89, however, the STA recorded an operating loss amounting to nearly £19 million (MSC, 1985; MSC, 1986; MSC, 1987b; Training Commission, 1988; NAO, 1991). The policy framework for privatising the STA skillcentres was now supported by the financial arguments for disposal of the skillcentres into the private sector. In late 1987, the Secretary of State for Employment set up a review of the STA following references by the National Audit Office and Public Accounts Committee to the need for the MSC to purchase its training on the basis of open competition between training providers (Training Commission, 1988).

These developments were consistent with the government's neo-liberal labour market strategy which had at its core an emphasis upon deregulation, privatisation and empowerment of employers within their local economy and local labour markets (King, 1993; Peck, 1994). As part of this policy, the Training and Enterprise Councils (TECs), were introduced in *'Employment for the 1990s'*, a government White Paper which included the view that the STA "would be in a better position...if it were to move into the private sector where it could adopt the best commercial practices" (Department of Employment, 1988, p. 37). The White Paper, although identifying within Britain's new industrial training system a continuing role at the national level, and a voluntary role at the industry level, envisaged most change at the local level where responsibility for training provision was to be vested within the private sector.

Employment for the 1990s proposed a radical deregulation and privatisation of the training system. While the influence of the state was to be drastically reduced at the sectoral/industry and national levels, new employer-led institutions of labour market governance and regulation were to be created at the local level. The view was that "localities [were] more likely to find solutions that work" (Department of Employment, 1988, p. 39). 1988 represented a point where the labour market institutions of the 'dependency culture' finally gave way to the regulatory mechanisms of the 'enterprise culture' (Coffield, 1990; Streeck, 1989) within a geographical context which now emphasised the local and not the national.

The first TECs were established in April 1990 and all were in place by October 1991 (Bennett, Wicks and McCoshan, 1994). Their creation and early development paralleled the privatisation of the skillcentres and their operation in the private sector. Both initiatives were examples of the 'localism' which underpinned the government's regulation of the labour market in the 1990s. The changes envisaged under the TEC programme were not simply to do with spatial scale and efficacy of delivery systems. The TECs, as a means of rebuilding the economy through local initiative, were concerned with empowerment, shifting responsibility from the national to the local and, from the public to the private sector. TECs were not intended to simply manage and deliver existing programmes at the local level. They were charged with assessing the economic and social needs of their locality, defining local strategies and allocating resources to stimulate local economic development (Coffield, 1990). They were to be "a new kind of organisation, locally based and born of the enterprise culture" (Training Agency, 1989, p. 4), revealing the logic of voluntarism and the radicalism of the market (Peck, 1992, 1995).

In reality, the TEC initiative shared many of the problems associated with the institutional change within the STA and the skillcentres including, the continuing dependency upon Department of Employment (DE) funds, with government placing social welfare objectives, particularly combating unemployment, as a restraint on 'market logic'; and, the imposition of the market logic onto a sphere of government activity, creating markets where arguably they were never present in a fully-functioning form.

The creation of a 'training market' by the government, and the requirement upon the STA to operate within an enterprise culture and a market logic, was also constrained by the employers. The failure of the private sector to take responsibility for skill formation has been an accepted feature of the British industrial system and a repeated basis for state intervention and labour regulation. The expectation that private funds would flow rapidly, and at an acceptable level, into the restructured STA and its skillcentres was ambitious within this context. Employers had traditionally been more concerned with the acquisition of skilled workers appropriate to their needs, rather than broader skill formation within the labour force.

Within this context, therefore, the sale of the Skills Training Agency, was the logical outcome of the failure of the reconstructed STA to fulfil the objectives set for it by the MSC and government. Deregulation in this instance meant privatisation, enabling the individual skillcentres to reflect the immediate skill needs of local employers, apparently unfettered by the requirement to respond to what was believed to be at that time an agenda of diminishing national importance, namely unemployment. The reconstructed STA had not 'empowered' local employers with any greater responsibility for training. The privatisation of the STA offered some scope for employer empowerment, regulated and controlled by the employer-led and directed TECs.

The government White Paper of late 1988 (Department of Employment, 1988) had introduced the intention to move the STA into the private sector. In March 1989, the Secretary of State for Employment informed Parliament of the decision to offer the Agency for sale by private tender, either as a whole or as a number of separate training businesses (NAO, 1991). The independent advisers on the feasibility of privatisation, reported that the STA, "reduced to a core of strategically located skillcentres", could be sold to the private sector as a viable training business (NAO, 1991, p. 1). In May 1990 the government completed the sale of 45 skillcentres to Astra Training Services Ltd (Astra), a company formed by three senior executives of the STA, achieving the first management buy-out within the Civil Service. The DE sold a further six skillcentres to three other organisations. The remaining nine skillcentres were to be closed.

The sale of the STA was contentious in terms of both the act of privatising the skillcentre network and the particulars of the sale to Astra. Twenty-six of the 33 organisations who had put in indicative offers were invited to submit final offers, which were received from sixteen organisations (NAO, 1991). Although this level of interest would suggest an appraisal, by a number of organisations, of the STA as a viable commercial opportunity, this is less evident from the breakdown of offers by skillcentre. From this perspective, other than offers from the network bidders, 26 centres were the subject of only one other bid and no further offers at all were made for 31 centres (NAO, 1991). In a significant number of cases it was apparent that there was in fact very little 'local' interest expressed in purchasing most of the skillcentres. Astra's successful bid, led to a net payment to Astra of £10.7 million, government allowing 'negative' bids to cover the cost of restructuring and rationalisation (NAO, 1991).

The Astra bid maintained the appearance of a national network in that it retained a presence in each of the STA regions and most of the major population centres throughout Britain. However, in July 1993, just over three years after the privatisation of the STA in May 1990, Astra Training Services, which had purchased 45 skillcentres, was placed in the hands of the receiver. Thirteen centres were sold to one businessman to form a new company, AST Training, intending to establish at least three regional divisions and seeking to expand in other areas. Other bids were received for other regional groupings but were rejected by the receiver. The collapse of Astra, the first collapse of a privatised government department agency, marked the end of any claims of a residual national network of skills training centres, even within the private sector. In June 1995, the AST group of companies, operating the last substantive group of skillcentres, albeit as separate regional companies, also went into voluntary liquidation.

The closure, following privatisation, of nearly all of the former STA skillcentres within a relatively short period of time, demonstrated the limitations of this policy of transferring responsibility for industrial training to the private sector, and following the imposition of the market logic, the vulnerability of that agency/company to market failure. Ironically, in a written submission to the House of Commons Employment Committee (1992), Astra, prior to its collapse, argued that "government-funded training suffered from the lack of a coherent long-term strategy". Most companies, they argued, in the context of recession and rising unemployment were cutting jobs. Astra concluded that many training providers were in financial difficulties and were leaving the training market, and that "this damage to the training infrastructure may take years to repair" (Employment Committee, 1992, p. 25).

In December 1988, when the privatisation of the STA was first proposed by government, the national level of unemployment had been falling for two years, dropping below 2 million. The recession began in the middle of 1990 and by early 1993 unemployment was again over 3 million. The unanticipated onset of recession led directly to declining government budgets, cutbacks in private training investment, and rising local unemployment. As Peck notes, regulatory environments with their 'market-led' policy initiatives have proved vulnerable to market failure and "a training system driven by the short-term needs of the market is self-evidently likely to produce under-investment in skill-formation" (Peck, 1992, p. 343). Establishment of the TECs, to which may be added the privatisation of the skillcentres was, "predicated on an expectation of tightening labour markets and falling unemployment" (Peck, 1994, p. 106). The TECs, designed to privatise the process of skill formation in tight labour markets, were now operating in a collapsed labour market with accelerating unemployment. The privatised skillcentres had been established within a framework which demanded an immediate response to expressed needs. In the context of recession, those expressed needs were for employers to cut costs and labour and not to buy-in external training services.

The parallel with the TECs, however, is not simply one of common experience within a local labour market and national economy situation; the performance of Astra was at this time closely related to the operation and funding of the TECs. The STA had been privatised in order to free the skillcentres from the inflexibility of the national training market, to allow them to be, within the context of a market situation, more responsive to local labour market needs, and to release them from a dependency upon DE/MSC, and consequently government funding. The recession from 1990, however, operated in a more complex manner than the expected reduction in private sector training investment. Increasing unemployment placed a major restriction on TEC budgets and spending priorities (Bennett, Wicks and McCoshan, 1994). The privatised skillcentres, far from breaking their link with government funding for training the unemployed, were just one more step removed from that source of funding as the local TECs, committed by government to supporting training for the unemployed, were now their largest customer. Astra, had sought to remind government of its policy by stating that they wanted the focus of government spending on training to be on those people who were most likely to benefit from it most rapidly. In 1992, however, just prior to its collapse, Astra was still predominately training unemployed people in the same basic craft skills that had been taught in the skillcentres and earlier Government Training Centres since the end of 1945, namely "building trades, engineering, welding and electronics" (Employment Committee, 1992, p. 27; Leonard, 1999b).

While Astra's decline was, therefore, in part attributable to the recession, it was its relationship with the local TECs, and with government which was critical. Astra's financial difficulties ultimately stemmed from changes in the structure and funding of government training programmes, particularly Employment Training (ET). ET was vulnerable to cuts by the

DE in 1992 as part of their response to the government's Public Expenditure Survey. Astra, as one of the largest providers of ET training, through its contracts with the local TECs, was particularly vulnerable to these cuts (Employment Committee, 1992). At the same time Astra was again caught, through the nature of its indirect customer relations with government, into low-skill training for the unemployed, with employer-funded training diminishing in the face of recession.

Within this context of national economic recession and cuts in public expenditure, both elements of Astra's income base, direct private sector contracts, and indirect public-sector funding through the TECs, reduced to a point where their business was not viable. The employer-led market forces approach had to accept that within a recession, the vulnerability of Astra to market failure was an acceptable consequence of the enterprise culture and free market competition. The scope for counter-cyclical skills training and the promotion of local economic growth, even during periods of economic recession, was not an option for a private company such as Astra.

Conclusion

The belief by government that the institutions of labour market regulation and governance would be more responsive to the needs of business and industry if they were employer-led, deregulated and privatised, has been seen in this instance and circumstance to be mis-directed and mis-placed. The notion of empowering business and focusing attention at the level of the locality, within a context of economic recession and increasing unemployment, simply served to subject formerly state-funded skills training provision to the harsh excesses of the competitive private market. Such an attempt to impose market logic within a situation where no real market for skills training existed was destined to early failure.

In addition, neo-liberal policy formation which aimed to remove the restrictive influence of the state and unleash market forces was an illusion. Although at the local level Astra and the TECs were the new private sector and employer-led institutions of local labour market regulation, the role of the state was still central and fundamental to their achievements, potential success and likely failure. Deregulation, marketisation and privatisation at the local level were coupled with a stronger and more powerful role for the state at the national level, a not unfamiliar outcome of neo-liberal policy formation. As recession deepened, the agencies of central government demanded, as the biggest purchasers of their services, a continued and extended counter-cyclical response from these privatised and employer-led bodies operating at the local level. The role of the state was not in these circumstances reduced, it became in fact reinforced and one-step removed to the national level where, given the immediate political objectives, policy was even less likely to reflect the expressed needs of business and industry operating in the circumstances and specificities of any particular local labour market and local economy. Policy formulation, aimed at promoting local economic growth, which seeks to remove the role of the state fails to acknowledge the fundamental interdependent relationship between the state and the economy. Understanding the nature of this intersection between the state and the economy at international, national and sub-national levels is fundamental to formulating policy to effectively support and promote local economic growth. The ideology of marketisation and privatisation, in this instance, had effectively the opposite effect. Business was not empowered, government became 'distant' and 'dislocated' from the needs of business at the local level and deregulation revealed the weaknesses of these new agencies and bodies to the worst excesses of the market.

In a period of economic growth, Astra may have had more success at adapting the skillcentre training provision into a flexible, modern, training provider reflecting the new-technology skills training needs of modern business and enterprise. However, even in these circumstances, the costs to the employer associated with purchasing this skills training would always be considerable compared with the effective subsidy offered by state provision. Industry has historically, and in general, never been willing to pay for skills training, in part due to the transferable nature of skills between employers. Astra may have survived as part of a 'boom' economy, but with the onset of 'bust' and economic recession, the euphemistic 'downsizing' of companies and corporations has frequently been paralleled by the reduction in company expenditure on formal skills training.

The national network of skillcentres could have been restructured and relaunched by government in order to meet the skills training needs of business and thereby have promoted local economic development. However, the concrete reality and historical specificity of the economic recession and increasing unemployment of the early 1990s ensured that the government's active neo-liberal policy strategy of meeting the needs of business and enterprise through deregulation, marketisation and privatisation was destined to reflect the worst outcomes of market competition rather than contribute to the empowerment and the promotion of local business, industry and enterprise.

References

Ainley, P. and Corney, M. (1990), *Training for the Future: The Rise and Fall of the Manpower Services Commission*, Cassell, London.

Bennett, R., Wicks, P. and McCoshan, A. (1994), *Local Empowerment and Business Services: Britain's Experiment with Training and Enterprise Councils,* UCL Press, London.

Block, F. (1994), 'The Roles of the State in the Economy', in N. Smelser and R. Swedberg (eds), *The Handbook of Economic Sociology*, Princeton University Press, Princeton NJ, pp. 691-710.

Coffield, F. (1990), 'From the Decade of the Enterprise Culture to the Decade of the TECs', *British Journal of Education and Work*, vol. 4, no. 1, pp. 59-78.

Department of Employment (1980a), 'Skillcentre Network to be Rationalised', *Department of Employment Gazette*, HMSO, London, p. 108.

Department of Employment (1980b), 'Changes in Skillcentre Network will Mean More Flexibility and Less Cost', *Department of Employment Gazette*, HMSO, London, p. 463.

Department of Employment (1984a), *Training for Jobs*, HMSO, London.

Department of Employment (1984b), 'Delayed Decision on Skillcentres and Larger Mobile Instructor Force', *Department of Employment Gazette*, HMSO, London, p. 524.

Department of Employment (1988), *Employment for the 1990s,* HMSO, London.

Employment Committee (1992), *Sale of the Skills Training Agency,* House of Commons, Session 1991-92, HMSO, London.

Employment Committee (1985), *Proposals for Changes in Skillcentres*, House of Commons, Second Report, Session 1984-85, HMSO, London.

Jones, B. (1996), 'The Social Constitution of Labour Markets: Why Skills Cannot be Commodities', in R. Crompton, D. Gallie and K. Purcell (eds), *Changing Forms of Employment: Organisations, Skills and Gender*, Routledge, London, pp. 109-132.

King, D. (1993), 'The Conservatives and Training Policy 1979-1992: from a Tripartite to a Neoliberal Regime', *Political Studies*, vol. 41, pp. 214-235.

Leonard, S. (1999a), *Quasi-Markets Within the Welfare State: Lessons from State-Funded Skills Training and the Skillcentre Programme 1984-93,* Paper presented at the annual conference of the Royal Geographical Society with the Institute of British Geographers, Leicester, England, January.

Leonard, S. (1999b), *Geographies of Labour Market Regulation: Industrial Training in Government Training Centres and Skillcentres in Britain and London 1917-93,* Unpublished PhD thesis, Department of Geography and Environment, London School of Economics, University of London.

Leonard, S. (2000), *Geographical Consequences of a Neo-Liberal State: Marketisation and Privatisation of British State-Funded Skills Training,* Paper presented at the annual conference of the Association of American Geographers, Pittsburgh, USA, April.

Likierman, A. (1984), *The Case for Shutting Skillcentres*, London Business School.

Manpower Services Commission (1980), *MSC Manpower Review 1980,* MSC, London.

Manpower Services Commission (1982), *Skillcentre Review,* MSC, London.

Manpower Services Commission (1983), *Annual Report 1982-83,* MSC, Sheffield.

Manpower Services Commission (1984), *Skillcentre Training Agency: Progress Report and Future Development,* MSC, Sheffield.

Manpower Services Commission (1985), *Annual Report 1984-85,* MSC, Sheffield.

Manpower Services Commission (1986), *Annual Report 1985-86,* MSC, Sheffield.

Manpower Services Commission (1987a), *The Funding of Vocational Education and Training: a Consultation Document,* MSC, Sheffield.

Manpower Services Commission (1987b), *Annual Report 1986-87,* MSC, Sheffield.

National Audit Office (1987), *Department of Employment and Manpower Services Commission: Adult Training Strategy,* HMSO, London.

National Audit Office (1991), *Sale of the Skills Training Agency* HMSO, London.

Peck, J. (1989), 'Reconceptualizing the Local Labour Market: Space, Segmentation and the State', *Progress in Human Geography*, vol. 13, pp. 42-61.

Peck, J. (1992), 'TECs and the Local Politics of Training', *Political Geography*, vol. 11, no. 4, pp. 335-354.

Peck, J. (1994), 'From Corporatism to Localism, from MSC to TECs: Developing Neoliberal Labour Regulation in Britain', *Economies et Societes, Serie Economie du Travail*, vol. 18, no. 8, pp. 99-119.

Peck, J. (1995), *Geographies of Governance: TECs and the Remaking of 'Community Interests',* Paper presented at the Institute of British Geographers annual conference, University of Northumbria at Newcastle, January.

Peck, J. (1996), *Work-Place: The Social Regulation of Labor Markets,* The Guilford Press, New York.

Public Accounts Committee (1980), *Department of Employment: Manpower Services Commission Skillcentres,* House of Commons, 12th Report, Session 1979-80, HMSO, London.

Streeck, W. (1989), 'Skills and the limits of neo-liberalism: the enterprise of the future as a place of learning', *Work, Employment and Society*, vol. 3, pp. 89-104.

Stringer, J. and Richardson, J. (1982), 'Policy Stability and Policy Change: Industrial Training 1964-1982', *Public Administration Bulletin*, vol. 39, pp. 22-39.

Training Agency (1989), *Training and Enterprise Councils: A Prospectus for the 1990s*, Training Agency, Sheffield.

Training Commission (1988), *Annual Report 1987-88,* Training Commission, Sheffield.

20 Rethinking Auckland: Local Response to Global Challenges

Richard Le Heron and Philip McDermott

Introduction: Participating in the Globalising Economy

After a century in which Auckland's population never declined there may be grounds for economic optimism. We argue otherwise, suggesting instead that Auckland's roles in the emerging international division of labour are by no means clear or assured. Furthermore, traditional models of economic growth based on the supply, quality and capacity of local resources, capital and labour perpetuate an inward-looking policy focus. This potentially undermines local initiatives for growth management.

While an emergent national policy emphasis on scientific and technical expertise may improve opportunities for the Region's businesses and labour to participate in a globalising economy, by itself this will not assure economic growth. Indeed, in the clamour to foster a knowledge economy in New Zealand, knowledge about the relationships among local and wider processes and the potential contribution and constraints of Auckland, the nation's largest urban centre, is remarkably sketchy.

New Zealand, more than most countries, has embraced globalisation through its willingness to lead the world in the reregulation of markets and borders (Le Heron and Pawson, 1996). The late 20th Century saw Auckland indisputably the main point of connection between New Zealand and the globalising world economy. Whether connection is enough, however, is a vital question that needs to be answered in part by addressing the role Auckland, and through it, New Zealand, plays in global circuits of capital. In this chapter we ask how Auckland might lead the challenge for

New Zealand to create a knowledge economy, given its position in global value chains?

Answering this question takes on some urgency, as the policy pendulum in New Zealand shifted recently towards a centrist mandate exercised by a newly elected (October 1999) centre-left government. The electoral success of the Labour-Alliance government suggests a willingness on the part of the electorate to veer away from 15 years of movement towards unimpeded market regulation. A question arising from the new mandate is how far economic intervention can take effect in an environment of globalisation that appears to concentrate national production at fewer sites. This question needs to be addressed in light of fundamental shifts in the landscape of *global connectivity and interaction* that are likely to shape Auckland's (and New Zealand's) future participation in the globalising world economy.

Economic globalisation involves, *inter alia*, the development of increasingly complex international production, trade and finance regimes (Allen, 1995; Dicken, Peck and Tickell, 1997; Pieterse, 1995). Recognising these international regimes, this chapter deals with the ways in which networks of interdependence and influence lock together economic agents and institutions to make particular regions 'work' within globalising value chains. Only in this way might industries, enterprises and labour in Auckland be valorised in global networks and circuits.

The term 'global value chains' relates to the ability to assemble knowledge and capacity from disparate organisations across cultures and localities to produce goods and services with *a world-wide market* (Gereffi and Korzeniwiez, 1994; Gereffi, 1996; Le Heron and Roche, 1996; Mol and Law, 1994; Whatmore and Thorne, 1997; Whitney, 1996). The accumulated value can be attributed to no single organisation or locality, and the disposition of the shares of value and the distribution of investment are subject to intra-chain power relations. Indeed, even local logistical arrangements for members of such groups may be mediated by the affiliation of a distant parent with a global service provider, like a freight forwarding company, software provider, or communications service. A value chain creates a web which, through strong horizontal as well as vertical linkages, effectively limits the autonomy of local business units regardless of the management model adopted (Dicken, 1998; Le Heron, 1998). Consequently, value chains are only partially represented in analyses of trade that do not capture the institutional context of transactions.

Localities may participate in both these non-contiguous production regimes (trade blocs and value chains) through the selective movement of local producers into international domains defining new relationships between areas of production and areas of consumption (Cooke, 1996; Fuellhart, 1999; Gordon, 1996; Morgan, 1996; Storper, 1992, 1997). Through this process of differential participation dissimilarity among places may reflect what they contribute within a series of value chains, and may increase as activities are realigned in response to chain development. Local specialisation can be defined in terms of contributions to international complexes of business and exchange rather than with reference to the concentration of a particular skill or production base at a particular locality (Asheim, 1996; Cooke *et al.*, 1998; Florida, 1995; Kenney and von Burg, 1999).

The Policy Challenge

In the past, one response of localities aspiring to participate in the global economy was to attract international and national investors by driving down local costs, through bounties, subsidies, tariffs and the like. Yet, today policy intervention that focuses on reducing the cost of doing business in a locality in order to increase the probability of attracting capital is intrinsically risky. It does not address the issue of local capacities (innovation, skills and leadership) that are increasingly the focus of area development policies. Even policies addressing these issues, though, may not be enough to sustain a presence in international circuits of capital where uncertainty is multiplied by rapid obsolescence, accelerating product cycles, tensions over the ownership of intellectual capital and highly mobile product and process innovation.

Another orthodox response, promoting local clusters of (high tech) industry, may no longer work, either. Successful innovation is associated with advances by companies that dominate global markets, in fields such as communications and electronics, materials and automation, transport and logistics (Grunsven and Van Egeraat, 1999; Malmberg and Solvell, 1995; Shefer and Frenkel, 1998; Tamasy, this volume). For example, in the currently high yielding realm of software and high tech stocks, returns reflect the capacity of entrepreneurs to sell into global capital circuits more than the intrinsic value or even the novelty of products *per se*. The development of local expertise represents an attempt to accumulate new intellectual capital in companies and this is always potentially relocatable across borders within organisations (Murphy, 2000).

At best, policy can entice international capital to a particular locality through a neo-Fordist response of driving down the costs of doing business, or encourage local businesses to develop attributes so they can to link into the global economy, a neo-liberal ideal that has few demonstration models.

The latter sits in a framework emphasising specialisation, flexibility and collaboration as a basis for creating value at particular sites. The aim is to make places "key staging points or centres of competitive advantage within respective global industrial filieres and value chains" (Amin, 1997, p. 133). Local economies that do not participate in the development of the linkages underlying global value chains are constrained by local capital, natural, and human resource capacities – the traditional endogenous economic model of growth – and risk being left on the periphery.

Capital risk is also mediated through non-governmental agencies rating the credit risk not just of companies, but also of places. This instrumental approach to investment undermines the capacity for places to nurture innovation, to commercialise local technology, and to capture associated production opportunities. The traditional relationship between place and capital founded on familiarity and proximity, which in the past shaped production complexes, is diminished. The reduction of a place's rating by a private credit agency may weigh heavily on the capacity of businesses there to raise capital. This is reinforced by the changing nature of private savings and investment funds, which no longer observe local loyalties and hedge risk internationally.

It follows that labour must also compete globally (Le Heron, 1994). If labour is mobile (typically more skilled and knowledge-rich labour (Callister and Rose, 1996)), the competition for employment stretches across localities, even though the site of employment may be fixed. If labour is immobile (typically unskilled and less-informed) competition is more likely to be based on lowering the price locally. This may be through modification of employment conditions and industrial law. Increasing the quality of labour through education and training policies is a response that can fulfil two roles. By lifting productivity it may reduce the costs of doing business at a locality. By increasing capacity of labour, it may facilitate a move across the divide between (price elastic) immobile and (price inelastic) mobile labour. However, while the differentiation of labour by price and productivity may offer an advantage to capital in a location temporarily, local institutional pressures and international business dynamics mean that this is likely to be short-lived, and the investment attracted ephemeral.

The preceding discussion suggests that global value chains encompass diverse spatial contexts and cross spatial scales. The outcomes for a particular area are indeterminate, contextually specific, attainable through multiple pathways and nonlinear in nature. We now turn to recent developments in New Zealand that are germane to comprehending Auckland's connections to global chains.

Auckland: A Peripheral Global City?

New Zealand's participation in international trade and production remains centred on land-based resource sectors. Smaller provincial centres either participate in the global value chains through specialist resource based sectors or else become marginalised. Most New Zealand provincial centres are indirectly, if somewhat tenuously, embedded in global value chains through servicing primary activities. Examples include Rotorua (population 30,000) embedded in the international forestry and tourism chains (Britton *et al.*, 1992; Schlotjes and Le Heron, 1997); New Plymouth (68,700), with oil and gas based petrochemicals and dairying-related services; Tauranga (78,318, forestry and horticulture); and Napier (54,224, forestry, horticulture and viticulture).

With the internationalisation of the resource sectors, even Auckland's contribution as a centre of international commerce, management and marketing, however, is reduced. Its primacy is still based on its capacity to serve national business and consumption as a centre of imports and import substitution and as a service centre (see Taylor and Le Heron (1977) and McDermott (1973) for a 1970s benchmark of Auckland's linkages). An analysis of economic structure at the end of the 1990s reveals continuing primacy, but a shift from an economy founded on production to one centred on services and consumption (Table 20.1). The production of tangible goods accounted for only 22% of the region's total employment in 1999. Consumption (including retailing), personal and community services – services geared towards final demand – accounted for 45%.

The level of employment in the research and scientific institute sector is low, both in absolute terms and in relation to the rest of the country. This reflects a national research agenda still geared largely towards primary production, even at the end of the 20th Century, and so centred on provincial sites specialising in pastoral farming, horticulture, and forestry, for example. Employment in the education sector also lies substantially below what might be expected, both on the basis of share of national population (33%) and the younger age structure of Auckland compared with the rest of the country.

Table 20.1 Economic profile, Auckland 1997, 1999 (full-time employment equivalents)

	Auckland	% of Regional Total	New Zealand	Auckland as % New Zealand
Primary	14,281	3.2	177,700	8.0
Primary Processing	5,658	1.3	45,248	12.5
Intermediate Manufacturing	23,528	5.3	67,666	34.8
Transport Equipment	12,800	2.9	34,406	37.2
Final Goods	40,688	9.2	89,674	45.4
Subtotal Tangibles	*82,674*	*22.0*	*236,994*	*34.9*
Utilities	1,454	0.3	6,972	20.9
Construction	32,634	7.4	102,367	31.9
Transport Services	23,890	5.4	64,231	37.2
Communications	10,725	2.4	27,836	38.5
Intermediate Services	*68,703*	*15.6*	*201,406*	*34.1*
Computer Bureaux	6,301	1.4	14,017	45.0
Banking, Investment & Insurance	16,817	3.8	45,718	36.8
Legal & Accounting Services	10,675	2.4	28,425	37.6
Real Estate	4,276	1.0	13,138	32.5
Other business services	34,522	7.8	80,361	43.0
Research and Scientific Institutes	624	0.1	5,551	11.2
Business Services	*73,215*	*16.6*	*187,210*	*39.1*
Sanitary and Cleaning Services	5,537	1.3	16,327	33.9
Central Govt	10,129	2.3	39,962	25.3
Local Govt	2,739	0.6	12,350	22.2
Government Services	*18,405*	*4.2*	*68,639*	*26.8*
Miscellaneous Personal	11,677	2.7	34,210	34.1
Medical, Dental, Other Health and Veterinary Services	25,168	5.7	83,539	30.1
Education Services	30,345	6.9	109,278	27.8
Welfare Services	11,275	2.6	46,041	24.5
Personal & Community Service	*78,465*	*17.8*	*273,068*	*28.7*
Wholesale & Retail	91,661	20.8	251,444	36.5
Catering & Accommodation	13,933	3.2	42,361	32.9
Entertainment & Cultural Services	12,976	2.9	36,517	35.5
Consumption Services	*118,570*	*26.9*	*330,322*	*35.9*
TOTAL	*440,032*	*100.0*	*1,297,639*	*33.9*

Sources: EcoLink Database (McDermott Fairgray Group, 1999b)
Business Directory, Statistics New Zealand

A key policy question for the new economy, then, is how far the dominant Auckland economy – geared towards consumption with minimal and declining production, and only limited investment in science and education – can sustain New Zealand's participation in a rapidly changing global value chains? Can this primate city supply the knowledge framework, the learning opportunities, the milieu for entrepreneurship and innovation, the setting for international relationships, and the quality of labour, necessary to take advantage of the anticipated benefits of global engagement?

Answering questions such as this requires recasting the way we interpret Auckland's current economic activity. Addressing the needs of the 'new urban economy' may call for rethinking the policy agenda for local development, and the research questions that underpin it. In the following section we explore five sectors to illustrate dimensions of connection and tensions of governance that pervade the Auckland scene.

Understanding the Connections

Despite the election of a left leaning government in 1999, the investment most appropriate for New Zealand and the role of central government in influencing this are, still under debate. The debate centres on the issue of New Zealand's international competitiveness, the role of international investment in enhancing competitiveness, and the commitment to a regime of free trade to regulate competition and innovation. Central government's role is pivotal but subject to contrasting views. The neoliberal position of minimalist government is seen as one way of encouraging international participation in the economy to make good deficiencies in local capital and skills. But, with the change in government, direct intervention in finance and labour markets has been resurrected as the alternative.

Despite this apparent shift, the debate is not resolved. It may not be unless we acknowledge the increasing inappropriateness of the idea of a distinctive New Zealand economy (Britton *et al.*, 1992). Thinking remains dominated by policies intended to encourage inward investment in productive assets, rather than encouraging outward connection through participation in global value chains. The latter stance would mean recognising and exploring the ability of local business to 'act at a distance' by co-opting (through networks) other actors, materials, systems, procedures, rules and knowledges (Becker and Novy, 1999; Hall and Hubbard, 1997; Horan and Jonas, 1998; Keil, 1998; Le Gales, 1998; Morgan, 1997. Adopting this view would be an important step in policy thinking. Instead of having to choose between a local or global stance, the notion of network allows us to think of highly connected global entities, which remain nevertheless locally linked.

Within Auckland itself, economic debate is taking place almost independently of the globalisation occurring around it. Local policy preoccupation reflects expectations of growth based on the experience of the late 20th Century. It is dominated by a regulatory paradigm defined by principles of environmental preservation, and manifest in policies focused on protecting the rural environment from urban encroachment (Auckland Regional Council, 1999). It does not confront the sources and prospects of future growth, nor address changing modes of production, distribution and consumption, nor consider international linkages as drivers of growth. Consequently, policy is formed on extrapolating recent experience, rather than any consideration of structural and institutional shifts occurring as a consequence of globalisation.

More generally, some of the issues that will influence how New Zealand and Auckland participate in a globalising economy remain largely unquestioned. There has been little debate over how the shift from a production-led to consumption-led economy is taking place, for example, and its consequences (McDermott, 1996). While there has been much talk about localised sector clusters as economic drivers (Ffowcs Williams, 1996), little consideration has been given to how they may relate to global value chains, and how they will become and remain internationally connected. Technology in general is emerging as a policy panacea, yet with scant consideration of the mechanisms through which supply side initiatives in education and research might enhance local engagement in the global economy. Indeed, a recent review of the contribution of a provincial university (Massey University) to regional development suggested that technological innovation and agreements not only bypassed the Manawatu Region, where it is situated, but also Auckland, in favour of direct international connection (McDermott and Le Heron, 1999).

The policy vacuum at central and local government levels left by *de*regulation has left economic actors to reorient from local and national space to world space without consideration of the potential benefits of coordination (and therefore governance arrangements) in global value chains. The identification of supply side policy options (emphasising local initiatives) ignores this prospect.

Three policy questions arise:

- What is needed to make Auckland's social and physical capital work in global value chains?
- What powers can be mobilised from involving Auckland?

- What 'spatiality' (socially produced space) is beginning to emerge from the new relationships and interactions of New Zealand's restructuring in the 1990s?

The following discussion examines several of the dimensions of Auckland's economic composition suggested by Table 20.1 from the perspective of participation in globalisation. The order chosen for discussion reflects, first, current concerns over infrastructure and, then, emerging interest in how New Zealand might grow a globally oriented knowledge economy. The section provides no answers but attempts to sketch a research agenda that has as its object the understanding of connectivity and the part that reflexive institution-building could play in developing a knowledge economy.

Trade and Physical Infrastructure

Preoccupation with infrastructure in Auckland revolves around the quality of the local environment – with some concern over the impacts of congestion or rehabilitation of ageing infrastructure on local costs (Tremaine, 1999). The key to the economic development of the city, however, may lie as much with the role and development of infrastructure that connects city, suburbs – and nation – to the global economy, as with local roads and underground services. Hence, Auckland's international airport and ports emerge as critical nodes in the physical chain between New Zealand producers and consumers with international suppliers and distributors. The airport is a critical focal point in networks of relationships and connections that cannot be satisfied simply through electronic and communications media. Internationally, airports are playing a practical and symbolic role in both sustaining growth and maintaining an international presence for cities and states.

Analysis of the total value of trade (FOB values for exports and CIF values for imports) confirms the rapidly growing role of Auckland's International Airport in the 1990s (Figure 20.1). In 1988 the Airport accounted for 29% of total international trade movements to and from the Region by value, and was up to 33% by 1999, based on an annual average growth rate of 7% (compared with 5% for the seaport). Put another way, at the beginning of the period Auckland Region handled 65% of New Zealand's total imports and 31% of its exports. By the end of the period, it handled 73% of imports and 38% of exports. This is despite the evidence that its dominance in the production of tangible goods might be starting to diminish.

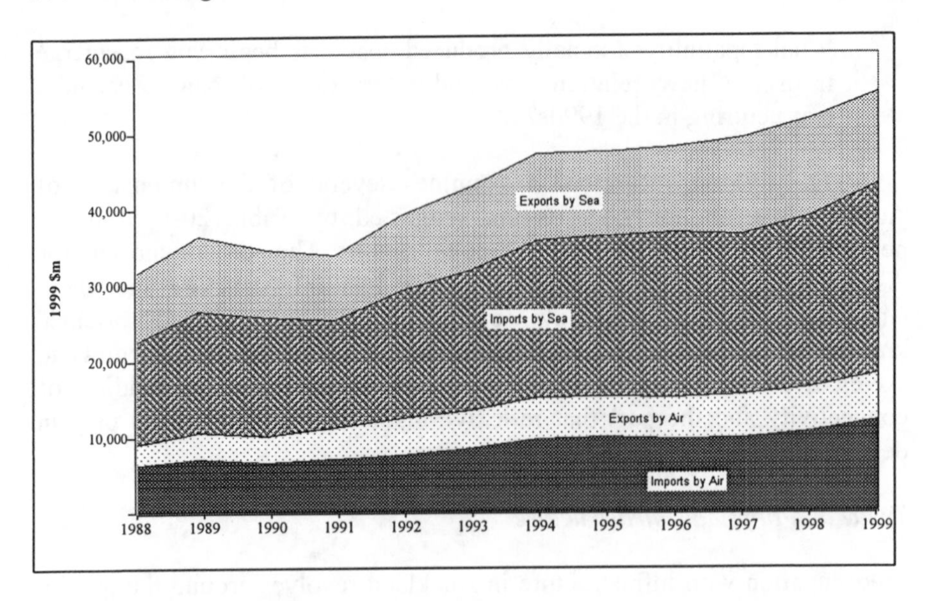

Figure 20.1 Trends in international trade ports of Auckland and Auckland International Airport, 1988-1999

Source: Statistics New Zealand

Some 84% of all New Zealand airfreight was handled through Auckland International Airport. This reflects the growth in higher value products traded, which favour airfreight. These freight trends highlight the role of Auckland as a centre of services, distribution and consumption ahead of its role as an exporter. Imports into the region expanded at an average of 6.3% per year from 1988 to 1999, compared with 4.5% for exports. In fact, applying the value chain approach to the question confirms the centrality of the Airport. In 1994, for example, Auckland International Airport was associated with activity that supported some 11% of the country's measured output (GDP), and 27% of Auckland Region's (McDermott Fairgray, 1995). A large share of this was attributed to the movement of international passengers, both tourists and business people.

This example reinforces the need to think about how a particular region might develop or exploit infrastructure as a basis for participating in globalisation. International capital recognises and acts on the opportunity to extract value from domestic infrastructure which has suffered from poor productivity and under-investment, particularly if it exists in a quasi-monopoly market (Ferenbach, 1996). The risk that productivity gains are captured almost entirely in financing costs and increased margins to the owners of capital suggests that within domestic markets, at least, increased competitive oversight (by as yet uncreated institutions) might be a necessary corollary to globalisation. Any move from a cost-focused trading regime to a focus on extracting value from value chains calls for extra vigilance in the ownership and operation of critical infrastructure.

A research agenda for the 'new economy' has to examine the circumstances of infrastructure planning and development carefully. For example, are there opportunities for extracting value within Auckland's strained infrastructure that may be realised by local markets, institutions and communities? Or will an absence of local liquidity see the necessary investment driven by international capital?

One local response has been to set up a quasi-local authority with explicit responsibility for investment in transport and stormwater infrastructure. Infrastructure Auckland is governed by a board appointed by an electoral college of local politicians. It has established its own criteria for evaluating competing infrastructure projects through which it assesses proposals by local councils, with its funding driven primarily by the profits of the Ports of Auckland (Lindsey, 1999). Its procedures for evaluation show little recognition of the role that transport infrastructure, in particular, might play in providing the milieu either for extracting value of for forging the linkages on which value might be created. Infrastructure Auckland favours, instead, an approach which uses a detailed process of disassembling and measuring individual elements to assess largely localised environmental benefits, despite the commitment of close to NZ$1 billion to leverage large scale regional transport and stormwater projects.

Telecommunications

The communications sector illustrates the role of infrastructure within value chains and the capacity of investors to extract value across borders from those chains (Larner, 1999). Telecommunications links places and catalyses travel between them. In many respects it is an aggregative force. By breaking down isolation and distance it facilitates economic concentration among key international centres, despite the hopes held for its capacity to support the dispersal of economic activity. How, then, might a place assume and maintain the role of a hub within the telecommunications field, rather than that of a spoke increasingly distant from the economic centre (Malecki, this volume)? At the most obvious level, the electronic integration of the distribution chain, first through Electronic Document Interchange (facilitating freight forwarding and customs clearance in international trade) and subsequently through integration via web-based systems, facilitates and accelerates trade (Forer, 1999). It reduces the need for widely dispersed services for handling trade and concentrates responsibility in the hands of a diminishing number of increasingly global logistics companies rather than a wide distribution of national carriers.

There may be potential to resist concentration by promoting the quality of local physical environments, given rapid convergence in the quality of communications among places. If a locality cannot offer advantages other than access to ubiquitous communications facilities, quality communications may act as a double-edged sword, though, supporting the shift of control functions, for example, to higher order or more attractive places.

At the same time, the emergence of specialist international communications complexes sees local communications supporting accumulation elsewhere. Indeed, this is recognised through the integration of communications, media and software, with a diminishing number of key players influencing where value accrues throughout this chain, a chain which has been reinforced by horizontal linkages and coalescence across formerly quite separate and disparate sectors.

Finance and Business Services

The finance sector has been undergoing restructuring in two directions. First, more diverse agencies are offering a greater range of often overlapping services. Traditional consumer-oriented savings banks are displaced transformed into trading banks, while trading banks broaden their service base in response to market erosion by specialist investment service providers and the extension of former insurance companies into a wider range of financial services.

Second, technology is leading to substantial productivity gains so that the sector, which sustained substantial employment growth in metropolitan centres during the 1980s and early 1990s, has begun to contract.

Table 20.2 indicates that Auckland is by no means immune to the contraction occurring within the financial sector. This is especially so as the Trans-Tasman 'rationalisation' of financial service undermines Auckland's role in servicing much of the national economy through loss of banking services to Australia. This raises questions that go well beyond the implications for regional employment and income. Coupled with a more instrumental approach to risk assessment, increasing distance may limit the capacity to finance operations that do not meet strict international standards, undermining the capacity to innovate locally. Moreover, it indicates an institutional network within which national borders are of diminishing relevance to the flow of funds. Local savings and capital formation in this context refer neither to the region nor the nation, but to a Trans-Tasman complex of investment agencies, which may be linked, to Auckland enterprise. Perhaps the questions reach beyond issues of local development to those of economic sovereignty, as the balance of financial flows and the capacity to manage monetary conditions move to a trans-national stage.

Table 20.2 The finance sector, Auckland 1997, 1999 (full-time employment equivalents)

	FTEs, 1999		% Shift 1997-1999		Ak % NZ
	Auckland	New Zealand	Auckland	NZ	1999
Reserve Bank and Trading Banks	5,651	17,740	-4.0	-7.8	31.9
Savings Banks	1,622	5,030	-4.0	-7.7	32.2
Other Financing	1,391	2,875	1.8	10.2	48.4
Investment	1,129	2,693	-31.7	-26.0	41.9
Services to Finance and Investment	1,824	4,605	16.2	17.6	39.6
Life Insurance	873	2,550	-16.6	-30.4	34.2
Medical Insurance	414	756	1.2	0.7	54.8
General Insurance	1,914	4,684	-10.1	-13.1	40.9
Services to Insurance and Superannuation	1,999	4,782	-1.9	-3.0	41.8
TOTAL	16,817	45,718	-5.5	-7.8	36.8

Source: EcoLink, McDermott Fairgray Group, 1999b

Wider access to capital may be advantageous for innovation and enterprises that can secure a place in global value chains; but we may find fewer enterprises occupying the higher order positions that enable them to access the emergent form of finance.

Research, Technology, Education and Training

Turning to the issue of developing a local environment that fosters innovation, the immediate question is how people can begin to exploit local knowledge in an environment in which the necessary supporting infrastructure – investment services in particular – are essentially 'disconnected'. The answer may lie in the establishment of individual relationships across borders, effected through a range of institutions including, but not limited to, individual employers.

In New Zealand, the government has supported and on occasion directed the development of the science and education sectors, in the near absence of private sources. Today, changing funding arrangements and structures are intended to break down the traditional focus on agriculture and associated research complexes around provincial universities (with the risk that in backing off agriculture a range of internationally linked and competitive activities will begin to languish). Two groups of players are especially important, the Crown Research Institutes (CRIs) and the Universities.

Several dimensions are pertinent. First, the universities and polytechnics, which until recently were focused on national goals (e.g. education and training for jobs in New Zealand) are actively re-orienting to become fuller participants in the global education and research complex (e.g. Universitas 21). This involves adopting benchmarking, performance indicators and best practice methodologies, in addition to overseas marketing missions (Cohen Wesley and Goe, 1994; Patel and Pavitt, 1994; Patchell, this volume).

Second, despite this, traditional assumptions about the composition and dynamics of the sector, stemming from a national framework, still guide thinking. The conjunction of a decline in school leavers, more than a decade of in-migration from parts of Asia and southern Africa, the rise in Pacific Island and Maori available to enter the sector and growing resistance by students to borrowing to fund education have resulted in difficulties for the sector. These include: falling first year enrolments, which constrain the flexibility to redesign curricula; a new block of students with high numeracy but low English language skills stretching the capacities of providers; and less enthusiasm for graduate study than anticipated in a knowledge-economy.

The CRIs, evolving from government departments, have been struggling to create international networks in keeping with those already associated with the universities. They have remained dependent on government funding and so continue to work within a centrally enunciated research framework, which has been both slow to change and procedurally constrained. Even recent attempts to establish a new national research and development agenda – the Foresight Project, for example – has involved substantial consultation with existing providers and so leans towards the application of new ideas to old fields and incremental rather than fundamental change. In any case, unstable central funding, subject to both political whim and bureaucratic procedures for contesting funds and monitoring research processes, inevitably engenders conservatism among research agencies.

Furthermore, there has been an absence of the institutional structure that might facilitate the movement from pure research to applications. With a small population, a thin manufacturing sector, no substantive military establishment, and a high degree of dependence on imports to meet intermediate and final demand, the local milieu is not conducive to the development of new products. Public sector support for commercial applications has, in the past, relied on the prospect of commercialisation of university-based research, or spin-offs from the CRIs. Only now is the government addressing the commercial environment for R & D, primarily by extending its central priorities and funding programmes to the private sector. For its part, the residual private sector in New Zealand, in particular, argues mainly for tax equivalence with competing localities, seeking both the autonomy and tax advantages available elsewhere.

Ultimately, the real risk might be that a centrally conceived and controlled programme in a globalising environment might create an individual capacity to undertake research and development, but not the institutional capacity to take advantage of it. The consequence will be the continued migration of the small, highly educated and skilled beneficiaries of research and development funding to localities with a more active enterprise culture, while local businesses struggle to connect to global value chains.

Entertainment, Leisure and Tourism

These sectors are defined less by consumption rather than production and illustrate how Auckland is re-emerging as a 'new' place in neoliberal space. They contribute to the transformation of urban landscapes to cater for purchasers rather than producers. The importance of tourism, which in New Zealand still accounts for a relatively small portion of GDP, is that it is expanding much more rapidly than the local economy, and provides a highly visible interface with a global economy. Consequently, cities are being reconfigured to cater for a growing leisure sector, with that reconfiguration defined by international tastes and expectations.

The question that is often implicit in this reconfiguration is whether it is for a domestic or global consumer? At least three developments are discernible. First, local initiatives are connecting into value chains by attracting global capital, people and knowledge. The most visible example is investment between 1998-2000 for the America's Cup international challenge match yacht racing on Auckland's Hauraki Gulf. Investment covered race facilities, marina and on-shore amenities for spectators, software development (Virtual Spectator, showing the race to the world), and media. A significant expansion in Auckland's inshore and ocean going boat industries accompanied this spectacle (Glass and Hayward, 2001).

Second, a reconfiguring of local space by local investors is occurring as part of conscious or unconscious international integration. Examples include: major changes in the character of Ponsonby, an area adjacent to Central Auckland through a mix of gentrifying, heritage and café-culture investment (Latham, 1999) and The Edge development, a revitalisation by private-public interests of a sizeable portion of central Auckland, and North Harbour Stadium. The last-mentioned is a second major sporting arena in the Region. It is of a size and design intended to claim a place on international sporting circuits, yet attracts little more than a sparse mix of local or Trans-Tasman interest rugby union, rugby league, and tennis fixtures, and occasional outdoor shows. Third, overseas initiatives to generate local value by embedding globally valorised entertainment into Auckland's networks include the Phantom of the Opera and Aida, and the re-running of selected shows in Auckland after the annual (and highly successful) International Festival of Arts in Wellington.

The question over Auckland is how far its leisure spaces and activities are configured to meet domestic needs and how far to reflect international demand and tastes? The possibility is that the two sit uncomfortably together at present, given the limited capacity of the local population to sustain the spectacle that might be required to claim international attention.

Conclusions

In the chapter we have explored what insight might come from rethinking Auckland in terms of its present and prospective relationships with global value chains. We did so cognisant of a growing belief in government and business circles that economic salvation for New Zealand and Auckland lies with the aggressive development of a 'knowledge economy'. Most rhetoric speaks of what would obtain were a knowledge economy achieved. Little analysis has tackled the vexing questions that must be posed and answered if we are to have any reasonable idea of the processes potentially involved in moving to a 'New Zealand knowledge economy'.

By grounding our analysis in the debates of the moment, we have attempted to show two crucial things. First, a key to unlocking the processes of a knowledge economy is recognising the connectivity of New Zealand to the many value chains that make up the evolving globalising world economy.

We argue that the neoliberal framework fails to unpack the critical connections that go into the formation of the market relationships of global value chains. The competitive model freezes the world into a mosaic of nations and misses the web like nature and dynamic of production to consumption networks and circuits into which producers of any territory insert themselves.

Moreover, governance by competition is potentially antithetical to fostering processes that might form part of a more knowledge-centred economy. The neoliberal agenda of favouring markets, deregulation and privatisation clouds the realities that economic processes require multiple forms of governance. Yet, in some areas, such as R & D, the centre has struggled to retain control, in part through ever more tortuous consultative processes that serve only to represent embedded relationships and limit the extent to which development resources might embrace the new. In others, such as leisure and entertainment, the global is increasingly reflected in the local landscape. However, this may be as much about opportunities for international capital to benefit from investment in local property and infrastructure as reflecting realities about the internationalisation of the New Zealand economy.

Second, policy preoccupation over infrastructure in Auckland may impede dialogue about the issues, risks and governance possibilities that are part of the drive for a knowledge economy. Localised concern for growth management is a weak response to global challenges. It takes for granted the capacity of key drivers – infrastructure itself, telecommunications, research and education, and leisure and entertainment – to ensure Auckland's continuing relevance on a global stage. Our analysis suggests that this confidence is misplaced. Thinking of Auckland as a knowledge node in a globalising world is groundless unless knowledge about and from social processes, human dimensions, performativity, culture and so on are actively enshrined in institutional practices.

Three conclusions can be drawn. First, the idea of a knowledge economy will take on different meanings according to a locale's position in relation to particular global value chains. Knowledge of this aspect is scant yet vitally strategic in developing local capacities to engage in globalising processes.

Second, New Zealanders must realise that other countries are approaching the issues and risks of globalisation in similar ways – other areas are also entangled in global value chains. All are facing similar structural pressures arising from capitalist dynamics, the world holds few secrets for long, and attempts to copy the successful models of others simply overlooks the geographically and historically specific conditions in which successful models have flourished. Gaining some understanding of the experiences and lessons of others who have sought to reposition companies and governments in regional or global value chains is therefore a high priority.

Third, the call for a knowledge economy comes precisely when urban sustainability is attracting local and central political attention. Auckland as a site for development, life and livelihood is produced through processes that may or may not be sustainable (Millar and Le Heron, 1999). The greatest barrier to a knowledge society and sustainability (both processes rather than outcomes) is the absence of institutional designs for defining and implementing learning and sustainable practices in local contexts. While there are signs that a new collaborative ethos is rapidly emerging amongst Auckland's territorial authorities, without an understanding of where the Region, its enterprise and its communities fit within chains of value connecting different places, the question remains one of collaboration towards what? The research agenda that might define Auckland's responses to growth must commence with an understanding of the global challenges to growth that the Region is facing.

Acknowledgment

Research relating to this chapter is an outgrowth of other projects. Richard Le Heron acknowledges funding from the 'Political Economy of APEC' project, UoA 509, supported by the Foundation for Research Science and Technology and the University of Auckland Research Fund to investigate infrastructure and regional competitiveness.

References

Allen, J. (1995), 'Crossing Borders: Footloose Multinationals', in J. Allen and C. Hamnett (eds), *A Shrinking World*, Oxford University Press, Oxford.

Amin, A. (1997), 'Placing Globalisation', *Theory, Culture and Society*, vol.14, no. 2, pp. 123-137.

Asheim, B. (1996), 'Industrial Districts as 'Learning Regions': a Condition for Prosperity?', *European Planning Studies*, vol. 4, pp. 379-400.

Auckland Regional Council, Auckland City Council, Waitakere City Council, Rodney District Council, Manukau City Council, North Shore City Council and Franklin District Council, (1999), *Auckland Region: Business and Economy 1999*. Available at: http://www.arc.govt.nz

Becker, L. and A. Novy, (1999), 'Divergence and Convergence of National and Local Regulation: the Case of Austria and Vienna', *European Urban and Regional Studies*, vol. 6, no. 2, pp. 127-143.

Britton, S., Le Heron, R. and Pawson, E. (eds) (1992), *Changing Places in New Zealand. A Geography of Restructuring*, New Zealand Geographical Society, Christchurch.

Callister, P. and Rose, D. (1996), 'Up-skilling and De-skilling', in R. Le Heron and E. Pawson (eds), *Changing Places: New Zealand in the Nineties*, Longman Paul, Auckland, pp. 105-108.

Cohen Wesley, R.F. and Goe, W.R. (1994), *University-Industry Research Centers in the United States*, Carnegie Mellon University, Pittsburgh.

Cooke, P. (1996), 'Reinventing the Region: Firms, Clusters and Networks in Economic Development', in P.W. Daniels and W.F. Lever (eds), *The Global Economy in Transition*, Longman, Harlow.

Cooke, P., Uranga, M.G. and Etxebarria, G. (1998), 'Regional Systems of Innovation: an Evolutionary Perspective', *Environment and Planning A,* vol. 30, no. 9, pp. 1563-1584.

Dicken, P. (1998), *Global Shift: Transforming the World Economy*, Third Edition, Paul Chapman, London.

Dicken, P., Peck, J. and Tickell, A. (1997), 'Unpacking the Global', in R. Lee and J. Willis (eds), *Geographies of Economy*, Arnold, London, pp. 158-167.

Ferenbach, C. - Berkshire Partners (1996), 'Investing in Rail: an International Perspective', Keynote Address, Proceedings of the Australasian Transport Research Forum, Auckland, 28-30 August.

Ffowcs Williams, I. (1996), *Report on 'Local Clusters Go-Bush'*, Tradenz, Wellington.

Florida, R. (1995), 'Towards the Learning Region', *Futures*, vol. 27, pp. 527-536.

Forer, P. (1999), 'Fabricating Space', in R. Le Heron, L. Murphy, P. Forer and M. Goldstone (eds), *Explorations in Human Geography: Encountering Place*, Oxford, Auckland, pp. 86-116.

Fuellhart, K. (1999), 'Localization and the Use of Information Sources: the Case of the Carpet Industry', *European Urban and Regional Studies*, vol. 6, no. 1, pp. 39-58.

Gereffi, G. (1996), 'Global Commodity Chains: New Forms of Coordination and Control among Nations and Firms in International Industries', *Competition and Change*, vol. 4, pp. 427-439.

Gereffi, G. and Korzicwiez, M. (eds) (1994), *Commodity Chains and Global Capitalism*, Greenwood Press, Westport Connecticut.

Glass, M.R. and Hayward, D.J. (2001), 'Innovation and Interdependencies in the New Zealand Custom Boat-building Industry', *International Journal of Urban and Regional Research*, forthcoming.

Gordon, R. (1995), 'Globalisation, New Production Systems and the Spatial Division of Labour', in W. Littek and T. Charles (eds), *The New Division of Labour, Emerging Forms of Work Organisation in International Perspective*, Walter de Gruyter, Berlin.

Grunsven, L.V. and Van Egeraat, C.V. (1999), 'Achievements of the Industrial 'High-road' and Clustering Strategies in Singapore and their Relevance to European Peripheral Economies', *European Planning Studies*, vol. 7, no. 2, pp. 145-173.

Hall, T. and Hubbard, P. (1997), *The Entrepreneurial City: Politics, Regime and Representation*, Wiley, London.

Horan, C. and Jonas, A.E.G. (1998), 'Governing Massachusetts: Uneven Development and Politics in Metropolitan Boston', *Economic Geography*, Special Issue, pp. 83-95.

Keil, R. (1998), 'Globalization Makes States: Perspectives of Local Governance in the Age of the World City', *Review of the International Political Economy*, vol. 5, no. 4, pp. 616-646.

Kenney, M. and von Burg, U. (1999), 'Technology, Entrepreneurship and Path Dependence: Industrial Clustering in Silicon Valley and Route 128', *Industrial and Corporate Change*, vol. 8, no. 1, pp. 67-103.

Krate, S. (1999), 'A Regulationist Approach to Regional Studies', *Environment and Planning A*, vol. 31, no. 4, pp. 683-704.

Larner, W. (1998), 'Hitching a Ride on a Tiger's back: Globalisation and Spatial Imaginaries in New Zealand', *Environment and Planning D: Society and Space*, vol. 16, pp. 599-614.

Le Gales, P. (1998), 'Regulations and Governance in European Cities', *International Journal of Urban and Regional Research*, vol. 22, no. 3, pp. 482-506.

Le Heron, R. (1994), 'Complexity, the Rural Problematic and Sustainability', *Proceedings 17th New Zealand Geography Conference*, Christchurch, pp. 283-288.

Le Heron R. (1998), 'Organising Industrial Spaces: A Commentary on the IGU Commission. State-of-the-Art: IGC 1996', *Tijdschrift voor Economische en Sociale Geografie*, vol. 89, no. 2, pp. 203-209.

Le Heron, R. and Pawson, E. (eds) (1996), *Changing Places: New Zealand in the Nineties*, Longman Paul, Auckland.

Latham, A. (1999), 'Heritage, Gentrification and the Good City: Reinventing an Inner-city Suburb in Auckland, New Zealand', Paper presented to the Association of American Geographers Conference, Honolulu, March.

Lindsey, D. (1999), Presentation to 'Region and Economy' class, Auckland Regional Council, July.

Malmberg, A. and Solvell, O. (1995), 'Spatial Clustering, Accumulation of Knowledge and Firm Competitiveness', paper presented to the 1995 Residential Conference of the IGU Commission on the Organisation of Industrial Space, Seoul, August.

McDermott Fairgray (1995), *The Role of Auckland International Airport in the Economy*, Report to Auckland International Airport Ltd, Auckland.

McDermott Fairgray (1999), *Auckland Regional Growth Strategy Gap Analysis,* Auckland Regional Council, Auckland.

McDermott, P. (1973), 'Spatial Margins and Industrial Location in New Zealand', *New Zealand Geographer,* vol. 29, no. 1, pp. 64-74.

McDermott, P. (1996), 'The Ports', in R. Le Heron and E. Pawson (eds), *Changing Places: New Zealand in the Nineties*, Longman Paul, Auckland.

McDermott, P. and Le Heron, R. (1999), 'Becoming Independent: the Local Implications of a Strategy of National Expansion by a Provincial University', unpublished manuscript.

Millar, C. and Le Heron, R. (1999), 'Economic Drivers and Relationships with Urban Environments', Proceedings of Workshop on Urban Sustainability sponsored by the Royal Society of New Zealand, the New Zealand National Commission for UNESCO and the Parliamentary Commissioner for the Environment, Royal Society of New Zealand, Miscellaneous Series 53, pp. 35-42.

Mol, A.M. and Law, J. (1994), 'Regions, Networks and Fluids: Anaemia and Social Topology', *Social Studies of Science*, vol. 24, pp. 641-671.

Morgan, K. (1996), 'The Learning Region, Institutions, Innovation and Regional Renewal', *Regional Studies*, vol. 31, no. 5, pp. 491-503.

Morgan, K. (1997), 'The Regional Animateur: Taking Stock of the Welsh Development Agency', *Regional and Federal Studies*, vol. 7, pp. 70-94.

Moricz, Z. and Murphy, L. (1997), 'Space Traders: Reregulation, Property Companies and Auckland's Office Market, 1975-1994', *International Journal of Urban and Rural Research*, vol. 21, no. 2, pp. 165-179.

Murphy, A. (2000), 'Retailing in Cyberspace: Implications of Online Grocery Shopping', Paper presented at the 96th Annual Meeting Association of American Geographers, Pittsburgh, April.

Pieterse, J. (1995), 'Globalisation as Hybridisation', *International Sociology*, vol. 9, no. 2, pp. 161-184.

Patel, P. and Pavitt, K. (1994), 'National Innovation Systems: Why they are Important and how they Might be Measured and Compared', *Economics of Innovation and New Technology*, vol. 3, pp. 77-95.

Perry, M. and McArthur, A. (1992), 'The rise and fall of venture capital in New Zealand', in M. Green (ed.), *Venture Capital: International Comparisons*, Routledge, London.

Scholtjes, R. and Le Heron, R. (1997), 'Internationalisation and private sector and institutional restructuring: The example of Rotorua, New Zealand', *Australasian Journal of Regional Studies,* vol. 2, pp. 1-20.

Shefer, D. and Frenkel, A. (1998), 'Local milieu and innovations: some empirical results', *Annals of Regional Science*, vol. 32, no. 1, pp. 185-200.

Storper, M. (1992), 'The limits to globalization: technology districts and international trade', *Economic Geography*, vol. 68, pp. 60-93.

Storper, M. (1997), *The Regional World. Territorial Development in a Global Economy*, Guilford Press, New York

Taylor, M.J. and Le Heron, R.B. (1977), 'Agglomeration, Location and Regional Growth in New Zealand: the Role of Auckland', in G. Bush and C. Scott (eds), *Auckland at Full Stretch,* Auckland City Council and University of Auckland, Auckland, pp. 188-199.

Tremaine, K. (1999), 'Successfully Managing Auckland's Growth in the Next Millennium: Issues of Governance', Second Annual Lecture of the Auckland Branch, *New Zealand Geographical Society*, 12 October.

Whatmore, S. and Thorne, L. (1997), 'Nourishing Networks: Alternative Geographies of Food', in D. Goodman and M. Watts (eds), *Globalising Food: Agrarian Questions and Global Restructuring*, Routledge, London.

Whitley, R. (1996), 'Business Systems and Global Commodity Chains: Competing or Complementary Forms of Economic Organisation', *Competition and Change*, vol. 4, pp. 411-425.

Index